AF365181

Engineering Mathematics-I

Engineering
Mathematics-I

Dr. R. N. Yadava
M.Sc. (BHU Varanasi), Ph.D. (IIT Bombay)
*Professor & Director of R&D, AISECT University
&
Former Director Gr. Sc. and Head Resources
Management and Modelling
CSIR-Advanced Material and Process Research
Institute Bhopal (M.P.)*

Dr. Anil Goyal
M.Sc., Ph.D.
*Professor of Mathematics
Dean, Student Welfare &
Dean Applied Sciences
R.G.P.V., Bhopal (M.P.)*

Dr. Ramakant Bhardwaj
M.Sc., Ph.D.
*Deputy Director (R & D)
TIT Group of Institutions Bhopal
& Head, Department of Mathematics
TIT & Science Bhopal (M.P.)*

Dr. Sarvesh Kumar Agrawal
M.Sc., M.Phil, Ph.D.
*Associate Professor
Department of Mathematics
TRUBA College of Science & Technology
Bhopal (M.P.)*

Dr. Neeta Tiwarl
M.Sc., Ph.D. (Mathematics)
*Dean Academics
Bansal Institutes of Research and Technology
Bhopal (M.P.)*

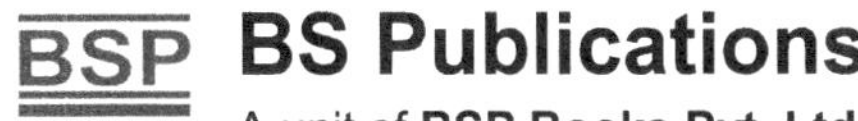 **BS Publications**
A unit of **BSP Books Pvt. Ltd.**

4-4-309/316, Giriraj Lane, Sultan Bazar,
Hyderabad - 500 095
Phone : 040 - 23445605, 23445688

Engineering Mathematics - I by *R. N. Yadava, Anil Goyal, Ramakant Bhardwaj, Sarvesh Kumar Agrawal and Neeta Tiwari.*

© 2016, *by Publisher*

Published by :

 **BS Publications**
A unit of **BSP Books Pvt. Ltd.**

4-4-309/316, Giriraj Lane, Sultan Bazar,
Hyderabad - 500 095
Phone : 040 - 23445605, 23445688
e-mail : info@bspbooks.net

ISBN : 978-93-5230-109-6 (HB)

Preface

The need for a textbook on Engineering Mathematics for Under Graduate students of Engineering as per the revised syllabus was felt by the UG students of Engineering. Taking into account the fact that there is no comprehensive text book for the Bachelor of Engineering students on Engineering Mathematics an attempt has been made to fill a niche that exists between more elementary books and books for advance studies.

Objective of this book on Engineering Mathematics is to provide to the students of Bachelor of Technology/Engineering a simple, clear and logical presentation of Basics of Differentiation, Rolle's and Lagrange's Theorem, Tangent and Normal, Indefinite and Definite Integral under Recapitulation of Mathematics as Unit - I. Further Expansion of Functions, Maxima and Minima, Curvature and Curve Tracing has been explained in logical manner under Ordinary Derivatives and Applications as Unit - II. Subsequently Partial Differentiation, Jacobian and Approximations of Errors are discussed in Unit - III and Definite Integral as a Limit of Sum, Beta and Gamma Functions are discussed in Unit - IV in simplest way for the benefit of the students. Also an attempt has been made to explain Multiple Integrals, Volume and Surface of Solid in best possible manner within the reach of students under Applications of Integral Calculus as Unit - V.

The book has been prepared according to the syllabus of various Technical Universities and is conceived as a text book for the course in Bachelor of Engineering of all branches. The contents of the book have been systematically organized and spread over fifteen chapters. Each chapter begins with the fundamentals and explained with different type of solved examples. The book not only covers the entire scope of the subject but explains the philosophy of the Basics of Engineering Mathematics and makes the understanding of the subject more interesting.

R. N. Yadava
Anil Goyal
Ramakant Bhardwaj
Sarvesh Kumar Agrawal
Neeta Tiwari

Contents

Chapter – 4: Indefinite and Definite Integral

Unit - II

Ordinary Derivatives and Applications

Chapter – 5: Expansion of Functions

Chapter – 6: Maxima and Minima

Chapter – 7: Curvature

Chapter – 8: Curve Tracing

Unit - III

Partial Derivatives and Applications

Chapter – 9: Partial Differentiation

Chapter – 10: Jacobian

Chapter – 11: Approximations and Errors

Unit - IV

Integral Calculus

Chapter – 12: Definite Integral as a Limit of Sum

Chapter – 13: Beta and Gamma Functions

Unit - V

Applications of Integral Calculus

Chapter – 14: Multiple Integrals

Chapter – 15: Volume and Surface of Solid

Unit – I

Recapitulation of Mathematics

- Basics of Differentiation
- Rolle's and Lagrange's Theorem
- Tangent and Normal
- Indefinite and Definite Integral

CHAPTER – 1

Basics of Differentiation

1.1 Introduction

If f(x) is a function of x, then $\underset{h \to 0}{\text{Lim}} \dfrac{f(a+h)-f(a)}{h}$ is defined as the differential coefficient of f(x) at $x = a$, provided this limit exists and is finite.

The differential coefficient of f(x) with respect to x is generally written as $\dfrac{d}{dx} f(x)$ or $f'(x)$ or Df(x).

Note: Generally, $\dfrac{d}{dx} f(x) = \underset{h \to 0}{\text{Lim}} \dfrac{f(x+h)-f(x)}{h}$ is known as the first principle of differential calculus.

1.2 Differential Coefficient of a Function at a Point

The value of the derivative of f(x) obtained by putting $x = a$ is called the differential coefficient of f(x) at $x = a$ and is denoted by $f'(a)$ or $\left(\dfrac{dy}{dx} \right)_{x=a}$.

List of Formulae

1. $\dfrac{d}{dx} x^n = nx^{n-1}$

2. $\dfrac{d}{dx} \log_e x = \dfrac{1}{x}$

3. $\dfrac{d}{dx} e^x = e^x$

4. $\dfrac{d}{dx} a^x = a^x \log_e a$

5. $\dfrac{d}{dx} x = 1$

6. $\dfrac{d}{dx} \sin x = \cos x$

7. $\dfrac{d}{dx} \cos x = -\sin x$

8. $\dfrac{d}{dx} \tan x = \sec^2 x$

9. $\dfrac{d}{dx} \cot x = -\text{cosec}^2 x$

10. $\dfrac{d}{dx} \sec x = \sec x \tan x$

11. $\dfrac{d}{dx} \text{cosec } x = -\text{cosec } x \cot x$

12. $\dfrac{d}{dx} C = 0$, where C is constant

13. $\dfrac{d}{dx} \log_a x = \dfrac{1}{x} \log_a e$

Example 1: Differentiate $\dfrac{lx^2 + mx + n}{\sqrt{x}}$ w.r.t.x.

Solution:

$$y = \frac{lx^2 + mx + n}{\sqrt{x}}$$

$$y = lx^{\frac{3}{2}} + mx^{\frac{1}{2}} + nx^{\frac{-1}{2}}$$

$$\frac{dy}{dx} = l\frac{d}{dx}\, x^{\frac{3}{2}} + m\frac{d}{dx}\, x^{\frac{1}{2}} + n\frac{d}{dx}\, x^{\frac{-1}{2}}$$

$$= l.\,\frac{3}{2}\, x^{\frac{3}{2}-1} + m.\,\frac{1}{2}\, x^{\frac{1}{2}-1} + n\left(\frac{-1}{2}\right)x^{\frac{-1}{2}-1}$$

$$\frac{dy}{dx} = \frac{3l}{2}\, x^{\frac{1}{2}} + \frac{m}{2}\, x^{\frac{-1}{2}} - \frac{n}{2}\, x^{\frac{-3}{2}}$$

1.3   Differentiation from First Principl

$$\frac{dy}{dx} = \lim_{h\to 0}\frac{f(x+h)-f(x)}{h}$$

Example 2: If $y = \sqrt{x}$, find $\dfrac{dy}{dx}$ by first principle

Solution: Given $f(x) = \sqrt{x}$

by definition $\dfrac{dy}{dx} = \lim_{h\to 0}\dfrac{f(x+h)-f(x)}{h}$

$$= \lim_{h\to 0}\frac{\sqrt{x+h}-\sqrt{x}}{h}$$

$$= \lim_{h\to 0}\frac{\sqrt{x+h}-\sqrt{x}}{h} \times \frac{(\sqrt{x+h}+\sqrt{x})}{(\sqrt{x+h}+\sqrt{x})}$$

$$= \lim_{h\to 0}\frac{x+h-x}{h(\sqrt{x+h}+\sqrt{x})}$$

$$= \lim_{h\to 0}\frac{1}{\sqrt{x+h}+\sqrt{x}} = \frac{1}{2\sqrt{x}}$$

Example 3: Differentiate $\sin^2 x$ from first principle

Solution: Given $f(x) = \sin^2 x$, $f(x+h) = \sin^2(x+h)$

by definition

$$\frac{dy}{dx} = \underset{h \to 0}{Lim} \frac{\sin^2(x+h) - \sin^2 x}{h}$$

$$= \underset{h \to 0}{Lim} \frac{\sin(x+h+x)\sin(x+h-x)}{h}$$

$$(\because \sin^2 A - \sin^2 B = \sin(A+B)\sin(A-B))$$

$$= \underset{h \to 0}{Lim} \frac{\sin(2x+h)\sin h}{h}$$

$$= \underset{h \to 0}{Lim} (\sin 2x + h). \underset{h \to 0}{Lim} \frac{\sin h}{h}$$

$$= \underset{h \to 0}{Lim} \sin(2x+h) = \sin 2x$$

(i)   Differential coefficient of the product of two functions   **(product rule)**

$$\frac{d}{dx}[f_1(x).f_2(x)] = f_1(x)\frac{d}{dx}f_2(x) + f_2(x)\frac{d}{dx}f_1(x)$$

(ii)   Differential coefficient of the quotient of two functions   **(quotient rule)**

$$\frac{d}{dx}\frac{f_1(x)}{f_2(x)} = \frac{f_2(x)\dfrac{d}{dx}f_1(x) - f_1(x)\dfrac{d}{dx}f_2(x)}{[f_2(x)]^2}$$

***Example* 4:** Differentiate $x^2 \log x$ w.r.t. x.

Solution: $\dfrac{d}{dx}(x^2. \log x) = x^2 \dfrac{d}{dx}\log x + \log x \dfrac{d}{dx}x^2$

$$= x^2.\frac{1}{x} + \log x. 2x$$

$$= x + 2x \log x$$

$$= x(1 + 2\log x)$$

***Example* 5:** If $y = \dfrac{x}{x+5}$, then prove that $x\dfrac{dy}{dx} = y(1-y)$.

Solution: $y = \dfrac{x}{x+5}$ then $\dfrac{dy}{dx} = \dfrac{(x+5)\dfrac{dy}{dx}x - x\dfrac{d}{dx}(x+5)}{(x+5)^2}$

$$= \frac{(x+5).1 - x(1+0)}{(x+5)^2} = \frac{(x+5)-x}{(x+5)^2}$$

$$\frac{dy}{dx} = \frac{5}{(x+5)^2} \text{ then } x\frac{dy}{dx} = \frac{5x}{(x+5)^2}$$

$$= \frac{x}{x+5} \cdot \frac{5}{x+5} = y\left(1 - \frac{x}{x+5}\right) \Rightarrow x\frac{dy}{dx} = y(1-y)$$

1.4   Differential Coefficient of a Function of Function

If $y = f(z)$, where $z = F(x)$, then by eliminating z, we get $y = f\{F(x)\}$

e.g., $\sin x^3$, $e^{\sin x}$, $\cos 2x$ are the function of function.

Again, let $\qquad y = f(t)$, where $t = g(x)$

Then $\qquad \dfrac{dy}{dx} = \dfrac{dy}{dt} \cdot \dfrac{dt}{dx}$, this is known as '*chain rule*'.

***Example* 6:** Differentiate $\log(\log \sin x)$ w.r.t. x.

***Solution*:** Let $\qquad y = \log(\log \sin x)$

Put $\qquad \log \sin x = t$

$$y = \log t$$

$$\frac{dy}{dx} = \frac{dy}{dt} \cdot \frac{dt}{dx}$$

$$= \frac{d}{dt}\log t \cdot \frac{d}{dx}\log \sin x$$

$$= \frac{1}{t}\frac{d}{dx}\log \sin x$$

Again, put $\sin x = u$

$$\frac{dy}{dx} = \frac{1}{t} \cdot \frac{d}{du}\log u \frac{du}{dx}$$

$$= \frac{1}{t} \cdot \frac{1}{u} \cdot \frac{d}{dx}\sin x = \frac{1}{\log \sin x} \cdot \frac{1}{\sin x} \cdot \cos x$$

$$= \frac{\cot x}{\log \sin x}$$

***Example* 7:** If $y = \log\sqrt{\dfrac{1 - \cos mx}{1 + \cos mx}}$, then find $\dfrac{dy}{dx}$.

***Solution*:** $\qquad\qquad y = \log\sqrt{\dfrac{1 - \cos mx}{1 + \cos mx}}$

$$\frac{dy}{dx} = \frac{d}{dx} \log \sqrt{\frac{2\sin^2 \frac{mx}{2}}{2\cos^2 \frac{mx}{2}}}$$

$$= \frac{d}{dx} \log \tan \frac{mx}{2}$$

Put $\qquad \tan \frac{mx}{2} = t$

$$\frac{dy}{dx} = \frac{d}{dt} \log t . \frac{d}{dx} \tan \frac{mx}{2}$$

$$= \frac{1}{t} . \frac{d}{dx} \tan \frac{mx}{2}$$

Again put $\quad \dfrac{mx}{2} = u$

$$\frac{dy}{dx} = \frac{1}{\tan \frac{mx}{2}} . \frac{d}{du} \tan u . \frac{d}{dx} \frac{mx}{2}$$

$$= \frac{1}{\tan \frac{mx}{2}} \sec^2 u . \frac{m}{2}$$

$$= \frac{m}{2} \frac{1}{\tan \frac{mx}{2}} . \sec^2 \frac{mx}{2}$$

$$= \frac{m}{2} . \frac{\cos \frac{mx}{2}}{\sin \frac{mx}{2}} . \frac{1}{\cos^2 \frac{mx}{2}}$$

$$= m \operatorname{cosec} mx$$

1.5 Differential Coefficient of Inverse Trigonometric Functions or Trigonometrical Transformation

Formula   1. $\dfrac{d}{dx} \sin^{-1} x = \dfrac{1}{\sqrt{1-x^2}}$ $\qquad$ 2. $\dfrac{d}{dx} \cos^{-1} x = \dfrac{-1}{\sqrt{1-x^2}}$

$\qquad$ 3. $\dfrac{d}{dx} \tan^{-1} x = \dfrac{1}{1+x^2}$ $\qquad$ 4. $\dfrac{d}{dx} \cot^{-1} x = \dfrac{-1}{1+x^2}$

5. $\dfrac{d}{dx}\sec^{-1}x = \dfrac{1}{x\sqrt{x^2-1}}$

6. $\dfrac{d}{dx}\operatorname{cosec}^{-1}x = \dfrac{-1}{x\sqrt{x^2-1}}$

***Example* 8:** Differentiate $\sin^{-1}\left(\dfrac{2x}{1+x^2}\right)$ w.r.t. x.

***Solution*:** Let

$$y = \sin^{-1}\left(\dfrac{2x}{1+x^2}\right)$$

Put

$$x = \tan\theta \Rightarrow \theta = \tan^{-1}x$$

$$y = \sin^{-1}\left(\dfrac{2\tan\theta}{1+\tan^2\theta}\right)$$

$$= \sin^{-1}(\sin 2\theta) \qquad \left(\because \sin 2\theta = \dfrac{2\tan\theta}{1+\tan^2\theta}\right)$$

$$y = 2\theta\,,\ y = 2\tan^{-1}x$$

$$\dfrac{dy}{dx} = 2\dfrac{d}{dx}\tan^{-1}x = \dfrac{2}{1+x^2}$$

***Example* 9:** If $y = \cot^{-1}\left(\dfrac{\sqrt{1+x^2}+1}{x}\right)$, then find $\dfrac{dy}{dx}$.

***Solution*:**

$$y = \cot^{-1}\left(\dfrac{\sqrt{1+x^2}+1}{x}\right)$$

Put

$$x = \tan\theta \Rightarrow \theta = \tan^{-1}x$$

$$y = \cot^{-1}\left(\dfrac{\sqrt{1+\tan^2\theta}+1}{\tan\theta}\right)$$

$$= \cot^{-1}\left(\dfrac{\sec\theta+1}{\tan\theta}\right)$$

$$= \cot^{-1}\left(\dfrac{\dfrac{1}{\cos\theta}+1}{\dfrac{\sin\theta}{\cos\theta}}\right)$$

$$= \cot^{-1}\left(\dfrac{1+\cos\theta}{\sin\theta}\right)$$

$$= \cot^{-1}\left(\frac{2\cos^2\dfrac{\theta}{2}}{2\sin\dfrac{\theta}{2}\cos\dfrac{\theta}{2}}\right)$$

$$= \cot^{-1}\cot\frac{\theta}{2}$$

$$y = \frac{\theta}{2} \implies \qquad y = \frac{1}{2}\tan^{-1}x$$

$$\frac{dy}{dx} = \frac{1}{2}\cdot\frac{1}{1+x^2}$$

***Example* 10:** Differentiate $\sin^{-1}\left(x\sqrt{1-x} - \sqrt{x}\sqrt{1-x^2}\right)$ w.r.t.x.

***Solution*:**   Let $y = \sin^{-1}\left(x\sqrt{1-x} - \sqrt{x}\sqrt{1-x^2}\right)$

Put $x = \sin A,\ \sqrt{x} = \sin B \quad \implies \quad A = \sin^{-1}x, \qquad B = \sin^{-1}\sqrt{x}$

$$y = \sin^{-1}\left(\sin A\sqrt{1-\sin^2 B} - \sin B\sqrt{1-\sin^2 A}\right)$$

$$= \sin^{-1}\left(\sin A\cos B - \cos A\sin B\right)$$

$$= \sin^{-1}\sin(A-B)$$

$$y = A - B$$

$$y = \sin^{-1}x - \sin^{-1}\sqrt{x}$$

$$\frac{dy}{dx} = \frac{d}{dx}\sin^{-1}x - \frac{d}{dx}\sin^{-1}\sqrt{x}$$

$$= \frac{1}{\sqrt{1-x^2}} - \frac{d}{dt}\sin^{-1}t\cdot\frac{dt}{dx} \qquad\qquad \text{put } \sqrt{x} = t$$

$$= \frac{1}{\sqrt{1-x^2}} - \frac{1}{\sqrt{1-t^2}}\cdot\frac{d}{dx}\sqrt{x}$$

$$= \frac{1}{\sqrt{1-x^2}} - \frac{1}{\sqrt{1-x}}\cdot\frac{1}{2\sqrt{x}}$$

$$= \frac{1}{\sqrt{1-x^2}} - \frac{1}{2\sqrt{x-x^2}}$$

1.6   Differentiation of Implicit Functions

There may be a relation between x and y when it is not possible to express it in the form of $y = f(x)$ 'or' to solve an equation for y in terms of x, relation is called implicit function.

For $\dfrac{dy}{dx}$ by differentiating the given relation w.r.t. x

***Example* 11:** Find $\dfrac{dy}{dx}$, when $x^2 + y^2 = a^2$

Solution: Given $x^2 + y^2 = a^2$

Differentiating both the sides w.r.t. x, we get

$$\frac{d}{dx}x^2 + \frac{d}{dx}y^2 = \frac{d}{dx}a^2$$

$$2x + 2y\,\frac{dy}{dx} = 0$$

$$\frac{dy}{dx} = \frac{-2x}{2y}$$

***Example* 12:** Find $\dfrac{dy}{dx}$, when $ax^2 + 2hxy + by^2 + 2gx + 2fy + c = 0$

Solution: Given $ax^2 + 2hxy + by^2 + 2gx + 2fy + c = 0$

Differentiating both the sides w.r.t. x, we get

$$a\frac{d}{dx}x^2 + 2h\frac{d}{dx}xy + b\frac{d}{dx}y^2 + 2g\frac{d}{dx}x + 2f\frac{d}{dx}y + \frac{d}{dx}c = 0$$

$$2ax + 2h\left(x\frac{dy}{dx} + y\frac{d}{dx}x\right) + 2by\,\frac{dy}{dx} + 2g + 2f\,\frac{dy}{dx} + 0 = 0$$

$$2ax + 2hx\frac{dy}{dx} + 2yh + 2by\,\frac{dy}{dx} + 2g + 2f\,\frac{dy}{dx} = 0$$

$$(2ax + 2hy + 2g) + (2hx + 2by + 2f)\frac{dy}{dx} = 0$$

$$2\,(hx + by + f)\frac{dy}{dx} = -2\,(ax + hy + g)$$

$$\frac{dy}{dx} = \frac{-(ax + hy + g)}{(hx + by + f)}$$

1.7   Logarithmic Differentiation

If the power of a function of x is also a function of x, then it would be convenient for us if we first take logarithms w.r.t the base e and then differentiate both sides w.r.t x.

Let us recall the formulae

(i) $\log m^n = n \log m$

(ii) $\log(mn) = \log m + \log n$

(iii) $\log \dfrac{m}{n} = \log m - \log n$

***Example* 13:** If $y = x^x$, then prove that $\dfrac{dy}{dx} = x^x (1 + \log x)$

***Solution*:**   Given $y = x^x$

Taking log on both the sides, we get

$$\log y = \log x^x$$

$$\log y = x \cdot \log x$$

Differentiating both the sides w.r.t x, we get

$$\frac{d}{dx}\log y = x\frac{d}{dx}\log x + \log x\frac{d}{dx}x$$

$$\frac{1}{y}\frac{dy}{dx} = x.\frac{1}{x} + \log x.1$$

$$\frac{1}{y}\frac{dy}{dx} = 1 + \log x$$

$$\frac{dy}{dx} = y(1 + \log x)$$

$$= x^x(1 + \log x)$$

1.8   Differentiation of Parametric Equation

When variables x and y are expressed in terms of a third variable like t or θ, then this third variable is called parameter.

e.g., $x = g(t)$ and $y = h(t)$, these are called parametric equations.

Formula $\dfrac{dy}{dx} = \dfrac{dy/dt}{dx/dt}$

***Example* 14:** If $x = at^2$, $y = 2at$, then find $\dfrac{dy}{dx}$.

***Solution*:**   Given $x = at^2$, $y = 2at$

$$\frac{dx}{dt} = 2at, \qquad \frac{dy}{dt} = 2a$$

$$\frac{dy}{dx} = \frac{dy/dt}{dx/dt}$$

$$= \frac{1}{t}$$

1.9   Differentiation of Infinite Series

When the function is given in infinite series. In this case we use the fact that if a term is deleted from an infinite series it remains uneffected.

***Example* 15:** If $y = x^{x^{x^{x^{x\ldots\ldots\infty}}}}$, find $\dfrac{dy}{dx}$.

***Solution*:**   Here $y = x^y$

$\Rightarrow$ $\log y = y \log x$

Differentiating w.r.t x, we have

$$\frac{1}{y}\frac{dy}{dx} = y\frac{d}{dx}\log x + \log x \frac{dy}{dx}$$

$$= \frac{y}{x} + \log x \frac{dy}{dx}$$

$$\frac{dy}{dx}\left(\frac{1}{y} - \log x\right) = \frac{y}{x}$$

$$\frac{dy}{dx}\left(\frac{1 - y\log x}{y}\right) = \frac{y}{x}$$

$$\frac{dy}{dx} = \frac{y^2}{x(1 - y\log x)}$$

***Example* 16:** If $y = \sqrt{\sin x + \sqrt{\sin x + \sqrt{\sin x + \ldots\ldots\infty}}}$, Prove that $\dfrac{dy}{dx} = \dfrac{\cos x}{2y - 1}$

***Solution*:**   Here $y = \sqrt{\sin x + y}$

$$y^2 = \sin x + y$$

$$2y\frac{dy}{dx} = \cos x + \frac{dy}{dx}$$

$$\frac{dy}{dx} = \frac{\cos x}{2y - 1}$$

***Example* 17:** If $y = e^{x + e^{x + e^x + \ldots\ldots \infty}}$, show that $\dfrac{dy}{dx} = \dfrac{y}{1-y}$

***Solution*:**   Here $y = e^{x+y}$

Taking log on both sides

$$\log y = (x + y) \log e \qquad\qquad (\because \log e = 1)$$

$$\log y = x + y$$

Differentiating w.r.t x, we get

$$\frac{1}{y}\frac{dy}{dx} = 1 + \frac{dy}{dx}$$

$$\frac{dy}{dx}\left(\frac{1}{y} - 1\right) = 1$$

$$\frac{dy}{dx}\left(\frac{1-y}{y}\right) = 1$$

$$\frac{dy}{dx} = \frac{y}{1-y}$$

***Example* 18:** If $y = \sqrt{x + \sqrt{x + \sqrt{x + \ldots\ldots \infty}}}$ then prove that $\dfrac{dy}{dx} = \dfrac{1}{2y-1}$

***Solution*:**   Here $y = \sqrt{x + y}$ .....(1)

Taking log on both sides

$$\log y = \frac{1}{2}\log(x + y)$$

Differentiating w.r.t x, we get

$$\frac{1}{y}\frac{dy}{dx} = \frac{1}{2}\cdot\frac{1}{x+y}\left(1 + \frac{dy}{dx}\right)$$

$$\left[\frac{1}{y} - \frac{1}{2(x+y)}\right]\frac{dy}{dx} = \frac{1}{2(x+y)} \qquad\qquad .....(2)$$

From eq. (1) $x + y = y^2$

Putting in eq. (2) we get

$$\left(\frac{1}{y} - \frac{1}{2y^2}\right)\frac{dy}{dx} = \frac{1}{2y^2}$$

$$\left(\frac{2y-1}{2y^2}\right)\frac{dy}{dx} = \frac{1}{2y^2}$$

$$\frac{dy}{dx} = \frac{1}{2y-1}$$

1.10   Successive Differentiation

Differentiating the first derivative of a function we obtain second derivative, differentiating the second derivative, we get the third derivative and so on. This process of finding derivatives of higher order is known as successive differentiation.

***Example* 19:** If $y = x + \tan x$, Prove that $\cos^2 x \dfrac{d^2y}{dx^2} + 2x = 2y$

***Solution*:**   Given $y = x + \tan x$

Differentiating w.r.t x, we get

$$\frac{dy}{dx} = 1 + \sec^2 x$$

Differentiating again, we get

$$\frac{d}{dx}\left(\frac{dy}{dx}\right) = 0 + 2\sec x \cdot \sec x \tan x$$

$$\frac{d^2y}{dx^2} = 2\sec^2 x \tan x$$

$$= \frac{2\tan x}{\cos^2 x}$$

$$\cos^2 x\,\frac{d^2y}{dx^2} = 2\tan x$$

$$\cos^2 x\,\frac{d^2y}{dx^2} = 2(y - x)$$

$$\cos^2 x\,\frac{d^2y}{dx^2} = 2y - 2x$$

$$\cos^2 x\,\frac{d^2y}{dx^2} + 2x = 2y$$

***Example* 20:** If $y = ae^{mx} + be^{-mx}$, show that $\dfrac{d^2y}{dx^2} = m^2 y$

***Solution*:**   Let $y = ae^{mx} + be^{-mx}$

differentiating w.r.t. x, we get

$$\frac{dy}{dx} = am\,e^{mx} - bm\,e^{-mx}$$

differentiating again w.r.t. x, we have

$$\frac{d^2y}{dx^2} = am^2 e^{mx} + bm^2 e^{-mx}$$

$$= m^2 (ae^{mx} + be^{-mx})$$

$$\frac{d^2y}{dx^2} = m^2y$$

***Example* 21:** If y = A cosnx + B sin nx, show that $\dfrac{d^2y}{dx^2} + n^2y = 0$

***Solution*:**   Let   y = A cosnx + B sin nx

differentiating w.r.t. x

$$\frac{dy}{dx} = -\,An\sin nx + Bn\cos nx$$

differentiating again w.r.t. x

$$\frac{d^2y}{dx^2} = -An^2 \cos nx - Bn^2 \sin nx$$

$$= -n^2 (A \cos nx + B \sin nx)$$

$$= -n^2 y$$

$$\frac{d^2y}{dx^2} + n^2y = 0$$

***Example* 22:** If y – A cos (log x) + B sin (log x) prove that $x^2 \dfrac{d^2y}{dx^2} + x\dfrac{dy}{dx} + y = 0$

***Solution*:**   Let   y = A cos (log x) + B sin (log x)

differentiating w.r.t. x, $\dfrac{dy}{dx} = -\,A \sin (\log x).\dfrac{1}{x} + B \cos (\log x).\dfrac{1}{x}$

$$x\,\frac{dy}{dx} = -\,A \sin (\log x) + B \cos (\log x)$$

differentiating again w.r.t. x, we have

$$x\,\frac{d^2y}{dx^2} + \frac{dy}{dx} = -\,A \cos(\log x).\frac{1}{x} - B \sin (\log x).\frac{1}{x}$$

$$x^2\frac{d^2y}{dx^2} + x\,\frac{dy}{dx} = -\,[A \cos(\log x) + B \sin (\log x)]$$

$$= -y$$

$$\therefore \qquad x^2 \frac{d^2y}{dx^2} + x\, \frac{dy}{dx} + y = 0$$

***Example* 23:** If $y = Ae^{px} + Be^{qx}$ show that $\dfrac{d^2y}{dx^2} - (p+q)\dfrac{dy}{dx} + pqy = 0$

***Solution*:**   Let $y = Ae^{px} + Be^{qx}$

differentiating w.r.t. x,

$$\frac{dy}{dx} = Ape^{px} + Bqe^{qx} \qquad\qquad \ldots\ldots(i)$$

differentiating again w.r.t. x, we have

$$\frac{d^2y}{dx^2} = Ap^2e^{px} + Bq^2e^{qx} \qquad\qquad \ldots\ldots(ii)$$

Now $\dfrac{d^2y}{dx^2} - (p+q)\,\dfrac{dy}{dx} + pqy$

$$= Ap^2e^{px} + Bq^2e^{qx} - (p+q)(Ape^{px} + Bqe^{qx}) + pq(Ae^{px} + Be^{qx})$$

$$= Ap^2e^{px} + Bq^2e^{qx} - Ap^2e^{px} - Bq^2e^{qx} - Apqe^{px} - Bpqe^{qx} + pqAe^{px} + pqBe^{qx} = 0$$

Practice Problems

Find the differential coefficient of the following

1. $\sqrt{x^{-3}},\ 9x^{-8}$

2. $5x^3 - 4x^2 + 3x + 1$

3. $\dfrac{(x-4)(x+2)}{x}$

4. $\dfrac{(3x-1)(4x^{\frac{1}{2}}+5)}{x^{\frac{1}{2}}}$

5. $\dfrac{2x^3 + 6x^2 + 7}{\sqrt{x}}$

6. $x^5 - 2\log x$

7. $e^x + x^n$

8. $(x\sin x + 7x^2 - 3)\,.\dfrac{1}{x}$

9. $\dfrac{a + b\sqrt{x}}{\sqrt[3]{x}}$

10. $\dfrac{x\,e^x - 1}{x}$

11. $\dfrac{x^2\log x - x + x^2.e^x}{x^2}$

12. $3e^x + 7\log x + x^{-3}$

13. $e^{ax}\sin bx$

14. $3\log x \sin x$

15. $2\sqrt{x}\log x$

16. $ae^x\sin x + bx^n\cos x$

17. $(x^2 + 1) (e^x + 2 \sin x)$

18. $\sqrt{x} \, (1 - x) (x^2 + 2)$

19. $(1 - 2 \tan x) (5 + 4 \sin x)$

20. $\dfrac{2x^3 + 6x + 9}{x^2 + 5}$

21. $\dfrac{x^n}{\log_a x}$

22. $\dfrac{\sin x - x \cos x}{x \sin x + \cos x}$

23. $\dfrac{x \cdot \cos x}{\log x}$

24. $\dfrac{1 - \tan x}{1 + \tan x}$

25. $\dfrac{\sqrt{x} + \sqrt{a}}{\sqrt{a} - \sqrt{x}}$

26. $\dfrac{e^x + \tan x}{\cot x - x^n}$

27. $\dfrac{e^x + 5x^3 \log x}{\log x}$

28. $\dfrac{1}{(x + a) (x + b) (x + c)}$

29. $\dfrac{1 + \log x}{1 - \log x}$

30. $\left(\sqrt{x} + \dfrac{1}{\sqrt{x}} \right)^2$

31. $(2x - 3) \sqrt{7 + x^2}$

32. $\sqrt{\dfrac{1 - x}{1 + x}}$

33. $\dfrac{1}{x + \sqrt{1 + x^2}}$

34. $\dfrac{a^2 + x^2}{\sqrt{a^2 - x^2}}$

35. $\tan x^3$

36. $\log \tan x$

37. $\sin (\tan x)$

38. $\sqrt{\log x}$

39. $e^{ax + b}$

40. $\log \sqrt{\sin e^x}$

41. $\log \left(\dfrac{ax + b}{px + q} \right)$

42. $\dfrac{\sqrt{\sin x}}{\sin \sqrt{x}}$

43. $\cot (x^2 \sin 2x)$

44. $\sqrt{\dfrac{1 + \sin x}{1 - \sin x}}$

45. $\log \sqrt{\dfrac{1 + \sin x}{1 - \sin x}}$

46. $\log \dfrac{x + \sqrt{x^2 - 1}}{x - \sqrt{x^2 - 1}}$

47. $\log \left(\sqrt{x + 1} + \sqrt{x - 1} \right)$

48. $\sqrt{\operatorname{cosec} x \, \cot x}$

49. $\log \log \log x$

50. $\dfrac{e^{\sin x}}{\sin x^n}$

51. $\sqrt{\sec \sqrt{x}}$

52. If $y = \tan^{-1} \dfrac{\sqrt{1 - x^2}}{x}$, find $\dfrac{dy}{dx}$

53. Differentiate $\sin^{-1}(3x - 4x^3)$

54. If $y = \tan^{-1}\sqrt{\dfrac{1-\cos x}{1+\cos x}}$, find $\dfrac{dy}{dx}$ differentiate the following:

55. $\sin^{-1}\dfrac{2x}{1+x^2}$

56. $\tan^{-1}\dfrac{x}{\sqrt{1+x^2}}$

57. $\sin^{-1}\left(\dfrac{1-x^2}{1+x^2}\right)$

58. $\cos^{-1}\left(\dfrac{1-x^2}{1+x^2}\right)$

59. $2\tan^{-1}\sqrt{\dfrac{1-x}{1+x}}$

60. $\sin^{-1}\left(2x\sqrt{1-x^2}\right)$

61. $\cos^{-1}(4x^3 - 3x^2)$

62. $\tan^{-1}\left(\dfrac{a+x}{1-ax}\right)$

63. $\cos^{-1}\left(\dfrac{x-x^{-1}}{x+x^{-1}}\right)$

64. $\sin^{-1}\left(x\sqrt{1-x} - \sqrt{x}\sqrt{1-x^2}\right)$

65. If $y = 1 + \dfrac{x}{1!} + \dfrac{x^2}{2!} + \dfrac{x^3}{3!} + \ldots$ Then show that $\dfrac{dy}{dx} = y$

66. $\dfrac{3^x}{x + \tan x}$

67. $\sin x^\circ$ $\left(\text{Hint}: x^\circ = \dfrac{\pi}{180}\right)$

Find the Differential coefficient of the following functions by First Principle

68. x^7

69. $\cos(ax+b)$

70. $\tan ax$

71. $\cos^2 x$

72. $\cos^{-1} x$

73. e^{3x}

74. $\sin x^2$

75. $\cos\sqrt{x}$

76. $\tan 3x$

77. $x^{\frac{-3}{2}}$

78. If $\sin y = x\sin(a+y)$, prove that $\dfrac{dy}{dx} = \dfrac{\sin^2(a+y)}{\sin a}$

79. If $\dfrac{1}{2}(e^y - e^{-y}) = x$, prove that $\dfrac{dy}{dx} = \dfrac{1}{\sqrt{1+x^2}}$

80. If $y = \sqrt{\sin x + \sqrt{\sin x + \sqrt{\sin x + \ldots\infty}}}$ prove that $(2y-1)\dfrac{dy}{dx} = \cos x$.

81. If $y\sqrt{1-x^2} + x\sqrt{1-y^2} = 1$, prove that $\dfrac{dy}{dx} + \sqrt{\dfrac{1-y^2}{1+x^2}} = 0$

Find $\dfrac{dy}{dx}$ for implicit differentiation when;

82. $xy = c^2$

83. $\dfrac{x^2}{a^2} + \dfrac{y^2}{b^2} = 1$

84. $x^{\frac{2}{3}} + y^{\frac{2}{3}} = a^{\frac{2}{3}}$

85. $xy^3 - yx^3 = x$

86. $y \sec x + \tan x + x^2 y = 0$

87. $\tan(x + y) + \tan(x - y) = 1$

88. If $(x + y)^{m+n} = x^m y^n$, prove that $\dfrac{dy}{dx} = \dfrac{y}{x}$

89. Find the value of $\dfrac{dy}{dx}$ at $(2, -2)$ for the curve $x^2 + xy + y^2 = 4$

90. If $y = \sqrt{\log x \sqrt{\log x + \sqrt{\log x + ...\infty}}}$ show that $(2y - 1)\dfrac{dy}{dx} = \dfrac{1}{x}$

91. If $y = a^{x+y}$, show that $\dfrac{dy}{dx} = \dfrac{y \log a}{1 - y \log a}$

92. If $\cos(xy) = x$, show that $\dfrac{dy}{dx} + \dfrac{1 + y \sin(xy)}{x \sin(xy)} = 0$

93. If $x = y \log(xy)$, find $\dfrac{dy}{dx}$

94. If $x^3 + y^3 = \sin(x + y)$, find $\dfrac{dy}{dx}$

95. If $x^3 y^2 = \log(x + y) + \sin e^x$, find $\dfrac{dy}{dx}$

96. If $x^a . y^b = (x + y)^{a+b}$, prove that $\dfrac{dy}{dx} = \dfrac{y}{x}$ provided that $ay \neq bx$

Find $\dfrac{dy}{dx}$ of the following functions;

97. $(\log x)^x$

98. $(2x + 3)^{x-5}$

99. $(\cos x)^{\log x}$

100. $e^{x^{x^x}}$

101. $(2 - x)\left(\dfrac{3 - x}{1 + x}\right)^{\frac{1}{2}}$

102. $x^2 e^x \sin x$

103. $x \sin^{-1} x + (\log x)^x$

104. $(\tan x)^{\cot x} + (\cot x)^{\sin x}$

105. $\dfrac{1}{(x + a)(x + b)(x + c)}$

106. $\tan x \tan 2x \tan 3x \tan 4x$

107. $x^{x^{x^x}}$

108. $y = x^{1 + x + x^2}$

109. $\dfrac{(x + 1)(x + 2)(x + 3)}{(x - 1)(x - 2)(x - 3)}$

110. $\dfrac{(x+2)^{\frac{5}{2}}}{(x+4)^{\frac{1}{3}}(x+1)^{\frac{3}{2}}}$ Find $\dfrac{dy}{dx}$ for the following parametric equations;

111. $x = a \sec^2\theta,\ y = b \tan^2\theta$

112. $x = a \sin^3\theta,\ y = a \cos^3\theta$

113. $x = \sqrt{\sin 2\theta},\ y = \sqrt{\cos 2\theta}$

114. $x = at^4,\ y = bt^3$

115. $x = \dfrac{3at}{1+t^3},\ y = \dfrac{3at^2}{1+t^3}$

116. $x = \sin^{-1}\dfrac{2u}{1+u^2},\ y = \tan^{-1}\dfrac{2u}{1-u^2}$

117. $x = \log t,\ y = e^t + \cos t$

Find the derivative of the following functions

118. $\sin x^2$ w.r.t. x^2

119. $x^{\sin x}$ w.r.t. $(\sin x)^x$

120. $e^{\tan x}$ w.r.t. $\tan x$

121. $\tan^{-1}\dfrac{\sqrt{1+x^2}-1}{x}$ w.r.t. $\tan^{-1} x$

122. If $x = e^{\cos 2t}$ and $y = e^{\sin 2t}$, prove that $\dfrac{dy}{dx} = \dfrac{-y \log x}{x \log y}$

123. If $x = ae^t (\sin t + \cos t),\ y = ae^t (\sin t - \cos t)$ prove that $\dfrac{dy}{dx} = \tan t$

124. If $x = a\left(t+\dfrac{1}{t}\right)$ and $y = a\left(t-\dfrac{1}{t}\right)$ prove that $\dfrac{dy}{dx} = \dfrac{x}{y}$.

125. Find second derivative of $ax^3 + bx^2 + cx + d$

126. If $y = a \cos px + b \sin px$, prove that $\dfrac{d^2y}{dx^2} + p^2 y = 0$

127. If $y = \log (1 + \cos x)$ then prove that $y_1 y_2 + y_3 = 0$

$$\left[\text{Hint}:\ y_1 = \dfrac{dy}{dx},\ y_2 = \dfrac{d^2y}{dx^2},\ y_3 = \dfrac{d^3y}{dx^3}\right]$$

128. If $y = \tan x + \sec x$ then prove that $\dfrac{d^2y}{dx^2} = \dfrac{\cos x}{(1-\sin x)^2}$

129. If $y = x^2 \log x$ then prove that $\dfrac{d^3y}{dx^3} = \dfrac{2}{x}$

130. If $y = e^{ax} \sin bx$ then prove that $y_2 - 2ay_1 + (a^2 + b^2)y = 0$

131. If $y = e^{m \cos^{-1} x}$ then prove that $(1-x^2) y_2 - xy_1 - m^2 y = 0$

132. If $y = \sin (\sin x)$ then prove that $y_2 + y_1 \tan x + y \cos^2 x = 0$

133.   If $y = t^2 - 1$, $x = 2t$ then prove that $\dfrac{d^2y}{dx^2} = \dfrac{1}{2}$

134.   If $y = a \cos (\log x) + b \sin (\log x)$ then prove that $x^2 y_2 + x y_1 + y = 0$

135.   If $y = a \sin (\log x)$ then prove that $x^2 y_2 + x y_1 + y = 0$

136.   If $y = (\sin^{-1} x)^2$ then prove that $(1-x^2)\dfrac{d^2y}{dx^2} - x\dfrac{dy}{dx} = 2$

137.   If $y = x + \tan x$ then prove that $\cos^2 x\, \dfrac{d^2y}{dx^2} - 2y + 2x = 0$

138.   If $y = \dfrac{\sin^{-1}x}{\sqrt{1-x^2}}$ then prove that $(1-x^2)\dfrac{d^2y}{dx^2} - 3x\dfrac{dy}{dx} - y = 0$

139.   If $y = Ae^{-kt}\cos (pt + c)$ then show that $\dfrac{d^2y}{dt^2} - 2k\dfrac{dy}{dt} - n^2 y = 0$,

         where $n^2 = p^2 + k^2$.

140.   If $y = e^{\tan^{-1} x}$ then prove that $(1 - x^2)\, y_2 - x y_1 - 2 = 0$

141.   If $y = \dfrac{\log x}{x}$, then prove that $\dfrac{d^2y}{dx^2} + \dfrac{2 \log x - 3}{x^3}$

142.   If $y = \tan^{-1} x$ then prove that $\dfrac{d^2y}{dx^2} + \dfrac{2x}{(1+x^2)^2} = 0$

143.   If $y = \sin (x + y)$ then prove that $\dfrac{d^2y}{dx^2} + y\left(1 + \dfrac{dy}{dx}\right)^3 = 0$

144.   If $\dfrac{x^2}{a^2} + \dfrac{y^2}{b^2} = 1$, then show that $\dfrac{d^2y}{dx^2} = \dfrac{-b^4}{a^2 y^3}$.

Answers

1. $\dfrac{-3}{2}x^{\frac{-5}{2}}$, $-72x^{-9}$ 2. $15x^2 - 8x + 3$

3. $1 + \dfrac{8}{x^2}$ 4. $12 + \dfrac{15}{2}x^{\frac{-1}{2}} + \dfrac{5}{2}x^{\frac{-3}{2}}$

5. $5x^{\frac{3}{2}} + 9x^{\frac{1}{2}} - \dfrac{7}{2x^{\frac{3}{2}}}$ 6. $5x^4 - \dfrac{2}{x}$

7. $e^x + nx^{n-1}$ 8. $\cos x + 7 + \dfrac{3}{x^2}$

9. $\dfrac{-3}{2}ax^{\frac{-5}{2}} - \dfrac{b}{x^2}$

10. $e^x + \dfrac{1}{x^2}$

11. $\dfrac{x + x^2 e^x + 1}{x^2}$

12. $3e^x + \dfrac{7}{x} - 3x^{-4}$

13. $e^{ax}(a \sin bx + b \cos bx)$

14. $3\left(\dfrac{1}{x}\sin x + \cos x \log x\right)$

15. $\dfrac{1}{\sqrt{x}}(2 + \log x)$,   16. $ae^x(\cos x + \sin x) + bx^{n-1}(-x \sin x + n \cos x)$

17. $e^x(x^2 + 2x + 1) + 2\cos x\,(x^2 + 1) + 4x \sin x$

18. $2x^{\frac{3}{2}}(1 - x) + \dfrac{1}{2\sqrt{x}}(1 - 3x)(x^2 + 2)$

19. $4(\cos x - 2 \sin x) - 2 \sec^2 x\,(5 + 4 \sin x)$

20. $\dfrac{2(x^4 + 12x^2 - 9x + 15)}{(x^2 + 5)^2}$

21. $\tan\dfrac{x}{2}\sec^2\dfrac{x}{2}$

22. $\dfrac{nx^{n-1}\log_a x - x^{n-1}\log_a e}{(\log_a x)^2}$

23. $\dfrac{x^2}{(x \sin x + \cos x)^2}$

24. $\dfrac{\log x\,(\cos x - x \sin x) - \cos x}{(\log x)^2}$

25. $\dfrac{-2}{1 + \sin 2x}$

26. $\dfrac{\sqrt{a}}{\sqrt{x}\left(\sqrt{a} - \sqrt{x}\right)^2}$

27. $\dfrac{(e^x + \sec x^2)(\cot x - x^n) + (\operatorname{cosec}^2 x + nx^{n-1})(e^x + \tan x)}{(\cot x - x^n)^2}$

28. $\dfrac{(e^x + 15x^2 \log x + 5x^2)\log x - x^{-1}(e^x + 5x^3 \log x)}{(\log x)^2}$

29. $\dfrac{3x^2 + 2(a + b + c)x + (ab + bc + ca)}{(x + a)^2(x + b)^2(x + c)^2}$

30. $\dfrac{2}{x(1 - \log x)^2}$

31. $\left(\sqrt{x} + \dfrac{1}{\sqrt{x}}\right)\left(\dfrac{1}{\sqrt{x}} - \dfrac{1}{x\sqrt{x}}\right)$

32. $x(2x - 3)(7 + x^2)^{\frac{-1}{2}} + 2\sqrt{7 + x^2}$

33. $-\dfrac{1}{(1 + x)(1 - x^2)^{\frac{1}{2}}}$

34. $-\dfrac{1}{(x + \sqrt{1 + x^2})\sqrt{1 + x^2}}$

35. $\dfrac{x(3a^2 - x^2)}{(a^2 - x^2)^{\frac{3}{2}}}$

36. $3x^2 \sec^2 x^3$

37. $\cot x . \sec^2 x$

38. $\cos(\tan x)\sec^2 x$

39. $\dfrac{1}{2x \sqrt{\log x}}$

40. ae^{ax+b}

41. $\dfrac{e^x \cose^x}{2 \sin e^x}$

42. $\dfrac{a}{ax+b} - \dfrac{p}{px+q}$

43. $\dfrac{\sqrt{x} \cos x \sin \sqrt{x} - \sin x \cos \sqrt{x}}{2\sqrt{x} \sqrt{\sin x} \sin^2 \sqrt{x}}$

44. $- \cosec^2 (x^2 \sin 2x) (2x \sin 2x + 2x^2 \cos 2x)$

45. $\dfrac{1}{1 - \sin x}$

46. $\sec x$

47. $\dfrac{2}{\sqrt{x^2 - 1}}$

48. $\dfrac{1}{2\sqrt{x^2 - 1}}$

49. $\dfrac{- \cosec x (\cosec^2 x + \cot^2 x)}{2\sqrt{\cosec x . \cot x}}$

50. $\dfrac{1}{x \log x . \log \log x}$

51. $\dfrac{\sin x^n . e^{\sin x} . \cos x - e^{\sin x} . \cos x^n . nx^{n-1}}{\sin^2 x^n}$

52. $\dfrac{\sec \sqrt{x} \tan \sqrt{x}}{4\sqrt{x} \sqrt{\sec \sqrt{x}}}$

53. $\dfrac{-1}{\sqrt{1 - x^2}}$

54. $\dfrac{3}{\sqrt{1 - x^2}}$

55. $\dfrac{1}{2}$

56. $\dfrac{2}{1 + x^2}$

57. $\dfrac{1}{(1 + 2x^2) \sqrt{1 + x^2}}$

58. $\dfrac{-2}{1 + x^2}$

59. $\dfrac{2}{1 + x^2}$

60. $\dfrac{-1}{\sqrt{1 - x^2}}$

61. $\dfrac{2}{\sqrt{1 + x^2}}$

62. $\dfrac{-3}{\sqrt{1 - x^2}}$

63. $\dfrac{1}{1 + x^2}$

64. $\dfrac{-2}{1 + x^2}$

65. $\dfrac{1}{\sqrt{1 - x^2}} - \dfrac{1}{2\sqrt{x - x^2}}$

67. $\dfrac{3^x \left[(x + \tan x) \log 3 - (1 + \sec^2 x) \right]}{(x + \tan x)^2}$

68. $\dfrac{\pi}{180} \cos x^\circ$

69. $7x^6$

70. $- a \sin (ax + b)$

71. $a \sec^2 ax$

72. $-\sin 2x$

73. $\dfrac{-1}{\sqrt{1-x^2}}$

74. $3e^{3x}$

75. $2x \cos x^2$

76. $\dfrac{-\sin \sqrt{x}}{2\sqrt{x}}$

77. $3 \sec^2 3x$

78. $\dfrac{-3}{2} x^{\frac{-5}{2}}$

83. $\dfrac{-y}{x}$

84. $\dfrac{-b^2}{a^2} \cdot \dfrac{x}{y}$

85. $-\left(\dfrac{y}{x}\right)^{\frac{1}{3}}$

86. $\dfrac{1+3x^2y-y^3}{3xy^2-x^3}$

87. $\dfrac{-y \sec x \tan x + \sec^2 x + 2xy}{(\sec x + x^2)}$

88. $\dfrac{\sec^2(x+y)+\sec^2(x-y)}{\sec^2(x-y)-\sec^2(x+y)}$

90. $\dfrac{-2x+y}{x+2y}, \ 1$

94. $\dfrac{x-y}{x(1+\log xy)}$

95. $\dfrac{\cos(x+y)-3x^2}{3y^2-\cos(x+y)}$

96. $\dfrac{\dfrac{1}{x+y}+e^x \cose^x - 3x^2y^2}{2x^3y - \dfrac{1}{x+y}}$

98. $(\log x)^x \left[\log \log x + \dfrac{1}{\log x} \right]$

99. $(2x+3)^{x-5} \left[\dfrac{2(x-5)}{2x+3} + \log(2x+3) \right]$

100. $(\cos x)^{\log x} \left[\dfrac{1}{x} . \log \cos x - \tan x \log x \right]$

101. $e^{x^x} . x^x (1+\log x)$

102. $\dfrac{x^2-7}{(1+x)^{\frac{3}{2}}\sqrt{3-x}}$

103. $x^2 e^x \sin x \left[\dfrac{2}{x} + 1 + \cot x \right]$

104. $x^{\sin^{-1}x} \left[\dfrac{\sin^{-1} x}{x} + \dfrac{\log x}{\sqrt{1-x^2}} \right] + (\log x)^x \left[\dfrac{1}{\log x} + \log \log x \right]$

105. $(\tan x)^{\cot x} \cosec^2 x (1 - \log \tan x) + (\cot x)^{\tan x} . \sec^2 x (\log \cot x - 1)$

106. $\dfrac{-1}{(x+a)(x+b)(x+c)} \left[\dfrac{1}{x+a} + \dfrac{1}{x+b} + \dfrac{1}{x+c} \right]$

107. $2 \tan x \tan 2x \tan 3x \tan 4x (\cosec 2x + 2 \cosec 4x + 3 \cosec 6x + 4 \cosec 8x)$

108. $x^{x^x}\left[x^x(1+\log x \log x + x^{x-1}\right]$

109. $x^{1+x+x^2}\left[(2x+1)\log x + \dfrac{1}{x} + 1 + x\right]$

110. $\dfrac{(x+1)(x+2)(x+3)}{(x-1)(x-2)(x-3)}\left[\dfrac{2}{1-x^2} + \dfrac{4}{4-x^2} + \dfrac{6}{9-x^2}\right]$

111. $\dfrac{(x+2)^{\frac{5}{2}}}{(x+4)^{\frac{1}{3}}(x+1)^{\frac{3}{2}}}\left[\dfrac{5}{2(x+2)} - \dfrac{1}{3(x+4)} - \dfrac{3}{2(x+1)}\right]$

112. $\dfrac{b}{a}$

113. $-\cot\theta$

114. $-\left(\tan 2\theta\right)^{\frac{3}{2}}$

115. $\dfrac{3b}{4at}$

116. $\dfrac{t(2-t^3)}{1-2t^3}$

117. 1

118. $t(e^t - \sin t)$

119. $\cos x^2$

120. $\dfrac{x^{\sin x}\left(\dfrac{\sin x}{x} + \cos x \log x\right)}{(\sin x)^x(x\cot x + \log \sin x)}$

121. $\dfrac{e^{\tan x}}{\cos^3 x}$

122. $\dfrac{1}{2}$

126. $6ax + 2$

CHAPTER – 2

Rolle's and Lagrange's Theorem

2.1 Rolle's Theorem

If $f(x)$ is (i) continuous in $[a, b]$, (ii) differentiable in (a, b) and (iii) $f(a) = f(b)$, then there exists $c \in (a, b)$ such that $f'(c) = 0$.

Proof**:** Suppose

(i) $f(x)$ is a constant function throughout the interval $[a,b]$, then $f'(x) = 0 \ \forall x \in (a,b)$. Hence theorem is proved.

(ii) $f(x)$ is not a constant function in $[a ,b]$. As $f(x)$ is continuous in $[a,b]$, there exists a maximum value say, at 'c' $(a \le c \le b)$ and a minimum value , say at 'd' $(a \le d \le b)$ for $f(x)$ in (a, b).

$f(c) \ne f(d)$ and at least one of them is different from $f(a) = f(b)$

Suppose $f(c) \ne f(a)$ and 'c + h' be a point in the neighborhood of 'c'

then $\qquad \dfrac{f(c + h) - f(c)}{h} \le 0$ when $h > 0$. $\hfill .....(1)$

and $\qquad \dfrac{f(c + h) - f(c)}{h} \ge 0$ when $h < 0$. $\hfill .....(2)$

further $f(x)$ is differentiable in (a , b).

$$\therefore f'(c) = \underset{h \to 0}{Lt} \frac{f(c + h) - f(c)}{h} . \text{ As } h \to 0 \text{ then from (1) and (2) we get}$$

$$\underset{h \to 0}{Lt} \frac{f(c + h) - f(c)}{h} \le 0 \quad \text{and} \quad \underset{h \to 0}{Lt} \frac{f(c + h) - f(c)}{h} \ge 0 \quad \text{respectively.}$$

i.e., $f'(c) \le 0$ and $f'(c) \ge 0$ simultaneously $\Rightarrow f'(c) = 0$

Similarly the theorem can be proved when $f(d) \ne f(a)$

2.2 Geometrical Interpretation of Rolle's Theorem

Let P and Q be two points on the curve y = f(x). AP = BQ ordinates f(a) = f(b) and the curve is continuous from P to Q. It can be shown that there is at least one point on the curve y = f(x) between x = a and x = b at which the tangent to the curve is parallel to x-axis.

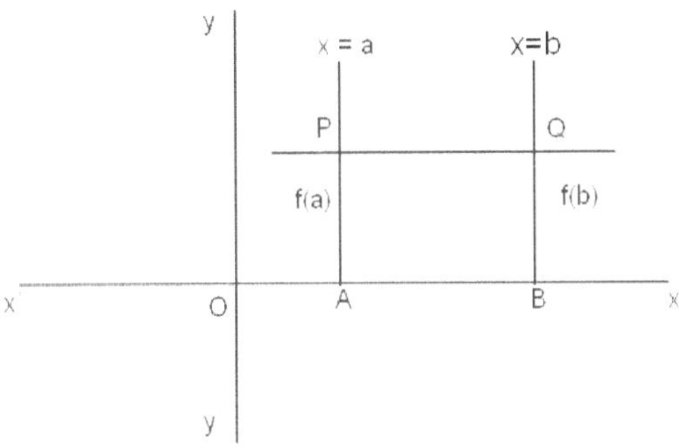

f(x) is a constant function

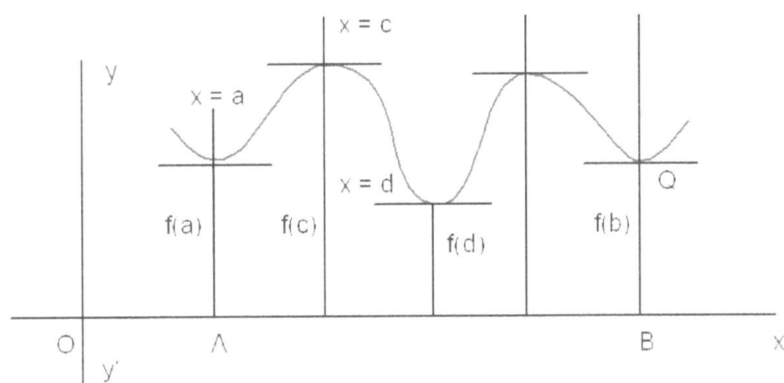

f(c) and f(d) are both different from f(a) = f(b)

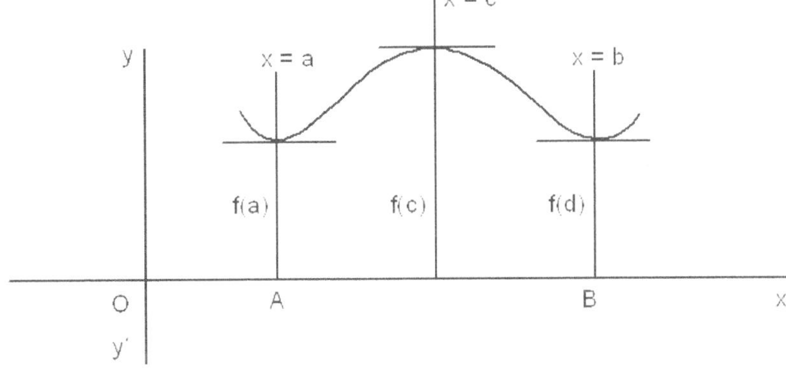

f(c) ≠ f(a) and f(d) = f(a) = f(b)

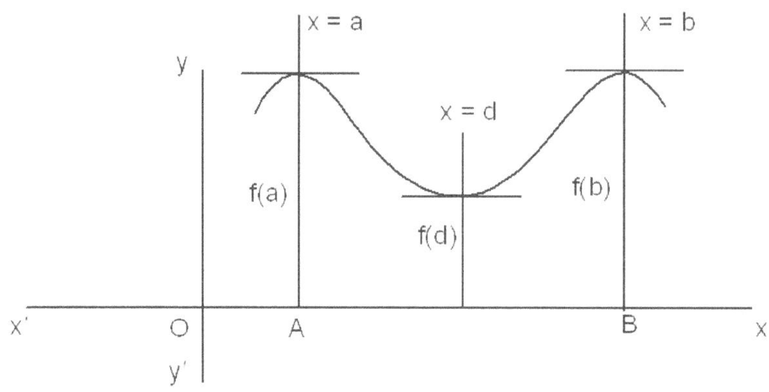

$f(d) \neq f(a)$ and $f(c) = f(a) = f(b)$

2.3   Algebraic Meaning of Rolle's Theorem

Let $f(t)$ be a polynomial with a and b are its roots. Since a polynomial function is everywhere continuous and differentiable a and b are roots of $f(t)$ therefore $f(a) = f(b) = 0$. So, $c \in (a,b)$ such that $f'(c) = 0$ that is $f'(t) = 0$ at $t = c$.

Hence in one sentence the algebraic meaning of Rolle's theorem is "Between any two roots of a polynomial $f(t)$ there is always a root of its derivative.

Some well known Results

1.   A polynomial function is everywhere continuous and differentiable.

2.   The exponential function, sine and cosine functions are everywhere continuous and differentiable.

3.   Logarithmic function is continuous and differentiable in its domain.

4. $\tan x$ is not continuous at $x = \pm\dfrac{\pi}{2}, \pm\dfrac{3\pi}{2}, \pm\dfrac{5\pi}{2} \dots \dots$

5. $|x|$ is not differentiable at $x = 0$.

6.   If $f'(x) = \pm\infty$ as $x \to k$, then $f(x)$ is not differentiable at $x = k$. For example   if $f(x) = (2x-1)^{1/2}$, then $f'(x) = \dfrac{1}{\sqrt{2x-1}}$ is such that $\underset{x \to \left(\frac{1}{2}\right)}{\mathrm{Lim}}{+}\, f'(x) = \infty$. So, $f(x)$ is not

   differentiable at $x = \dfrac{1}{2}$.

7.   The sum, difference, product and quotient of continuous (differentiable) functions is continuous (differentiable).

Illustrative Examples

***Example* 1:** Verify Rolle's theorem for $f(x) = \log\left\{\dfrac{x^2 + ab}{x(a+b)}\right\}$ in the interval (a, b).

Solution: Consider the function $f(x) = \log\left\{\dfrac{x^2 + ab}{x(a+b)}\right\}$ in the interval (a, b)

$$f'(x) = \underset{h \to 0\pm}{\text{Lt}} \frac{f(x+h) - f(x)}{h} = \underset{h \to 0\pm}{\text{Lt}} \frac{1}{h}\left[\log\frac{(x+h)^2 + ab}{(a+b)(x+h)} - \log\frac{x^2 + ab}{(a+b)x}\right]$$

$$= \underset{h \to 0\pm}{\text{Lt}} \frac{1}{h}\left[\log\frac{(x^2 + ab) + 2xh + h^2}{(x^2 + ab)} - \log\frac{x+h}{x}\right]$$

$$= \underset{h \to 0\pm}{\text{Lt}} \frac{1}{h}\left[\log\left\{1 + \frac{2xh + h^2}{x^2 + ab}\right\} - \log\left\{1 + \frac{h}{x}\right\}\right] = \underset{h \to 0\pm}{\text{Lt}} \left[\frac{2x}{x^2 + ab} - \frac{1}{x} + 0(h)\right]$$

where $0(h)$ indicates terms of order h and higher powers of h.

$$f'(x) = \frac{2x}{x^2 + ab} - \frac{1}{x}$$

which indicates that $f(x)$ is differentiable in (a, b) and hence continuous also.

Further $f(a) = \log\dfrac{\left(a^2 + ab\right)}{a(a+b)} = \log 1 = 0$

$f(b) = \log\dfrac{\left(b^2 + ab\right)}{b(a+b)} = 0$.   Thus $f(a) = 0 = f(b)$

All the conditions of Roll's theorem are satisfied. Hence $\exists c(a < c < b))$ such that $f'(c) = 0$.

$$f'(c) = 0 \text{ gives } \frac{2c}{c^2 + ab} - \frac{1}{c} = 0 \Rightarrow c^2 = ab \text{ or } c = \pm\sqrt{ab}$$

Clearly $c = +\sqrt{ab}$ is the G.M of a and b; $\sqrt{ab} \in (a, b)$

The theorem is thus verified.

***Example* 2:** Verify Rolle's Theorem for the function $f(x) = \dfrac{\sin x}{e^x}$ in $(0, \pi)$

Solution: $f(x) = \dfrac{\sin x}{e^x}$ in $(0, \pi)$

$$\underset{h \to 0\pm}{\text{Lt}} \frac{f(x+h) - f(x)}{h} = \underset{h \to 0\pm}{\text{Lt}} \frac{1}{h}\left[\frac{\sin(x+h)}{e^{(x+h)}} - \frac{\sin x}{e^x}\right]$$

$$= \underset{h \to 0\pm}{\text{Lt}} \frac{e^x}{h}\left[\frac{\sin(x+h) - e^h \sin x}{e^{x+h} \cdot e^x}\right]$$

$$\underset{h \to 0\pm}{Lt} \frac{1}{h} \left[\frac{\sin(x+h) - \sin x \left\{ 1 + \dfrac{h}{1!} + \dfrac{h^2}{2!} + \dots \right\}}{e^{x+h} \cdot e^x} \right]$$

$$\underset{h \to 0\pm}{Lt} \frac{1}{h} \left[\frac{2\cos\left(x + \dfrac{h}{2}\right).\sin\left\{\dfrac{h}{2}\right\} - h\left\{\sin x + 0(h)\right\}}{e^{x+h}} \right]$$

$$\underset{h \to 0\pm}{Lt} \frac{1}{h} \left[2\cos\left(x + \dfrac{h}{2}\right).\frac{\sin\left\{\dfrac{h}{2}\right\}}{\dfrac{h}{2}} - \left\{\sin x + 0(h)\right\} \right]$$

$$f'(x) = \frac{\cos x - \sin x}{e^x}$$

Thus f(x) is differentiable in $(0, \pi)$ and hence continuous there.

Further $f(0) = 0 = f(\pi)$. All the conditions of Rolle's theorem are satisfied.

$\therefore$ $\quad\quad\quad\quad \exists\, c\,(0 < c < \pi)$ such that $f'(c) = 0$

i.e., $\quad\quad \dfrac{\cos\ c - \sin\ c}{e^c} = 0 \Rightarrow c = \dfrac{\pi}{4}$ (principal value)

and clearly $c = \dfrac{\pi}{4} \in (0, \pi)$. Hence the theorem is verified.

***Example* 3:** Verify Rolle's Theorem for the function $f(x) = |x|$ in $(-1, 1)$.

***Solution*:** Here

$$\begin{aligned}
f(x) &= -x &&\text{for} &&-1 \le x \le 0 \\
&= 0 &&\text{for} &&x = 0 \\
&= x &&\text{for} &&0 \le x \le 1 \\
f(-1) &= 1, &&f(1) = 1;
\end{aligned}$$

Hence $\quad\quad f(-1) = f(1)$

$$\begin{aligned}
f'(x) &= -1 &&\text{for} &&-1 \le x \le 0 \\
f'(x) &= 1 &&\text{for} &&0 \le x \le 1
\end{aligned}$$

$\therefore$ $f'(x)$ does not exist at $x = 0$ and hence f(x) is not differentiable in $(-1, 1)$

$\therefore$ Rolle's theorem is not applicable to the function $f(x) = |x|$ in $(-1, 1)$.

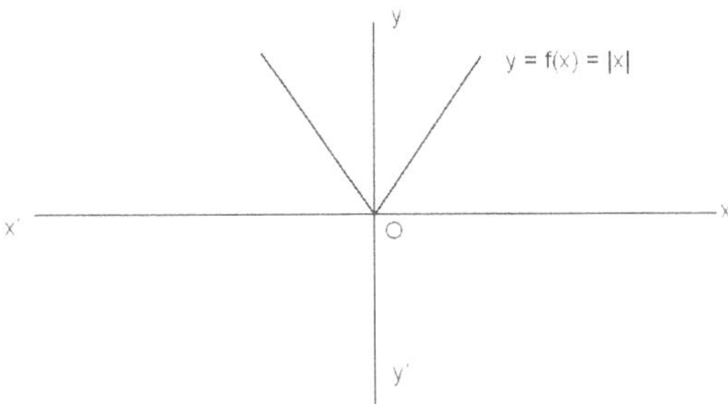

***Example* 4:** Verify rolle's theorem for $f(x) = (x + 2)^3 (x - 3)^4$ in $(-2, 3)$

***Solution*:** Given $f(x)$ is a polynomial in x and given interval is finite hence $f(x)$ is continuous in closed interval $(-2, 3)$

$$f'(x) = 3(x + 2)^2 (x - 3)^4 + (x + 2)^3 . 4(x - 3)^3 \qquad \qquad ......(1)$$

$$f'(x) = (x + 2)^2 (x - 3)^3 (7x - 1)$$

$f(x)$ is differentiable in the open integral $(-2, 3)$

$f(-2) = 0$ and $f(3) = 0$

∴ $\qquad f(-2) = f(3)$

Thus, all the three conditions of Rolle's theorem are satisfied. Hence there exists atleast one point $x = c$ in $(-2, 3)$

such that $\quad f'(c) = 0; \quad f'(c) = (c + 2)^2 (c - 3)^2 (7c - 1) = 0$

$$c = -2 \text{ or } 3 \text{ or } \frac{1}{7}$$

∴ $\qquad c = \dfrac{1}{7} \in (-2, 3)$

Hence Rolle's theorem verified.

***Example* 5:** Verify Rolle's Theorem for $f(x) = \tan x$ in $(0, \pi)$

***Solution*:** Given $f(x) = \tan(x)$ and the interval is $(0, \pi)$

$$f\left(\frac{\pi}{2}\right) = \tan\frac{\pi}{2} = \infty, \ \frac{\pi}{2} \in (0, \pi)$$

∴ $\qquad f(x) = \tan x$ is discontinuous at $x = \dfrac{\pi}{2}$

∴ $\qquad$ The condition of Rolle's theorem not satisfying

Hence we cannot apply Rolle's theorem.

***Example* 6:** Verify Rolle's Theorem for $f(x) = x^3$ in the interval $(1, 3)$

***Solution*:** Given $f(x) = x^3$ in the interval $(1, 3)$

f(x) is an algebraic polynomial in x and given interval (1, 3) is finite

$\therefore$ f(x) is continuous in the closed interval (1, 3) and differentiable in the open interval (1, 3)

f(1) = 1 and f(3) = 27

$\therefore$ f(1) ≠ f(3)

f(x) not satisfying the conditions of Rolle's theorem

Hence Rolle's theorem is not applicable to the function $f(x) = x^3$ in (1, 3)

***Example* 7:** Verify Rolle's theorem for the function $f(x) = x^2 - 5x + 6$ on the interval [2,3].

***Solution*:** Since a polynomial function is everywhere differentiable and so continuous also. Therefore, (i) f(x) is continuous on [2,3] and (ii) f(x) is differentiable on (2,3).

Also, $f(2) = 2^2 - 5 \times 2 + 6 = 0$ and $f(3) = 3^3 - 5 \times 3 + 6 = 0$

$f(2) = f(3)$

Thus all the conditions of Rolle's Theorem are satisfied. Now we have to show that there exists some c $\in$ (2, 3) such that f '(c) = 0.

For this we proceed as follows:

We have $f(x) = x^2 - 5x + 6 \Rightarrow f'(x) = 2x - 5$

$f'(x) = 2x - 5 = 0 \Rightarrow x = 2.5$

Thus, c = 2.5∈(2, 3) such that f '(c) = 0. Hence, Rolle's theorem is verified.

***Example* 8:** Verify Rolle's theorem for the function $f(x) = x^3 - 6x^2 + 11x - 6$ on the interval [1,3]

***Solution*:** Since a polynomial function is everywhere continuous and differentiable, therefore f(x) is continuous on [1,3] and differentiable on (1,3).

Also, $f(1) = 1^3 - 6 \times 1^2 + 11 \times 1 - 6 = 0$

and $f(3) = 3^3 - 6 \times 3^2 + 11 \times 3 - 6 = 0$

$\therefore$ $f(1) = f(3)$

Thus, all the three conditions of Rolle's theorem are satisfied. Now we have to show that there exists $c \in (1, 3)$ such that $f'(c) = 0$.

We have $f(x) = x^3 - 6x^2 + 11x - 6 \Rightarrow f'(x) = 3x^2 - 12x + 11$

$\therefore$ $f'(x) = 0 \Rightarrow 3x^2 - 12x + 11 = 0 \Rightarrow x = \dfrac{12 \pm \sqrt{144 - 132}}{6} \Rightarrow x = 2 \pm \dfrac{1}{\sqrt{3}}$

Clearly, both the values of x lie in the interval (1,3)

Thus, c = $2 \pm \dfrac{1}{\sqrt{3}} \in$ (2,3) such that f '(c) = 0. Hence, Rolle's theorem is verified.

***Example* 9:** Verify Rolle's theorem for the function $f(x) = \sqrt{4 - x^2}$ on [−2,2]

Solution: Clearly, $f(x)$ is defined for all $x \in [-2, 2]$ and has a unique value for each $x \in [-2, 2]$. So, at each point of $[-2, 2]$, the limit of $f(x)$ is equal to the value of the function. Therefore, $f(x)$ is continuous on $[-2, 2]$.

Also, $f(-2) = f(2) = 0$.

Thus, all the three conditions of Rolle's theorem are satisfied. Now we have to show that there exists $c \in (-2, 2)$ such that $f'(c) = 0$.

We have $f(x) = \sqrt{4 - x^2} \Rightarrow f'(x) = \dfrac{-x}{\sqrt{4-x^2}}$

$\therefore \qquad f'(x) = 0 \Rightarrow \dfrac{-x}{\sqrt{4-x^2}} = 0 \Rightarrow x = 0$

Since $c = 0 \in (-2, 2)$ such that $f'(c) = 0$. Hence, Rolle's theorem is verified.

***Example* 10:** Verify Rolle's theorem for each of the following functions on indicated intervals:

(a) $f(x) = \sin^2 x$ on $0 \leq x \leq \pi$

> ***Solution:*** Since $\sin x$ is everywhere continuous and differentiable and the product of two continuous (differentiable) functions is continuous (differentiable), therefore, $f(x) = \sin^2 x = \sin x . \sin x$ is continuous on $[0, \pi]$ and differentiable on $(0, \pi)$.
>
> Also, $f(0) = \sin^2 0 = 0$ and $f(\pi) = \sin^2 \pi = 0$
>
> $\therefore f(0) = f(\pi)$
>
> Thus, $f(x)$ satisfies all the three conditions of Rolle's theorem.
>
> Now, we have to show that there exists $c \in (0, \pi)$ such that $f'(c) = 0$
>
> We have $f(x) = \sin^2 x \Rightarrow f'(x) = 2\sin x \cos x = \sin 2x$
>
> $\therefore f'(x) = 0 \Rightarrow \sin 2x = 0 \Rightarrow 2x = \pi \Rightarrow x = \pi/2$
>
> Since $c = \dfrac{\pi}{2} \in (0, \pi)$ such that $f'(c) = 0$. Hence, Rolle's theorem is verified.

(b) $f(x) = \sin x + \cos x - 1$ on $[0, \frac{\pi}{2}]$

> ***Solution:*** Since $\sin x$ and $\cos x$ are everywhere continuous and differentiable, therefore $f(x) = \sin x + \cos x - 1$ is continuous on $[0, \frac{\pi}{2}]$ and differentiable on $\left(0, \frac{\pi}{2}\right)$.
>
> Also, $f(0) = \sin 0 + \cos 0 - 1 = 0$ and $f\left(\dfrac{\pi}{2}\right) = \sin\dfrac{\pi}{2} + \cos\dfrac{\pi}{2} - 1 = 1 - 1 = 0$
>
> $\therefore f(0) = f\left(\dfrac{\pi}{2}\right)$.
>
> Thus, $f(x)$ satisfies conditions of Rolle's Theorem on $\left[0, \dfrac{\pi}{2}\right]$. Therefore, there exists $c \in (0, \dfrac{\pi}{2})$ such that $f'(c) = 0$.
>
> Now, $f(x) = \sin x + \cos x - 1 \Rightarrow f'(x) = \cos x - \sin x$
>
> $\therefore f'(x) = 0 \Rightarrow \cos x - \sin x = 0 \Rightarrow \sin x = \cos x \Rightarrow \tan x = 1$
>
> $$\Rightarrow x = \pi/4$$

Thus, $c = \frac{\pi}{4} \in \left(0, \frac{\pi}{2}\right)$ such that $f'(x) = 0$. Hence, Rolle's theorem is verified.

(c) $f(x) = \sin x - \sin 2x$ on $[0, \pi]$

Solution: Since sine function is everywhere continuous and differentiable, therefore so are $\sin x$ and $\sin 2x$. Consequently $f(x) = \sin x - \sin 2x$ is continuous on $[0, \pi]$.

Also, $f(0) = \sin 0 - \sin 0 = 0$ and $f(\pi) = \sin \pi - \sin 2\pi = 0$

$\therefore f(0) = f(\pi)$

Thus, $f(x)$ satisfies all the three conditions of Rolle's theorem on $[0, \pi]$. Consequently thereexists $c \in (0, \pi)$ such that $f'(c) = 0$.

Now, $f(x) = \sin x - \sin 2x \Rightarrow f'(x) = \cos x - 2 \cos 2x$

$\therefore f'(x) = 0 \Rightarrow \cos 2x - 2 \cos x = 0 \Rightarrow \cos x - 2(2\cos^2 x - 1) = 0$

$$\Rightarrow 4\cos^2 x - \cos x + 2 = 0 \Rightarrow \cos x = \frac{1 \pm \sqrt{33}}{8} = 0.8431 \text{ or } -0.5931$$

$$\Rightarrow x = \cos^{-1}(0.8431) \text{ or } \cos^{-1}(-0.5931)$$

$$\Rightarrow x = \cos^{-1}(0.8431) \text{ or } 180° - \cos^{-1}(0.5931)$$

$$[\because \cos^{-1}(-x) = \pi - \cos^{-1} x]$$

$$\Rightarrow x = 32°\, 32' \text{ or } x = 126°\, 23'$$

Thus, $c = 32°32'; 126°23' \in (0,\ \pi)$ such that $f'(c) = 0$.

Hence, Rolle's theorem is verified.

Example **11:** Verify Rolle's theorem for each of the following functions on the indicated interval:

(i) $f(x) = x(x+3)e^{-x/2}$ on $[-3,\ 0]$.

(ii) $f(x) = e^x(\sin x - \cos x)$ on $\left[\frac{\pi}{4}, \frac{5\pi}{4}\right]$.

Solution: (i) Since a polynomial function and an exponential function are everywhere continuous and differentiable, therefore, $f(x)$, being product of these two, is continuous on $[-3,\ 0]$ and differentiable on $(-3,\ 0)$.

Also, $f(-3) = -3(-3+3)e^{3/2} = 0$ and $f(0) = 0$

$\therefore f(-3) = f(0)$

Thus, $f(x)$ satisfies all the three conditions of Rolle's theorem on $[-3, 0]$. Consequently there exists $c \in (-3, 0)$ such that $f'(c) = 0$.

Now, $f(x) = x(x+3)e^{-\frac{x}{2}}$

$$\Rightarrow f'(x) = (2x+3)e^{-x/2} + (x^2 + 3x)\left(-\frac{1}{2}\right)e^{-x/2} = e^{-x/2}\left\{\frac{-x^2 + x + 6}{2}\right\}$$

$$\therefore f'(x) = 0 \Rightarrow e^{-x/2}\left\{\frac{-x^2 + x + 6}{2}\right\} = 0 \Rightarrow -x^2 + x + 6 = 0$$

$$\Rightarrow x^2 - x - 6 = 0 \Rightarrow (x - 3)(x + 2) = 0 \Rightarrow x = -2, 3$$

Thus, $c \in (-3, 0)$ such that $f'(c) = 0$. Hence, theorem is verified.

(ii) Since an exponential function and sine and cosine functions are everywhere continuous and differentiable, therefore $f(x)$ is continuous on $\left[\frac{\pi}{4}, \frac{5\pi}{4}\right]$ and differentiable on $\left(\frac{\pi}{4}, \frac{5\pi}{4}\right)$

$$\text{Also, } f\left(\frac{\pi}{4}\right) = e^{\frac{\pi}{4}}\left(\sin\frac{\pi}{4} - \cos\frac{\pi}{4}\right) = e^{\frac{\pi}{4}}\left(\frac{1}{\sqrt{2}} - \frac{1}{\sqrt{2}}\right) = 0$$

$$f\left(\frac{5\pi}{4}\right) = e^{\frac{5\pi}{4}}\left(\sin\frac{5\pi}{4} - \cos\frac{5\pi}{4}\right) = e^{\frac{5\pi}{4}}\left(-\sin\frac{\pi}{4} + \cos\frac{\pi}{4}\right)$$

$$= e^{\frac{5\pi}{4}}\left(-\frac{1}{\sqrt{2}} + \frac{1}{\sqrt{2}}\right) = 0$$

$$\therefore f\left(\frac{\pi}{4}\right) = f\left(\frac{5\pi}{4}\right)$$

Thus, $f(x)$ satisfies all the three conditions of Rolle's theorem on $\left[\frac{\pi}{4}, \frac{5\pi}{4}\right]$ such that $f'(c) = 0$.

Now, $f(x) = e^x(\sin x - \cos x)$

$$\Rightarrow f'(x) = e^x(\sin x - \cos x) + e^x(\cos x + \sin x) = 2e^x \sin x$$

$$\therefore \qquad f'(x) = 0 \Rightarrow 2e^x \sin x = 0 \Rightarrow \sin x = 0 \qquad\qquad [\because e^x \neq 0]$$

$$\Rightarrow x = \pi$$

Thus, $c = \pi \in \left(\frac{\pi}{4}, \frac{5\pi}{4}\right)$ such that $f'(c) = 0$. Hence, Rolle's theorem is verified.

***Example* 12:** Discuss the applicability of Rolle's theorem for the following functions on the indicated intervals:

  (i) $f = |x|$ on $[-1, 1]$

  (ii) $f(x) = 3 + (x - 2)^{2/3}$ on $[1, 3]$

***Solution*:**

(i)   We have $f(x) = |x| = \begin{cases} -x, & \text{when } -1 \leq x < 0 \\ x, & \text{when } 0 \leq x \leq 1 \end{cases}$

Since a polynomial function is everywhere continuous and differentiable, therefore $f(x)$ is continuous and differentiable for all $x < 0$ and for all $x > 0$ except possible at $x = 0$.

So, consider the point $x = 0$.

We have $\lim\limits_{x \to 0^-} f(x) = \lim\limits_{x \to 0} -x = 0$ and $\lim\limits_{x \to 0^+} f(x) = \lim\limits_{x \to 0} x = 0$

$$\therefore \qquad \lim\limits_{x \to 0^-} f(x) = \text{ and } \lim\limits_{x \to 0^+} f(x) = f(0)$$

Thus, $f(x)$ is continuous at $x = 0$. Hence $f(x)$ is continuous on $[-1, 1]$.

Now, (LHD at $x = 0$) $= \lim\limits_{x \to 0^-} \dfrac{f(x) - f(0)}{x - 0}$

$$= \lim\limits_{x \to 0} \dfrac{-x - 0}{x}$$

$$= \lim\limits_{x \to 0} \dfrac{-x}{x} = \lim\limits_{x \to 0} -1 = -1$$

and (RHD at $x = 0$) $= \lim\limits_{x \to 0} \dfrac{f(x) - f(0)}{x - 0} = \lim\limits_{x \to 0} \dfrac{x - 0}{x}$

$$= \lim\limits_{x \to 0} \dfrac{x}{x} = \lim\limits_{x \to 0} 1 = 1$$

$\therefore$ (LHD at $x = 0$) $\neq$ RHD at $x = 0$

This shows that $f(x)$ is not differentiable at $x = 0 \in (-1, 1)$. Thus, the condition of differentiability at each point of $(-1, 1)$ is not satisfied.

Hence, Rolle's theorem is not applicable to $f(x) = |x| \text{ on } [-1, 1]$.

(ii) We have $f(x) = 3 + (x - 2)^{2/3}, x \in [2, 3]$

$\therefore f'(x) = \left(\dfrac{2}{3}\right)(x - 2)^{\frac{1}{3}}$

Clearly, $\lim\limits_{x \to 2} f'(x) = \infty$. So, $f(x)$ is not differentiable at $x = \in (1, 3)$.

Hence, Rolle's theorem is not applicable to $f(x) = 3 + (x - 2)^{2/3}$ on the interval $[1, 3]$.

***Example* 13:** Discuss the applicability of Rolle'stheorem on the function

$$f(x) = \begin{cases} x^2 + 1, & \text{when } 0 \le x \le 1 \\ 3 - x, & \text{when } 1 < x \le 2 \end{cases}$$

Solution: Since a polynomial function is everywhere continuous and differentiable, therefore $f(x)$is continuous and differentiable at all points except possibly at $x = 1$.

Now, we consider the differentiability of $f(x)$ at $x = 1$.

We have (LHD at $x = 1$) $= \lim\limits_{x \to 1^-} \dfrac{f(x) - f(1)}{x - 1}$

$$= \lim\limits_{x \to 1} \dfrac{(x^2 + 1) - (1 + 1)}{x - 1} \qquad\qquad [\because f(x) = x^2 + 1 \text{ for } 0 \le x \le 1]$$

$$= \lim\limits_{x \to 1} \dfrac{x^2 - 1}{x - 1} = \lim\limits_{x \to 1}(x + 1) = 2$$

and (RHD at $x = 1$) $\lim\limits_{x \to 0^+} \dfrac{f(x) - f(1)}{x - 1} = \lim\limits_{x \to 1} \dfrac{(3 - x) - (1 + 1)}{x - 1}$

$$= \lim\limits_{x \to 1} \dfrac{-(x - 1)}{x - 1} = -1$$

$\therefore$ (LHD at $x = 1$) $\neq$ (RHD at $x = 1$). So, $f(x)$ is not differentiable at $x = 1$.

Thus, the condition of differentiability at each point of the given interval in not satisfied.

Hence, Rolle's theorem is not applicable to the given function on the interval $[0, 2]$.

Practice Problems

1. Check the functions whether Rolle's theorem applying or not

   (i) $f(x) = 3 + (x - 2)^{2/3}$ or $[1,3]$

   (ii) $f(x) = \sin\frac{1}{x}$ for $-1 \le x \le 1$.

   (iii) $f(x) = 2x^2 - 5x + 3$ on $[1,3]$

   (iv) $f(x) = \begin{cases} -4x + 5, & 0 \le x \le 1 \\ 2x - 3, & 1 < x \le 2 \end{cases}$

2. Verify Rolle's theorem for the function $f(x) = x(x - 4)^2$ on the interval $[0, 4]$.

3. Verify Rolle's theorem for the function $f(x) = x(x - 2)^2$ on the interval $[0, 2]$.

4. Verify Rolle's theorem for the function $f(x) = x^2 + 5x + 6$ on the interval $[-3, -2]$.

5. Verify Rolle's theorem for each of the following functions on the indicated intervals:

   (i) $f(x) = x^2 - 8x + 12$ on $[2,6]$

   (ii) $f(x) = x^2 - 4x + 3$ on $[1,3]$

   (iii) $f(x) = (x - 1)(x - 2)^2$ on $[1,2]$

   (iv) $f(x) = x(x - 1)^2$ on $[0,1]$

6. Using Rolle's theorem, find point on the curve $y = 16 - x^2, x \in [-1, 1]$ where tangent is parallel to x-axis.   [**Ans.** $(0,16)$]

7. Verify Rolle's theorem and find c

   1. $f(x) = x(x + 3)e^{-x/2}$ in $(-3. \ 0)$ [**Ans.** $c = -2$]

   2. $f(x) = \sin x$ in $(0. \ \pi)$ [**Ans.** $c = \pi/2$]

   3. $f(x) = c^x(\sin x - \cos x)$ in $\left(\frac{\pi}{4}. \ \frac{5\pi}{4}\right)$ [**Ans.** $c = \pi$]

   4. $f(x) = \log\left(\frac{x^2+6}{5x}\right)$ in $(2, 3)$ [**Ans.** $c = \sqrt{6}$]

   5. $f(x) = (x - a)^m(x - b)^n$ in $a, b)$, $m > 0, \ n > 0$ [**Ans.** $c = \frac{(mb+na)}{m+n}$]

   6. $f(x) = x^2 - 5x + 7$ in $(2, 3)$ [**Ans.** $c = 5/2$]

2.4 Lagrange's Mean Value Theorem (First mean Value Theorem)

If f(x) is (i) continuous in [a, b] and (ii) differentiable in (a, b) then $\exists$ at least one value 'c' in

(a, b) such that $f'(c) = \dfrac{f(b) - f(a)}{b - a}$

Proof: Define a new function $\phi(x) = f(x) + A\,x$ .....(1)

where A is a constant $\exists \, \phi(a) = \phi(b)$

i.e., $f(a) + A\,a = f(b) + Ab$

i.e., $A = -\left[\dfrac{f(b) - f(a)}{b - a}\right]$ .....(2)

$f(x)$ and Ax are continuous in $[a, b]$

Hence $\phi(x)$ is continuous in $[a, b]$ .....(3)

$f(x)$ and $A\,x$ are differential in (a, b), ...(4)

Hence $\phi(x)$ is differential in (a, b) .....(5)

and $\phi(a) = \phi(b)$

(3), (4), (5) show that $\phi(x)$ satisfies all the conditions of Rolle's theorem.

$\therefore \, \exists$ atleast one value c in (a, b) $i.e.\,\phi'\,(c) = 0$

$$\phi'\,(x) = f'\,(x) + A$$

$$\phi'\,(c) = f'\,(c) + A = 0$$

$$\Rightarrow A = -\,f'\,(c)$$

From (2) and (6) it follows that

$$= f'(c) = \frac{f(b) - f(a)}{b - a}$$

2.5 Geometrical Interpretation of Lagrange's Mean Value Theorem

P and Q are two points on the continuous curve $y = f(x)$ corresponding to $x = a$ and $x = b$ respectively.

$\therefore$ $P[a, \, f(a)], Q\,[b, \, f(b)]$ are two points on the curve.

Slope on the line joining the points P and Q is $\dfrac{f(b) - f(a)}{b - a}$,

R is a point on the curve between P and Q corresponding to $x = c$, so that $f'(c)$ is the slope of the tangent line at R $[c, \, f(c)]$

$f'(c) = \dfrac{f(b) - f(a)}{b - a}$ means that the tangent at R is parallel to the chord P Q

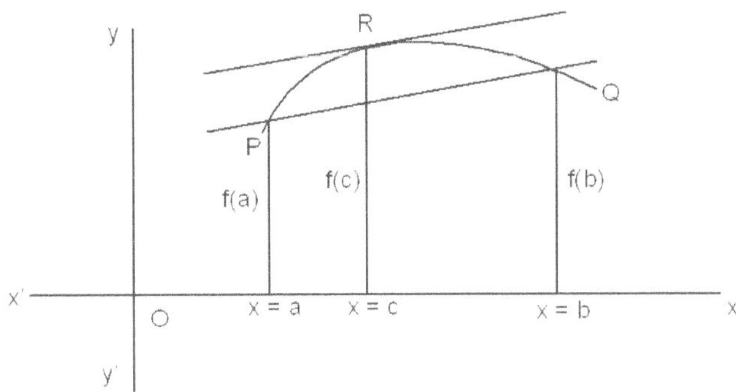

Hence this theorem tells that there is at least one point R on the curve PQ where the tangent to the curve is parallel to the chord PQ.

Illustrative Examples

***Example* 14:** Verify Lagrange's theorem for the function $f(x) = (x-1)(x-2)(x-3)$ in $(0, 4)$.

Solution:

$$f(x) = (x-1)(x-2)(x-3) \text{ and } a = 0, b = 4$$

$$f(x) = x^3 - 6x^2 + 11x - 6 \text{ is an algebric polynomial and } (0, 4) \text{ is a finite interval.}$$

Hence $f(x)$ is differentiable in $(0, 4)$ and is continuous in $[0, 4]$ showing that the conditions of Lagrange's Mean Value theorem are satisfied.

$\therefore \quad \exists$ atleast one value 'c' in $(0, 4)$, such that

$$f'(c) = \frac{f(b) - f(a)}{b - a} \qquad \qquad .....(i)$$

$$f(0) = -6, \quad f(4) = 6$$

$$f'(x) = 3x^2 - 12x + 11$$

$$f'(c) = 3c^2 - 12c + 11$$

From (1) $3c^2 - 12c + 11 = \dfrac{6 - (-6)}{4 - 0}$

$$\Rightarrow 3c^2 - 12c + 8 = 0$$

$$\therefore \quad c = \frac{6 \pm 2\sqrt{3}}{3}$$

The point $c = \dfrac{6 \pm 2\sqrt{3}}{3}$ which clearly lie in $(0, 4)$.

***Example* 15:** Verify Langrange's Mean Value theorem for the function $f(x) = e^x$ in $(0, 1)$

Solution: $f(x) = e^x$ is differentiable in $(0, 1)$ and continuous in $[0, 1]$

$\therefore$ $\exists$ atleast one value 'c' in (0, 1) such that

$$\ni f'(c) = \frac{f(b) - f(a)}{b - a} \qquad\qquad .....(i)$$

$$\ni f'(c) = \frac{f(1) - f(0)}{1 - 0}$$

$$f(0) = e^0 = 1, \ f(1) = e$$

$$f'(x) = e^x \text{ gives } f'(c) = e^c$$

From (1) $e^c = \dfrac{e - 1}{1 - 0}$

$\therefore$ $c = \log (e - 1)$, this $c \in (0, 1)$

***Example* 16:** Verify Langrange's Mean Value theorem for the function $f(x) = 5x^2 + 7x + 6$ in $(3, 4)$.

***Solution*:** $f(x)$ is an algebraic polynomial and the interval $(3, 4)$ is finite, $f(x)$ is differentiable in $(3, 4)$ and continuous in $[3, 4]$.

$\therefore$ $\exists$ atleast one value $c \in (3, 4)$ such that

$$f'(c) = \frac{f(4) - f(3)}{4 - 3} \qquad\qquad .....(i)$$

$$\ni f'(c) = \frac{f(b) - f(a)}{b - a}$$

$$f(3) = 72, \ f(4) = 114, \ \ f'(x) = 10x + 7$$

$$10\,c + 7 = \frac{114 - 72}{4 - 3}$$

$$c = 3.5 \in, (3, 4)$$

***Example* 17:** Verify Lagrange's Mean values theorem for $f(x) = x^{1/3}$ in $(-1, 1)$

***Solution*:** $f(x) = x^{1/3}$ is continuous in the closed interval $[-1, 1]$

$f(x) = x^{1/3}$ is not differentiable at $x = 0$ in the open interval $(-1, 1)$

Hence Lagrange's theorem is not applicable to the $f(x) = x^{1/3}$ in $(-1, 1)$

***Example* 18:** If $a < b$ prove that $\dfrac{b - a}{1 + b^2} < \tan^{-1} b - \tan^{-1} a < \dfrac{b - a}{1 + a^2}$ using Lagrange's Mean value theorem, hence deduce the following:

(i) $\dfrac{5\pi + 4}{20} < \tan^{-1} 2 < \dfrac{\pi + 2}{4}$
(ii) $\dfrac{25\pi + 12}{100} < \tan^{-1}\left(\dfrac{4}{3}\right) < \dfrac{3\pi + 2}{12}$

Solution: Given $f(x) = \tan^{-1}x$, interval is (a, b), $0 < a < b < 1$

$f(x)$ is continuous in the closed interval [a, b] and differentiable in the open interval (a, b). Hence there exists atleast one point $x = c$ in the open interval (a, b) such that

$$f'(c) = \frac{f(b) - f(a)}{b - a} \qquad \text{(Lagrange's Mean value theorem)}$$

Now $\qquad f'(x) = \dfrac{1}{1 + x^2} \quad \Rightarrow \quad f'(c) = \dfrac{1}{1 + c^2}$

$\therefore \qquad \dfrac{1}{1 + c^2} = \dfrac{\tan^{-1}(b) - \tan^{-1}(a)}{b - a} \qquad \qquad \text{.....(i)}$

we have $\qquad 1 + a^2 < 1 + c^2 < 1 + b^2$

$\therefore \qquad \dfrac{1}{1 + a^2} > \dfrac{1}{1 + c^2} > \dfrac{1}{1 + b^2} \qquad \qquad \text{.....(ii)}$

From (1) and (2) $\qquad \dfrac{1}{1 + a^2} > \dfrac{\tan^{-1}(b) - \tan^{-1}(a)}{b - a} > \dfrac{1}{1 + b^2}$

$\therefore \qquad \dfrac{b - a}{1 + b^2} < \tan^{-1}b - \tan^{-1}a < \dfrac{b - a}{1 + a^2} \qquad \qquad \text{.....(iii)}$

(i) substituting in (3) b = 2, a = 1

$$\frac{2 - 1}{1 + 2^2} < \tan^{-1}2 - \tan^{-1}1 < \frac{2 - 1}{1 + 1^2}$$

$$\frac{1}{5} < \tan^{-1}2 - \frac{\pi}{4} < \frac{1}{2}$$

$$\frac{1}{5} + \frac{\pi}{4} < \tan^{-1}2 < \frac{1}{2} + \frac{\pi}{4}$$

$$\frac{5\pi + 4}{20} < \tan^{-1}2 < \frac{\pi + 2}{4}$$

(ii) Substituting $b = \dfrac{4}{3}$ a = 1 in (3)

$$\frac{\frac{4}{3} - 1}{1 + \left(\frac{4}{3}\right)^2} < \tan^{-1}\frac{4}{3} - \tan^{-1}1 < \frac{\frac{4}{3} - 1}{1 + 1}$$

$$\frac{3}{25} < \tan^{-1}\frac{4}{3} - \frac{\pi}{4} < \frac{1}{6}$$

$$\frac{3}{25} + \frac{\pi}{4} < \tan^{-1}\frac{4}{3} < \frac{1}{6} + \frac{\pi}{4}$$

$$\frac{25\pi + 12}{100} < \tan^{-1}\frac{4}{3} < \frac{3\pi + 2}{12}$$

***Example* 19:** Show that for any $x > 0$, $1 + x < e^x < 1 + x\,e^x$

***Solution*:** Given $f(x) = e^x$ and $(0, x)$

$f(x)$ is continuous on $[0, x]$ and differentiable in the open interval $(0, x)$

From Langrange's theorem, there exists $x = c \in (0, x)$ such that

$$f'(c) = \frac{f(x) - f(0)}{x - 0}$$

$$\frac{e^x - e^0}{x} = e^c \qquad\qquad .....\text{(i)}$$

and $0 < c < x$

$$e^0 < e^c < e^x \qquad\qquad .....\text{(ii)}$$

(e^x is an increasing function)

From (1) and (2), we get

$$1 < \frac{e^x - 1}{x} < e^x$$

$$x < e^x - 1 < x\,e^x$$

$\therefore\qquad 1 + x < e^x < 1 + x\,e^x$

Practice Problems

I.   Verify Langrange's Mean Value theorem for the following functions:

1. $f(x) = x(x - 1)(x - 2)$ in $(0, \frac{1}{2})$ $\qquad$ [**Ans.** $1 \pm \sqrt{1/2}$]

2. $f(x) = \log x$ in $(1, e)$ $\qquad$ [**Ans.** e-1]

3. $f(x) = x^2 - 3x - 1$ in $\left(\frac{-11}{7}, \frac{13}{7}\right)$

$\qquad$ [**Ans.** 1/7]

4. $f(x) = 2x^2 - 7x + 10$ in $(2, 5)$ $\qquad$ [**Ans.** 3.5]

2.6   Cauchy's Mean Value Theorem

If two functions $f(x)$ and $g(x)$ are (i) continuous in $[a, b]$ (ii) differential in (a, b) and (iii) $g'(c) \neq 0$ in (a, b) then $\exists$ atleast one value 'c' in (a, b) such that;

$$\frac{f'(c)}{g'(c)} = \frac{f(b) - f(a)}{g(b) - g(a)}$$

Proof: Define a new function $\phi(x) = f(x) + A\, g(x)$ .....(1)

Where A is a constant such that $\phi(a) = \phi(b)$

$$f(a) + A\, g(a) = f(b) + A\, g(b) \qquad .....(2)$$

$$A = -\frac{[f(b) - f(a)]}{[g(b) - g(a)]} \qquad .....(3)$$

$f(x), A\, g(x)$ are continuous in $[a, b]$

Hence $\phi(x)$ is continuous in $[a, b]$ .....(4)

$f(x), A\, g(x)$ are differentiable in (a, b)

Hence $\phi(x)$ is differentiable in (a, b) .....(5)

and $\phi(a) = \phi(b)$ .....(6)

(4), (5), (6) show that $\phi(x)$ satisfies all the conditions of Rolle's theorem

$\therefore$ $\exists$ atleast one value c in (a, b) $\exists$ $\phi'(c) = 0$

But $\phi'(x) = f'(x) + Ag'(x)$

$\therefore$ $\phi'(c) = f'(c) + Ag'(c) = 0$

$$\Rightarrow A = -\frac{f'(c)}{g'(c)}$$

From (3), (7) we get $\dfrac{f'(c)}{g'(c)} = \dfrac{f(b) - f(a)}{g(b) - f(a)}$

Illustrative Examples

***Example* 20**: Verify Cauchy's Mean Value theorem for $f(x) = \dfrac{1}{x^2}$ and $g(x) = \dfrac{1}{x}$ in (a, b)

Solution: $f(x), g(x)$ are differentiable in (a, b) and continuous in $[a, b]$

$\therefore$ $\exists$ atleast one value 'c' in (a, b) such that

$$\frac{f'(c)}{g'(c)} = \frac{f(b) - f(a)}{g(b) - f(a)}$$

Here $f'(x) = \dfrac{-2}{x^3}$

$\qquad g'(x) = \dfrac{-1}{x^2}$

$$\therefore \qquad \frac{\left(\dfrac{-2}{c^3}\right)}{-\dfrac{1}{c^2}} = \frac{\dfrac{1}{b^2} - \dfrac{1}{a^2}}{\dfrac{1}{b} - \dfrac{1}{a}} \Rightarrow \frac{2}{c} = \frac{a+b}{ab}$$

$$c = \frac{2ab}{a+b} \quad \text{which is the Harmonic Mean of 'a' and 'b'}$$

$$c \; \varepsilon \; (a, b)$$

***Example* 21:** Verify Cauchy's Mean Valve Theorem for $f(x) = e^x$, $g(x) = e^{-x}$ in $(3, 7)$

Solution: $f(x)$, $g(x)$ are differentiable in $(3, 7)$ and continuous in $[3, 7]$

$\therefore \; \exists$ atleast one value 'c' in $(a, b) \ni$

$$\frac{f'(c)}{g'(c)} = \frac{f(b) - f(a)}{g(b) - f(a)}$$

Here $\quad f'(x) = e^x$

$$g'(x) = -e^{-x}$$

$$\frac{e^c}{-e^c} = \frac{e^7 - e^3}{e^{-7} - e^{-3}} \Rightarrow + e^{2c} = + e^{10}$$

$$\therefore \quad c = 5 \; \varepsilon \; (3, 7)$$

Practice Problems

1. Considering the functions $f(x) = x^2$, $g(x) = (x)$ in Cauchy's Mean value theorem for (a, b) prove that 'c' is the arithmetic mean between a and b.

2. Verify Cauchy's Mean Value theorem for $f(x) = \sin x$, $g(x) = \cos x$ in (a, b).

3. Verify Cauchy's Mean Value theorem for $f(x) = \sqrt{x}$, $g(x) = \dfrac{1}{\sqrt{x}}$ in (a, b).

4. Verify Cauchy's Mean Value theorem for (i) $f(x) = x^2$, $g(x) = x^3$ in $(1, 2)$ (ii) $f(x) = e^x$, $g(x) = e^{-x}$ in (a, b).

CHAPTER – 3

Tangent and Normal

3.1 Slopes of the Tangent and Normal

Slope of the tangent: Let $y = f(x)$ be a continuous curve, and let $P(x_1, y_1)$ be a point on it.

$\left(\dfrac{dy}{dx}\right)_p$ is the slope of the tangent to the curve $y = f(x)$ at point P i.e., $\left(\dfrac{dy}{dx}\right)_p = \tan \Psi = $ slope of the tangent at P, where Ψ is the angle which the tangent at $P(x_1, y_1)$ makes with the positive direction of x − axis.

Note 1: If the tangent at P is parallel to x − axis, then

$$\Psi = 0 \Rightarrow \tan \Psi = 0 \Rightarrow \text{slope} = 0 \Rightarrow \left(\dfrac{dy}{dx}\right)_p = 0.$$

Note 2: If the tangent at P is perpendicular to $axis$, or parallel to y − axis then

$$\Psi = \dfrac{\pi}{2} \Rightarrow \cot \Psi = 0 \Rightarrow \dfrac{1}{\tan \Psi} = 0 \Rightarrow \left(\dfrac{dx}{dy}\right)_p = 0.$$

Slope of the Normal: The normal to a curve at $P(x_1, y_1)$ is a line perpendicular to the tangent at P and passing through P.

$$\therefore \text{ slope of the normal at P} = -\dfrac{1}{\text{slope of the tangent at P}} = -\dfrac{1}{\left(\frac{dy}{dx}\right)_p} = -\left(\dfrac{dx}{dy}\right)_p.$$

Illustrative Examples

***Example* 1:** Show that the tangents to the curve $y = 2x^3 - 3$ at the points where $x = 2$ and $x = -2$ are parallel.

Solution: The equation of the curve is $y = 2x^3 - 3$ ......(i)

Differentiating w. r. t. x, we get $\dfrac{dy}{dx} = 6x^2$

Now, $m_1 = $ slope of the tangent at $x = 2 \Rightarrow \left(\dfrac{dy}{dx}\right)_{x=2} = 6(2)^2 = 24,$

And, $m_2 = $ slope of the tangent at $x = -2 \Rightarrow \left(\dfrac{dy}{dx}\right)_{x=-2} = 6(-2)^2 = 24.$

Clearly, $m_1 = m_2$. Thus, the tangents to the given curve at the points where $x = 2$ and $x = -2$ are parallel.

Example 2: Prove that the tangents to the curve $y = x^2 - 5x + 6$ at the points $(2,0)$ and $(3,0)$ are at right angles.

Solution: The equation of the curve is $y = x^2 - 5x + 6$

Differentiating w. r. t. x, we get $\dfrac{dy}{dx} = 2x - 5$.

Now, $m_1 = $ slope of the tangent at $(2,0) = \left(\dfrac{dy}{dx}\right)_{(2,0)} = 2 \times 2 - 5 = -1$

And, $m_2 = $ slope of the tangent at $(3,0) = \left(\dfrac{dy}{dx}\right)_{(3,0)} = 2 \times 3 - 5 = 1$.

$$\Rightarrow m_1 m_2 = -1 \times 1 = -1.$$

Thus, the tangents to the given curve at $(2, 0)$ and $(3, 0)$ are at right angles.

Example 3: Find the points on the curve $y = x^3 - 2x^2 - x$ at which the tangent lines are parallel to the line $y = 3x - 2$.

Solution: Let $P(x_1, y_1)$ be the required point. The given curve is

$$y = x^3 - 2x^2 - x \qquad\qquad\qquad ....(i)$$

$$\Rightarrow \dfrac{dy}{dx} = 3x^2 - 4x - 1 \Rightarrow \left(\dfrac{dy}{dx}\right)_{(x_1,y_1)} = 3x_1^2 - 4x_1 - 1.$$

Since the tangent at (x_1, y_1) is parallel to the line $y = 3x - 2$.

$\therefore$ slope of the tangent at $(x_1, y_1) = $ slope of the line $y = 3x - 2$.

$$\Rightarrow \left(\dfrac{dy}{dx}\right)_{(x_1, y_1)} = 3 \Rightarrow 3x_1^2 - 4x_1 - 1 = 3.$$

$$\Rightarrow 3x_1^2 - 4x_1 - 4 = 0 \Rightarrow (x_1 - 2)(3x_1 + 2) = 0 \Rightarrow x_1 = 2, \dfrac{-2}{3}.$$

Since, (x_1, y_1) lies on (i), therefore $y_1 = x_1^3 - 2x_1^2 - x_1$.

When $x_1 = 2$; $y_1 = 2^3 - 2(2)^2 - 2 = -2.$

When $x_1 = \dfrac{-2}{3}$; $y_1 = \left(\dfrac{-2}{3}\right)^3 - 2\left(-\dfrac{2}{3}\right)^2 + \dfrac{2}{3} = \dfrac{-14}{27}.$

Thus, the required points are $(2, -2)$ and $\left(\dfrac{-2}{3}, \dfrac{-14}{27}\right)$

Example 4: Find the point on the curve $y = 2x^2 - 6x - 4$ at which the tangent is parallel to the $x - $ axis.

Solution: Let the required point be $P(x_1, y_1)$. The given curve is

$$y = 2x^2 - 6x - 4 \qquad\qquad\qquad ....(i)$$

$$\Rightarrow \dfrac{dy}{dx} = 4x - 6 \Rightarrow \left(\dfrac{dy}{dx}\right)_{(x_1, y_1)} = 4x_1 - 6$$

Since the tangent at (x_1, y_1) lies on (i), therefore $y_1 = 2x_1^2 - 6x_1 - 4$

When $x = \frac{3}{2}$, $y_1 = 2\left(\frac{3}{2}\right)^2 - 6\left(\frac{3}{2}\right) - 4 = -\frac{17}{2}$

So, the required point is $(3/12, -17/2)$.

***Example* 5:** Find a point on the curve $y = (x - 3)^2$, where the tangent is parallel to the line joining $(4, 1)$ and $(3, 0)$.

***Solution*:** Let the required point be be $P(x_1, y_1)$. The equation of the given curve is

$$y = (x - 3)^2 \qquad \qquad \text{...(i)}$$

$$\Rightarrow \frac{dy}{dx} = 2(x - 3) \Rightarrow \left(\frac{dy}{dx}\right)_{(x_1,\ y_1)} = 2(x_1 - 3)$$

Since the tangent at P is parallel to the line joining $(4, 1)$ and $(3, 1)$. Therefore, Slope of the tangent at P = slope of the line joining $(4, 1)$ and $(3, 0)$

$$\Rightarrow \left(\frac{dy}{dx}\right)_{(x_1,\ y_1)} = \frac{0-1}{3-4} \Rightarrow 2(x_1 - 3) = 1 \Rightarrow x_1 = 7/2$$

Since the point $P(x_1, y_1)$ lies on (i), therefore $y_1 = (x_1 - 3)^2$

When $x_1 = \frac{7}{2}$; $y_1 = \left(\frac{7}{2} - 3\right)^2 = \frac{1}{4}$

Thus, the required point is $(7/2, 1/4)$.

Practice Problems

1. Find the slopes of the tangent and normal to the following curves at the indicated points:

   (i) $y = \sqrt{x^3}$ at $x = 4$

   (ii) $y = x^3 - x$ at $x = 2$

   (iii) $x = a(\theta - \sin\theta), y = a(1 - \cos\theta)$ at $\theta = \pi/2$

   (iv) $x^2 + 3y + y^2 = 5$ at $(1, 1)$

   (v) $xy = 6$ at $(1, 6)$

2. Find a point on the curve $y = x^3 - 3x$ where the tangent is parallel to the chord joining $(1, -2)$ and $(2, 2)$.

3. Find the points on the curve $y = x^3 - 2x^2 - 2x$ at which the tangent lines are parallel to the line $y = 2x - 3$.

4. Find the point on the curve $y = x^2$ where the slope of the tangent is equal to the x − coordinate of the point.

5. At what points on the circle $x^2 + y^2 - 2x - 4y + 1 = 0$, the tangent is parallel to x-axis.

6. At what point of the curve $y = x^2$ does the tangent make an angle of $45°$ with the x-axis?

7. Find the point on the curve $y = 3x^2 + 4$ at which the tangent is perpendicular to the line whose slope is $-\dfrac{1}{6}$.

8. At what points on the curve $y = x^2 - 4x + 5$ is the tangent perpendicular to the line $2y + x = 7$?

3.2 Equations of the Tangent and Normal

It is well known that the equation of a line passing through a point (x_1, y_1) and having slope m is

$$y - y_1 = m(x - x_1)$$

Since the slopes of the tangent and the normal to the curve $y = f(x)$ at a point $P(x_1, y_1)$ are $\left(\dfrac{dy}{dx}\right)_P$ and $-\dfrac{1}{\left(\dfrac{dy}{dx}\right)_P}$ respectively.

Therefore the equation of the tangent at $P(x_1, y_1)$ to the curve $y = f(x)$ is

$$y - y_1 = \left(\frac{dy}{dx}\right)_P (x - x_1) \qquad \qquad \text{...(1)}$$

Since the normal at $P(x_1, y_1)$ passes through P and has slope $-\dfrac{1}{\left(\dfrac{dy}{dx}\right)_P}$,

Therefore the equation of the normal at $P(x_1, y_1)$ to the curve $y(f(x)$ is

$$y - y_1 = -\frac{1}{\left(\frac{dy}{dx}\right)_P} (x - x_1) \qquad \qquad \text{...(2)}$$

Note: If $\left(\dfrac{dy}{dx}\right)_P = \infty$, then the tangent at (x_1, y_1) is parallel to y-axis and its equations is $x = x_1$.

Note: If $\left(\dfrac{dy}{dx}\right)_P = 0$, then the normal at (x_1, y_1) is parallel to y-axis and its equation is $x = x_1$.

Method for finding the equations of tangent and normal to the curve $y = f(x)$ at the point (x_1, y_1)

Step 1: Find $\dfrac{dy}{dx}$ from the given equation $y = f(x)$.

Step 2: Find the value of $\dfrac{dy}{dx}$ at the given point $P(x_1, y_1)$.

Step 3: If $\left(\dfrac{dy}{dx}\right)_{(x_1, y_1)}$ is a non- zero finite number, then obtain the equations of tangent and normal at (x_1, y_1) by using the formulae

$$y - y_1 = \left(\frac{dy}{dx}\right)_{(x_1, y_1)} (x - x_1) \text{ and } y - y_1 = -\frac{1}{\left(\frac{dy}{dx}\right)_{(x_1, y_1)}} (x - x_1)$$

Respectively. Otherwise go to step 4.

Step 4: If $\left(\dfrac{dy}{dx}\right)_{(x_1,\ y_1)} = 0$, then the equations of the tangent and normal at $(x_1,\ y_1)$ are $y - y_1 = 0$ and $x - x_1 = 0$ repectively.

If $\left(\dfrac{dy}{dx}\right)_{(x_1,\ y_1)} = \pm \infty$, then the equations of the tangent and normal at $(x_1,\ y_1)$ are $x - x_1 = 0$ and $y - y_1 = 0$ respectively.

Illustrative Examples

Example 6: Find the equations of the tangent and normal to the parabola $y^2 = 4a\,x$ at the point $(at^2, 2at)$.

Solution: The equation of the given curve is $y^2 = 4ax$

Differentiating (i) w. r. t. x, we get

$$2y\frac{dy}{dx} = 4a \Rightarrow \frac{dy}{dx} = \frac{2a}{y} \Rightarrow \left(\frac{dy}{dx}\right)_{(at^2,\ 2at)} = \frac{2a}{2at} = \frac{1}{t}$$

So, the equation of the tangent at $(at^2, 2at)$ is

$$y - 2at = -\frac{1}{\left(\frac{dy}{dx}\right)_{(at^2,\ 2at)}}(x - at^2) \Rightarrow y - 2at = \frac{1}{\frac{1}{2}}(x - at^2)$$

$$\Rightarrow y - 2at = -t((x - at^2) \Rightarrow y + tx = 2at + at^3$$

Example 7: Find the equation of the normal to the curve $y = 2x^2 + 3\sin x$ at $x = 0$.

Solution: The equation of the given curve is curve $y = 2x^2 + 3\sin x$ $\qquad\qquad$ ....(i)

Putting $x = 0$ in (i), we get $y = 0$. So, the point of contact is $(0,0)$.

Now, $y = 2x^2 + 3\sin x$

$$\Rightarrow \frac{dy}{dx} = 4x + 3\cos x \Rightarrow \left(\frac{dy}{dx}\right)_{(0,0)} = 4 \times 0 + 3\cos 0^\circ = 3$$

So, the equation of the normal at $(0, 0)$ is

$$y - 0 = -\frac{1}{3}(x - 0) \ \left[\text{using: } y - y_1 = \frac{-1}{\left(\frac{dy}{dx}\right)_{(x_1, y_1)}}\right]$$

$$\Rightarrow x + 3y = 0$$

Example 8: Find the equations of the tangent and the normal to $\dfrac{x^2}{9} + \dfrac{y^2}{16} = 1$ $(x_1,\ y_1)$ where $x_1 = 2$ and $y_1 > 0$.

Solution: The equation of the given curve is $\dfrac{x^2}{9} + \dfrac{y^2}{16} = 1$ $\qquad\qquad$ .....(i)

Since $(x_1,\ y_1)$ lies on (i), therefore

$$16x_1^2 + 9y_1^2 = 144 \Rightarrow 16(2)^2 + 9y_1^2 = 144 \Rightarrow y_1^2 = \frac{80}{9} \Rightarrow y_1 = \frac{4\sqrt{5}}{3} \qquad [\because y_1 > 0]$$

So, the given point is $\left(2, \frac{4\sqrt{5}}{3}\right)$.

Now, $\frac{x^2}{9} + \frac{y^2}{16} = 1 \Rightarrow 32x + 10y\frac{dy}{dx} = 0$

$$\Rightarrow \frac{dy}{dx} = \frac{-16x}{9y} \Rightarrow \left(\frac{dy}{dx}\right)_{\left(2,\frac{4\sqrt{5}}{3}\right)} = -\frac{16\times2}{9\times\frac{4\sqrt{5}}{3}} = -\frac{8}{3\sqrt{5}}$$

So, the equation of the tangent at $\left(2, \frac{4\sqrt{5}}{3}\right)$ is

$$y - \frac{4\sqrt{5}}{3} = \left(\frac{dy}{dx}\right)_{\left(2,\frac{4\sqrt{5}}{3}\right)}(x-2) \Rightarrow y - \frac{4\sqrt{5}}{3} = -\frac{8}{3\sqrt{5}}(x-2)$$

$$\Rightarrow 8x + 3\sqrt{5}y - 36 = 0$$

And, the equation of the normal at $\left(2, \frac{4\sqrt{5}}{3}\right)$ is

$$y - \frac{4\sqrt{5}}{3} = -\frac{1}{\left(\frac{dy}{dx}\right)_{\left(2,\frac{4\sqrt{5}}{3}\right)}}(x-2) \Rightarrow y - \frac{4\sqrt{5}}{3} = \frac{-1}{\frac{-8}{3\sqrt{5}}}(x-2)$$

$$\Rightarrow 9\sqrt{5}x - 24y + 14\sqrt{5} = 0$$

Example 9: Find the equations of the tangent and the normal at the point 't' on the curve $x = a\sin^3 t, y = b\cos^3 t$.

Solution: We have $x = a\sin^3 t, y = b\cos^3 t$.

$$\therefore \frac{dx}{dy} = 3a\sin^2 t\cos t; \frac{dy}{dx} = -3b\cos^2 t\sin t$$

$$\Rightarrow \frac{dy}{dx} = \frac{dy/dt}{dx/dt} = \frac{3b\cos^2 t\sin t}{3a\sin^2 t\cos t} = -\frac{b}{a}\cdot\frac{\cos t}{\sin t}$$

So, the equation of the tangent at the point 't' is $y - y_1 = \left(\frac{dy}{dx}\right)_{(x_1,y_1)}(x - x_1)$

$$y - b\cos^3 t = -\frac{b\cos t}{a\sin t}(x - a\sin^3 t)$$

$$\Rightarrow bx\cos t + ay\sin t = ab\sin t\cos t$$

The equation of the normal at the point 't' is $y - y_1 = \frac{-1}{\left(\frac{dy}{dx}\right)_{(x_1,y_1)}}(x - x_1)$

$$y - b\cos^3 t = -\frac{1}{\frac{-b\cos t}{a\sin t}}(x - a\sin^3 t)$$

$$\Rightarrow ax\sin t - by\cos t = a^2\sin^4 t - b^2\cos^4 t$$

Example 10: Find the equation of the tangent line to the curve $y = \sqrt{5x - 3} - 2$ which is parallel to the line $4x - 2y + 3 = 0$.

Solution: Let the point of contact of the tangent line parallel to the given line be $P(x_1, y_1)$.

The equation of the curve is $y = \sqrt{5x - 3} - 2$.

Differentiating both sides w. r. t. x, we get

$$\frac{dy}{dx} = \frac{5/2}{\sqrt{5x-3}} \Rightarrow \left(\frac{dy}{dx}\right)_{(x_1, y_1)} = \frac{5/2}{\sqrt{5x_1-3}}$$

Since the tangent at (x_1, y_1) is parallel to the line $4x - 2y + 3 = 0$. therefore,

$$\left(\frac{dy}{dx}\right)_{(x_1, y_1)} = \text{slope of the line } 4x - 2y + 3 = 0$$

$$\Rightarrow \frac{5}{2\sqrt{5x_1-3}} = \frac{-4}{-2} \Rightarrow 4\sqrt{5x_1 - 3} = 5 \Rightarrow 16(5x_1 - 3) = 25 \Rightarrow x_1 = \frac{73}{80}$$

Since (x_1, y_1) lies on $y = \sqrt{5x_1 - 2}$. Therefore,

$$y_1 = \sqrt{5x_1 - 3} - 2 \Rightarrow y_1 = \sqrt{5 \times \frac{73}{80} - 3} - 2$$

$$\left[\because x_1 = \frac{73}{80}\right]$$

$$\Rightarrow y_1 = -\frac{3}{4}$$

So, the point of contact is $\left(\frac{73}{80}, \frac{3}{4}\right)$

Hence, the required equation of the tangent is

$$y - \left(-\frac{3}{4}\right) = 2\left(x - \frac{73}{80}\right)$$

$$\left[\because \left(\frac{dy}{dx}\right)_{(x_1, y_1)} = 2\right]$$

$$\Rightarrow 80x - 40y - 103 = 0$$

***Example* 11:** Find the equation of tangent line to $y = 2x^2 + 7$ which is parallel to the line $4x - y + 3 = 0$.

***Solution*:** Let the point of contact of the required tangent line be (x_1, y_1).

The equation of the given curve is $y = 2x^2 + 7$.

Differentiating both sides w. r. t. x, we get

$$\frac{dy}{dx} = 4x \Rightarrow \left(\frac{dy}{dx}\right)_{(x_1, y_1)} = 4x_1$$

Since the line $4x - y + 3 = 0$ is parallel to the tangent at (x_1, y_1).

$\therefore$ slope of the tangent at (x_1, y_1) =slope of the line $4x - y + 3 = 0$

$$\Rightarrow \left(\frac{dy}{dx}\right)_{(x_1, y_1)} = \frac{-4}{-1} \qquad \left[\because \text{slope} = -\frac{\text{coeff. of x}}{\text{coeff. of y}}\right]$$

$$\Rightarrow 4x_1 = 4 \Rightarrow x_1 = 1$$

Now, (x_1, y_1) lies on $y = 2x^2 + 7$.

$$\therefore \quad y_1 = 2x_1^2 + 7 \Rightarrow y_1 = 2 + 7 = 9 \ [\because x_1 = 1]$$

So, the coordinates of the point of contact are (1, 9). Hence, the required equation of the tangent line is

$$y - 9 = 4(x - 1) \Rightarrow 4x - y + 5 = 0$$

***Example* 12:** For the curve $y = 4x^3 - 2x^5$ find all points at which the tangent passes through the origin.

***Solution*:** Let (x_1, y_1) be the required point on $y = 4x^3 - 2x^5$. Then

$$y_1 = 4x_1^3 - 2x_1^5 \qquad\qquad ....(i)$$

The equation of the given curve is on $y = 4x^3 - 2x^5$. Differentiating w. r. t. x, we get

$$\frac{dy}{dx} = 12x^2 - 10x^4 \Rightarrow \left(\frac{dy}{dx}\right)_{(x_1, y_1)} = 12x_1^2 - 10x_1^4$$

So, the equation of the tangent at (x_1, y_1) is

$$y - y_1 = \left(\frac{dy}{dx}\right)_{(x_1, y_1)} (x - x_1) \Rightarrow y - y_1 = (12x_1^2 - 10x_1^4)(x - x_1)$$

This passes through the origin, therefore

$$0 - y_1 = (12x_1^2 - 10x_1^4)(0 - x_1) \Rightarrow y_1 = 12x_1^3 - 10x_1^5 \qquad\qquad ...(ii)$$

Subtracting (ii) from (i), we get

$$0 = -8x_1^3 + 8x_1^5 \Rightarrow 8x_1^3(x_1^2 - 1) = 0 \Rightarrow x_1 = 0 \text{ or } x_1 = \pm 1$$

When $x_1 = 0$, $y_1 = 0$

When $x_1 = 1$, $y_1 = 12 - 10 = 2$

When $x_1 = -1$, $y_1 = -12 + 10 = -2$

Hence, the required points are (0, 0), (1, 2) and (–1, –2).

Practice Problems

1. Find the equation of the normal to $y = 2x^3 - x^2 + 3$ at (1,4).   **[Ans.** $x = y = 0$]

2. Find the equations of the tangent and the normal to the following curves;

   (i) $y = x^2$ at $(0, 0)$ **[Ans.** $x = y = 0$]

   (ii) $y = 2x^2 - 3x - 1$ at $(1, -2)$

   (iii) $\frac{x^2}{a^2} + \frac{y^2}{b^2} = 1$ at $(a \cos\theta, b \sin\theta$

   [Ans. $\frac{x\cos\theta}{a} + \frac{y\sin\theta}{b} = 1, ax\sec\theta - by\cosec\theta = a^2 - b^2$]

   (iv) $\frac{x^2}{a^2} - \frac{y^2}{b^2} = 1$ at $(a \sec\theta, b \tan\theta$

   [Ans. $\frac{x\sec\theta}{a} - \frac{y\tan\theta}{b} = 1, ax\cos\theta + by\cot\theta = a^2 + b^2$]

(v) $y^2 = 4ax$ at $\left(\dfrac{a}{m^2}, \dfrac{2a}{m}\right)$

$$[\textbf{Ans.} m^2 x - my + a = 0, m^2 x + m^3 y - 2am^2 - a = 0]$$

3.   Find the equation of the tangent to the curve $x = \theta + \sin\theta, y = 1 + \cos\theta$ at $\theta = \dfrac{\pi}{4}$.

$$[\textbf{Ans. } y - 1 - \tfrac{1}{\sqrt{2}} = \left(1 - \sqrt{2}\right)\left(x - \tfrac{\pi}{4} - \tfrac{1}{\sqrt{2}}\right) \dfrac{a^2 x}{x_0} + \dfrac{b^2 y}{y_0} = a^2 + b^2]$$

4.   Find the equation of the normal to the curve $x^2 + 2y^2 - 4x - 6y + 8 = 0$ at the point whose abscissa is 2. $[\textbf{Ans. } x = 2]$

5.   The equation of the tangent at $(2, 3)$ on the curve $y^2 = ax^3 + b$ is $y = 4x - 5$. Find the values of a and b. $[\textbf{Ans. } a = 2, b = -7]$

6.   Determine the equation(s) of tangent(s) line to the curve $y = 4x^3 - 3x + 5$ which are perpendicular to the line $9y + x + 3 = 0$. $[\textbf{Ans. } 9x - y - 3 = 0, 9x - y + 13 = 0]$

CHAPTER – 4

Indefinite and Definite Integral

4.1   Introduction

Integration of a function is the inverse process of differentiation. If $F(x)$ is the integral of $f(x)$ then we write $\int f(x)dx = F(x)$ 'or' $\int f(x)dx = F(x) + c$.

Here, the symbol $\int$ stands for integral and dx after $f(x)$ indicates that x is the variable of integration. The function $f(x)$ which is to be integrated is called the integrand and C is an arbitrary constant.

Standard Integral Formulae

1. $\int x^n dx = \dfrac{x^{n+1}}{n+1}$

2. $\int \dfrac{1}{x}dx = \log_e x$

3. $\int e^x dx = e^x$

4. $\int a^x dx = \dfrac{a^x}{\log_e a}$

5. $\int \sin x\, dx = -\cos x$

6. $\int \cos x\, dx = \sin x$

7. $\int \sec^2 x\, dx = \tan x$

8. $\int \operatorname{cosec}^2 x\, dx = -\cot x$

9. $\int \sec x \tan x\, dx = \sec x$

10. $\int \operatorname{cosec} x \cot x\, dx = -\operatorname{cosec} x$

11. $\int 1\, dx = x$

12. $\int 0\, dx = c$

13. $\int \dfrac{1}{\sqrt{1-x^2}}dx = \sin^{-1} x$

14. $\int \dfrac{1}{1+x^2}dx = \tan^{-1} x$

15. $\int \dfrac{1}{x\sqrt{x^2-1}}dx = \sec^{-1} x$

Integral of Trigonometrical Functions

***Example* 1:** Evaluate $\int \left(\dfrac{\sec x}{\sec x - \tan x} \right) dx$.

Solution:

$$\int \frac{\sec x}{\sec x - \tan x}dx = \int \frac{\sec x}{\sec x - \tan x} \times \frac{\sec x + \tan x}{\sec x + \tan x}dx$$

$$= \int \frac{\sec^2 x + \sec x \tan x}{\sec^2 x - \tan^2 x}dx$$

$$= \int \frac{\sec^2 x + \sec x \tan x}{1} \, dx$$

$$= \int \sec^2 x \, dx + \int \sec x \tan x \, dx$$

$$= \tan x + \sec x$$

***Example* 2:** Evaluate $\displaystyle\int \frac{\sin^2 x - \cos^2 x}{\sin^2 x \cos^2 x} \, dx$.

Solution:

$$\int \frac{\sin^2 x - \cos^2 x}{\sin^2 x \cos^2 x} \, dx = \int \left(\frac{\sin^2 x}{\sin^2 x \cos^2 x} - \frac{\cos^2 x}{\sin^2 x \cos^2 x} \right) dx$$

$$= \int \left(\frac{1}{\cos^2 x} - \frac{1}{\sin^2 x} \right) dx$$

$$= \int \sec^2 x \, dx - \int \cosec^2 x \, dx$$

$$= \tan x + \cot x$$

4.2   Integration by Substitution

When we have to integrate such functions which are not in any standard form, then we substitute a proper function of variable t for the variable x.

Let $\qquad I = \int f(x) dx$

Put $\qquad x = \phi(t)$

$\therefore \qquad \dfrac{dx}{dt} = \phi'(t)$

$\therefore \qquad I = \int f(x)\,\phi'(t)\,dt$

$\therefore \qquad \int f(x)\,dx = \int f\big[\phi(t)\big]\phi'(t)\,dt$

Some Important Results

(i) $\displaystyle\int \sin(ax+b)\,dx = \frac{-1}{a}\cos(ax+b)$
(ii) $\displaystyle\int \cos(ax+b)\,dx = \frac{1}{a}\sin(ax+b)$

(iii) $\displaystyle\int \sec^2(ax+b)\,dx = \frac{1}{a}\tan(ax+b)$
(iv) $\displaystyle\int e^{ax}\,dx = \frac{1}{a}e^{ax}$

(v) $\displaystyle\int \frac{1}{ax+b} = \frac{1}{a}\log(ax+b)$
(vi) $\displaystyle\int (ax+b)^n\,dx + \frac{1}{a(n+1)}(ax+b)^{n+1}$

(vii) $\int \sec(ax+b)\tan(ax+b)\,dx = \dfrac{1}{a}\sec(ax+b)$

(viii) $\int \sin ax\,dx = \dfrac{-1}{a}\cos ax$

(ix) $\int \cos ax\,dx = \dfrac{1}{a}\sin ax$ $\qquad\qquad$ (x) $\int \sec^2 ax\,dx = \dfrac{1}{a}\tan ax$

(xi) $\int \sec ax\,\tan ax\,dx = \dfrac{1}{a}\sec ax$ $\qquad$ (xii) $\int \operatorname{cosec} ax\,\cot ax\,dx = \dfrac{-1}{a}\operatorname{cosec} ax$

***Example* 3:** Evaluate $\int \cos x\,\cos 2x\,\cos 3x\,dx$

***Solution*:** Let $\qquad\qquad I = \int \cos x\,\cos 2x\,\cos 3x\,dx$

$$\text{Now, } \cos x\,\cos 2x\,\cos 3x = \frac{1}{2}\cdot 2\cos 3x\,\cos x\,\cos 2x$$

$$= \frac{1}{2}\,[\cos 4x + \cos 2x]\cos 2x$$

$$= \frac{1}{2}\,[\cos 4x\,\cos 2x + \cos^2 2x]$$

$$= \frac{1}{4}\,[\cos 6x + \cos 2x] + \frac{1}{2}\left[\frac{1+\cos 4x}{2}\right]$$

$$= \frac{1}{4}\,[\cos 6x + \cos 2x + \cos 4x + 1]$$

$$\therefore \qquad I = \frac{1}{4}\int [\cos 6x + \cos 4x + \cos 2x + 1]\,dx$$

$$= \frac{1}{4}\left[\frac{\sin 6x}{6} + \frac{\sin 4x}{4} + \frac{\sin 2x}{2} + x\right]$$

***Example* 4:** Evaluate $\int x^2 \sin x^3\,dx$

***Solution*:** Let $\qquad\qquad I = \int x^2 \sin x^3\,dx$

$$\text{Put} \qquad x^3 = t \Rightarrow 3x^2\,dx = dt \Rightarrow dx = \frac{1}{3x^2}\,dt$$

$$\therefore \qquad I = \int x^2 \sin t \cdot \frac{1}{3x^2}\,dt \;= \frac{1}{3}\int \sin t\,dt$$

$$= \frac{-1}{3}\cos t \;= \frac{-1}{3}\cos x^3$$

4.3   Indefinite Integrals of Additional Standard Form

Formula

(i) $\displaystyle\int \frac{dx}{\sqrt{a^2 - x^2}} = \sin^{-1}\frac{x}{a}$

(ii) $\displaystyle\int \frac{dx}{a^2 + x^2} = \frac{1}{a}\tan^{-1}\frac{x}{a}$

(iii) $\displaystyle\int \frac{dx}{x^2 - a^2} = \frac{1}{2a}\log\frac{x-a}{x+a}, x > a$

(iv) $\displaystyle\int \frac{dx}{a^2 - x^2} = \frac{1}{2a}\log\frac{a+x}{a-x}, x < a$

(v) $\displaystyle\int \frac{dx}{x\sqrt{x^2 - a^2}} = \frac{1}{a}\sec^{-1}\frac{x}{a}$

(vi) $\displaystyle\int \frac{dx}{\sqrt{x^2 + a^2}} = \log(x + \sqrt{x^2 + a^2})$

(vii) $\displaystyle\int \sqrt{a^2 - x^2}\,dx = \frac{1}{2}\left[x\sqrt{a^2 - x^2} + a^2 \sin^{-1}\frac{x}{a}\right]$

(viii) $\displaystyle\int \sqrt{a^2 + x^2}\,dx = \frac{1}{2}\left[x\sqrt{a^2 + x^2} + a^2 \log(x + \sqrt{a^2 + x^2})\right]$

(ix) $\displaystyle\int \sqrt{x^2 - a^2}\,dx = \frac{1}{2}\left[x\sqrt{x^2 - a^2} - a^2 \log(x + \sqrt{x^2 - a^2})\right]$

(x) $\displaystyle\int \tan x\, dx = \log(\sec x) = -\log(\cos x)$

(xi) $\displaystyle\int \cot x\, dx = \log(\sin x)$

(xiii) $\displaystyle\int \operatorname{cosec} x\, dx = \log(\operatorname{cosec} x - \cot x)$

(xiii) $\displaystyle\int \sec x\, dx = \log(\sec x + \tan x)$

Example 5: Evaluate $\displaystyle\int \frac{dx}{\sqrt{9 - 25x^2}}$

Solution: $\displaystyle\int \frac{dx}{\sqrt{9 - 25x^2}} = \frac{1}{5}\int \frac{dx}{\sqrt{\dfrac{9}{25} - x^2}}$

$$= \frac{1}{5}\int \frac{dx}{\sqrt{\left(\dfrac{3}{5}\right)^2 - x^2}} \qquad = \frac{1}{5}\sin^{-1}\frac{5}{3}x$$

Example 6: Evaluate $\displaystyle\int \sqrt{2 - 3x^2}\,dx$

Solution: $\displaystyle\int \sqrt{2 - 3x^2}\,dx = \sqrt{3}\int \sqrt{\frac{2}{3} - x^2}\,dx$

$$= \sqrt{3}\int \sqrt{\left(\frac{\sqrt{2}}{\sqrt{3}}\right)^2 - x^2}\,dx$$

$$= \sqrt{3}\cdot\frac{1}{2}\left[x\sqrt{\frac{2}{3}-x^2} + \left(\frac{\sqrt{2}}{\sqrt{3}}\right)^2 \sin^{-1}\frac{x}{\frac{\sqrt{2}}{\sqrt{3}}}\right]$$

$$\left(\text{by } \int\sqrt{a^2-x^2}\,dx\right)$$

$$= \frac{\sqrt{3}}{2}\left[x\sqrt{\frac{2}{3}-x^2} + \frac{2}{3}\sin^{-1}\frac{\sqrt{3}x}{\sqrt{2}}\right]$$

4.4   Integration by Parts

The integration of a product of two functions can be easily evaluated by the rule of integration by parts.

Method: $\int u.v\, dx = u\int v\, dx - \int\left[\frac{du}{dx}\int v\, dx\right]dx$

***Example* 7**: Solve $\int x\cos^2 x\, dx$

Solution: Let $\qquad I = \int x\cos^2 x\, dx$

$$= \int x\left(\frac{1+\cos 2x}{2}\right)dx$$

$$= \frac{1}{2}\int x(1+\cos x)dx$$

$$= \frac{1}{2}\int x\,dx + \frac{1}{2}\int x\cos x\, dx$$

Integrating by parts, we get

$$I = \frac{1}{2}\cdot\frac{x^2}{2} + \frac{1}{2}\left\{ x\int\cos x\,dx - \int\left[\frac{d}{dx}x\int\cos x\, dx\right]dx\right\}$$

$$= \frac{x^2}{4} + \frac{1}{2}\left\{ x\sin x - \int\sin x\, dx\right\}$$

$$= \frac{x^2}{4} + \frac{1}{2}\left\{ x\sin x + \cos x\right\}$$

***Example* 8**: Solve $\int\log_e x\, dx$

Solution: Let $\qquad I = \int\log_e x\, dx$

$$= \int\log_e x \,.\, 1\,dx$$

Integrating by parts, we get

$$I = \log_e x \int 1\, dx - \int \left[\frac{d}{dx} \log_e x \int 1\, dx \right] dx$$

$$= \log_e x.x - \int \frac{1}{x}.x\, dx \quad = x \log_e x - \int 1\, dx$$

$$= x \log_e x - x \quad = x(\log_e x - 1)$$

Formulae

(i) $\displaystyle \int e^{ax} \sin bx\, dx = \frac{e^{ax}}{a^2 + b^2}(a \sin bx - b \cos bx)$

(ii) $\displaystyle \int e^{ax} \sin bx\, dx = \frac{e^{ax}}{\sqrt{a^2 + b^2}} \sin(bx - \tan^{-1}\frac{b}{a})$

(iii) $\displaystyle \int e^{ax} \cos bx\, dx = \frac{e^{ax}}{a^2 + b^2}(a \cos bx + b \sin bx)$

(iv) $\displaystyle \int e^{ax} \cos bx\, dx = \frac{e^{ax}}{\sqrt{a^2 + b^2}} \cos(bx - \tan^{-1}\frac{b}{a})$

***Example* 9:** Solve $\displaystyle \int e^{3x} \sin 4x\, dx$

***Solution*:** Let $\qquad I = \displaystyle\int e^{3x} \sin 4x\, dx$

by compare $\displaystyle\int e^{ax} \sin bx\, dx$, $\quad a = 3, b = 4$

by using formula $\displaystyle\int e^{ax} \sin bx\, dx = \frac{e^{ax}}{a^2 + b^2}(a \sin bx - b \cos bx)$

$\therefore \qquad I = \displaystyle\int e^{3x} \sin 4x\, dx = \frac{e^{3x}}{9 + 16}(3 \sin 4x - 4 \cos 4x)$

$$= \frac{e^{3x}}{25}(3 \sin 4x - 4 \cos 4x)$$

4.5   Integration by Partial Fraction

Let $y = \dfrac{f(x)}{g(x)}$, where f(x) and g(x) are polynomial functions. If the degree of g(x) is more than that of f(x), we call such a function a partial fraction.

Rules

1. When the denominator of given function contains non repeated linear factor. Then,

$$\frac{f(x)}{(x+a)(x+b)(x+c)} = \frac{A}{x+a} + \frac{B}{x+b} + \frac{C}{x+c}$$

2. When the denominator contains repeated linear factors. Then,

$$\frac{f(x)}{(x+a)(x+b)^n} = \frac{A}{x+a} + \frac{B}{x+b} + \frac{C}{(x+b)^2} + ... + \frac{N}{(x+b)^n}$$

3. When the denominator contains a quadratic factor. Then,

$$\frac{f(x)}{(x+a)(x^2+b^2)} = \frac{A}{x+a} + \frac{Bx+C}{x^2+b^2}$$

4. When the denominator contains repeated quadratic factors. Then

$$\frac{f(x)}{(x^2+a^2)^2(x+b)} = \frac{Ax+B}{x^2+a^2} + \frac{Cx+D}{(x^2+a^2)^2} + \frac{E}{x+b}$$

***Example* 10:** Find the value of $\displaystyle\int \frac{x^2}{(x+1)(x+2)(x+3)}dx$

***Solution*:** Let

$$I = \frac{x^2}{(x+1)(x-2)(x+3)} = \frac{A}{x+1} + \frac{B}{x-2} + \frac{C}{x+3}$$

$$x^2 = A(x-2)(x+3) + B(x+1)(x+3) + (x+1)(x-2) \qquad .....(i)$$

Put $\qquad x+1=0 \Rightarrow x=-1$, in (1), we get

$$1 = A(-3)2 \quad \Rightarrow \quad 1 = -6A \qquad \therefore A = \frac{-1}{6}$$

Put $\qquad x-2=0 \qquad \Rightarrow \qquad x=2 \quad$ in (1)

$$4 = B(3)(5) \quad \Rightarrow \quad B = \frac{4}{15}$$

Put $\qquad x+3=0 \qquad \Rightarrow \qquad x=-3 \quad$ in (1)

$$9 = c(-2)(-5) \Rightarrow \quad c = \frac{9}{10}$$

$$\therefore \int \frac{x^2}{(x+1)(x-2)(x+3)}dx = \frac{-1}{6}\int\frac{1}{x+1}dx + \frac{4}{15}\int\frac{1}{x-2}dx + \frac{9}{10}\int\frac{1}{x+3}dx$$

$$= \frac{-1}{6}\log(x+1) + \frac{4}{15}\log(x-2) + \frac{9}{10}\log(x+3)$$

4.6   Definite Integral

If
$$\int f(x)dx = F(x)$$

Then
$$\int_a^b f(x)dx = F(b) - F(a)$$

Integral $\int_a^b f(x)dx$ is called the definite integral of f(x) between the limits a and b, where a is called lower limit and b is upper limit. The interval (a, b) is called the range of integration.

***Example* 11:** Evaluate the following

$$(i) \quad \int_1^{\sqrt{3}} \frac{dx}{1+x^2} \qquad\qquad (ii) \quad \int_0^a \frac{dx}{\sqrt{a^2 - x^2}}$$

***Solution*:** (i) $\displaystyle\int_1^{\sqrt{3}} \frac{dx}{1+x^2} = \left[\tan^{-1} x\right]_1^{\sqrt{3}}$

$$= \tan^{-1}\sqrt{3} - \tan^{-1} 1$$

$$= \frac{\pi}{3} - \frac{\pi}{4} = \frac{\pi}{12}$$

(ii) $\displaystyle\int_0^a \frac{dx}{\sqrt{a^2 - x^2}} = \left[\sin^{-1} \frac{x}{a}\right]_0^a$

$$= \sin^{-1}\frac{a}{a} - \sin^{-1}\frac{0}{a}$$

$$= \sin^{-1} 1 - \sin^{-1} 0$$

$$= \frac{\pi}{2} - 0 = \frac{\pi}{2}$$

Properties of Definite Integrals

1. $\displaystyle\int_a^b f(x)\, dx = \int_a^b f(t)\, dt$

2. $\displaystyle\int_a^b f(x)\, dx = -\int_b^a f(x)\, dx$

3. $\displaystyle\int_a^b f(x)\, dx = -\int_a^c f(x)\, dx + \int_c^b f(x)\, dx$

4. $\displaystyle\int_0^a f(x)\, dx = \int_0^a f(a - x)\, dx$

5. $\displaystyle\int_{-a}^{a} f(x)\,dx = 2\int_{0}^{a} f(x)\,dx$, if $f(x)$ is even function of x.

6. $\displaystyle\int_{-a}^{a} f(x)\,dx = 0$, if $f(x)$ is odd function of x.

7. $\displaystyle\int_{a}^{2a} f(x)\,dx = 2\int_{0}^{a} f(x)\,dx$,   if $f(2a-x)\,f(x)$

8. $\displaystyle\int_{a}^{2a} f(x)\,dx = 0$,   if $f(2a-x) = -f(x)$

***Example* 12:** Evaluate $\displaystyle\int_{-\frac{\pi}{2}}^{\frac{\pi}{2}} \sin^5 x\,dx$

***Solution*:** Here $f(x) = \sin^5 x$

$\Rightarrow$ $f(-x) = \sin^5(-x) = -\sin^5 x$

Since function $f(x)$ is odd function, then by property 6

$$\int_{-\frac{\pi}{2}}^{\frac{\pi}{2}} \sin^5 x\,dx = 0 \qquad\qquad \left(\because \int_{-a}^{a} f(x)\,dx = 0, f(x) \text{ is odd}\right)$$

***Example* 13:** Prove that $\displaystyle\int_{0}^{\frac{\pi}{2}} \frac{\sqrt{\sin x}}{\sqrt{\cos x}+\sqrt{\sin x}}\,dx = \frac{\pi}{4}$

***Solution*:** Let $\displaystyle I = \int_{0}^{\frac{\pi}{2}} \frac{\sqrt{\sin x}}{\sqrt{\cos x}+\sqrt{\sin x}}\,dx$ ....(i)

$$= \int_{0}^{\frac{\pi}{2}} \frac{\sqrt{\sin\left(\frac{\pi}{2}-x\right)}}{\sqrt{\cos(\frac{\pi}{2}-x)}+\sqrt{\sin(\frac{\pi}{2}-x)}}\,dx$$

$$\left(\because \int_{0}^{a} f(x)\,dx = \int_{0}^{a} f(a-x)\,dx\right)$$

$$I = \int_{0}^{\frac{\pi}{2}} \frac{\sqrt{\cos x}}{\sqrt{\sin x} + \sqrt{\cos x}} \, dx \qquad\qquad\qquad .....(ii)$$

Adding (1) and (2), we get

$$2I = \int_{0}^{\frac{\pi}{2}} \frac{\sqrt{\sin x} + \sqrt{\cos x}}{\sqrt{\sin x} + \sqrt{\cos x}} \, dx$$

$$= \int_{0}^{\frac{\pi}{2}} 1 \, dx = \left[x \right]_{0}^{\frac{\pi}{2}}, \quad I = \frac{\pi}{4} \quad \text{proved}$$

***Example* 14:** Evaluate

(i) $\displaystyle\int \frac{dx}{\sqrt{x+a} + \sqrt{x+b}}$ 　　　 (ii) $\displaystyle\int (3\cosec^2 x + 2\sin 3x)\, dx$

(iii) $\displaystyle\int \sin 2x \cos 3x \, dx$

***Solution*:** (i) 　　$\displaystyle\int \frac{dx}{\sqrt{x+a} + \sqrt{x+b}} = \int \frac{\sqrt{x+a} - \sqrt{x+b}}{\left(\sqrt{x+a} + \sqrt{x+b}\right)\left(\sqrt{x+a} - \sqrt{x+b}\right)} \, dx$

$$= \int \frac{\sqrt{x+a} - \sqrt{x+b}}{x+a-x-b} \, dx$$

$$= \frac{1}{a-b} \int \sqrt{x+a} \, dx - \frac{1}{a-b} \int \sqrt{x+b} \, dx$$

$$= \frac{1}{a-b} \left[\frac{(x+a)^{\frac{1}{2}+1}}{\frac{1}{2}+1} - \frac{(x+b)^{\frac{1}{2}+1}}{\frac{1}{2}+1} \right]$$

$$= \frac{1}{a-b} \left[\frac{(x+a)^{\frac{3}{2}}}{\frac{3}{2}} - \frac{(x+b)^{\frac{3}{2}}}{\frac{3}{2}} \right]$$

$$= \frac{2}{3(a-b)} \left[(x+a)^{\frac{3}{2}} - (x+b)^{\frac{3}{2}} \right]$$

(ii) 　　$\displaystyle\int (3\cosec^2 x + 2\sin 3x)\, dx = 3\int \cosec^2 x \, dx + 2\int \sin 3x \, dx$

$$= -3\cot x - 2\,\frac{\cos 3x}{3}$$

$$= -3\cot x - \frac{2}{3}\cos 3x$$

(iii) $\displaystyle\int \sin 2x \cos 3x \, dx = \frac{1}{2}\int 2\sin 2x \cos 3x \, dx$

$$= \frac{1}{2}\int [\, \sin(2x+3x) + \sin(2x-3x)\,]\, dx$$

$$= \frac{1}{2}\int [\, \sin 5x - \sin x\,]\, dx$$

$$= \frac{1}{2}\left[\int \sin 5x \, dx - \int \sin x \, dx \right]$$

$$= \frac{1}{2}\left[\frac{-\cos 5x}{5} + \cos x\right]$$

Example 15: Evaluate

(i) $\displaystyle\int \sin^3 x \, dx$ (ii) $\displaystyle\int \frac{\cos 2\theta - \cos 2\alpha}{\cos\theta - \cos\alpha}\, d\theta$

Solution: (i) $\displaystyle\int \sin^3 x \, dx = \int \frac{1}{4}(3\sin x - \sin 3x)\, dx$

$$(\because \sin^3 A = 3\sin A - 4\sin^3 A)$$

$$= \frac{1}{4}\left[3\int \sin x \, dx - \int \sin 3x \, dx\right]$$

$$= \frac{-3}{4}\cos x + \frac{1}{4}\times \frac{\cos 3x}{3}$$

$$= \frac{1}{12}\cos 3x - \frac{3}{4}\cos x$$

(ii) $\displaystyle\int \frac{\cos 2\theta - \cos 2\alpha}{\cos\theta - \cos\alpha}\, d\theta$

$$= \int \frac{2\cos^2\theta - 1 - 2\cos^2\alpha + 1}{\cos\theta - \cos\alpha}\, d\theta$$

$$= 2\int \frac{\cos^2\theta - \cos^2\alpha}{\cos\theta - \cos\alpha}\, d\theta$$

$$= 2\int \frac{(\cos\theta + \cos\alpha)(\cos\theta - \cos\alpha)}{\cos\theta - \cos\alpha}\, d\theta$$

$$= 2\int (\cos\theta + \cos\alpha)\, d\theta$$

$$= 2\,[\sin\theta + \theta\cos\alpha]$$

Example 16: Evaluate

$$\text{(i)} \quad \int \frac{x^2 \tan^{-1} x^3}{1+x^6}\,dx \qquad \text{(ii)} \quad \int x^{n-1}\sin x^n\,dx$$

Solution: (i)

$$\int \frac{x^2 \tan^{-1} x^3}{1+x^6}\,dx$$

Put $x^3 = t \quad \Rightarrow \quad 3x^2\,dx = dt$

$$\therefore \quad \int \frac{x^2 \tan^{-1} x^3}{1+(x^3)^2}\,dx = \frac{1}{3}\int \frac{\tan^{-1} t}{1+t^2}\,dt$$

Put $\tan^{-1} t = 4 \quad \Rightarrow \quad \dfrac{1\,dt}{1+t^2} = du$

$$\therefore \quad \frac{1}{3}\int \frac{\tan^{-1} t}{1+t^2}\,dt = \frac{1}{3}\int u\,du$$

$$= \frac{1}{3} \times \frac{4^2}{2}$$

$$= \frac{4^2}{6}$$

$$-\frac{1}{6}\left(\tan^{-1} t\right)^2$$

(ii) $\displaystyle \int x^{n-1}\sin x^n\,dx$

Let $x^n = t \quad \Rightarrow \quad nx^{n-1}\,dx = dt$

$$\therefore \quad \int x^{n-1}\sin x^n\,dx = \frac{1}{n}\int \sin t\,dt$$

$$= \frac{-1}{n}\cos t$$

$$= \frac{-1}{n}\cos x^n$$

Example 17: Evaluate

$$\text{(i)} \quad \int \frac{1}{x}\log x\,dx \qquad\qquad \text{(ii)} \quad \int \cos^2 x\,dx$$

$$(iii) \quad \int \frac{\sin(\tan^{-1} x)}{1+x^2} dx$$

Solution: (i)
$$\int \frac{1}{x} \log x \, dx$$

Let $\log x = t \Rightarrow \dfrac{1}{x} dx = dt$

$$\therefore \quad \int \frac{1}{x} \log x \, dx = \int t \, dt = \frac{t^2}{2} = \frac{1}{2} (\log x)^2$$

(ii)
$$\int \cos^2 x \, dx = \int \frac{1 + \cos 2x}{2} dx$$
$$= \frac{1}{2} \left[\int 1 dx + \int \cos 2x \, dx \right]$$
$$= \frac{1}{2} \left[x + \frac{\sin 2x}{2} \right]$$

(iii)
$$\int \frac{\sin(\tan^{-1} x)}{1+x^2} dx$$

Let $\tan^{-1} x = t \Rightarrow \dfrac{1}{1+x^2} dx = dt$

$$\therefore \quad \int \frac{\sin(\tan^{-1} x)}{1+x^2} dx = \int \sin t \, dt$$
$$= -\cos t$$
$$= -\cos(\tan^{-1} x)$$

Example 18: Evaluate $\displaystyle \int \frac{e^x - e^{-x}}{e^x + e^{-x}} dx$

Solution:
$$\int \frac{e^x - e^{-x}}{e^x + e^{-x}} dx$$

Let $\quad e^x + e^{-x} = t \Rightarrow (e^x - e^{-x}) \, dx = dt$

$$\therefore \quad \int \frac{e^x - e^{-x}}{e^x + e^{-x}} dx = \int \frac{1}{t} dt$$
$$= \log t = \log (e^x + e^{-x})$$

Example 19: Integrate

$$(i) \quad \int \frac{1}{x \log x} dx \qquad (ii) \quad \int \sin(e^x).e^x dx$$

Solution: (i) $\displaystyle\int \frac{1}{x \log x} dx$

Let $\log x = t \;\Rightarrow\; \dfrac{1}{x} dx = dt$

$\therefore \quad \displaystyle\int \frac{1}{x \log x} dx = \int \frac{1}{t} dt$

$$= \log t = \log(\log x)$$

(ii) $\displaystyle\int \sin(e^x).e^x dx$

Let $e^x = t \;\Rightarrow\; e^x dx = dt$

$\therefore \quad \displaystyle\int \sin(e^x).e^x dx = \int \sin t \, dt$

$$= -\cos t = -\cos(e^x)$$

***Example* 20:** Integrate

(i) $\displaystyle\int \frac{1 + \cos x}{x + \sin x} dx$ (ii) $\displaystyle\int \frac{\sec x}{\log(\sec x + \tan x)} dx$

(iii) $\displaystyle\int \frac{e^x - \sin x}{e^x + \cos x} dx$ (iv) $\displaystyle\int \frac{dx}{\sqrt{1 + \cos 2x}}$

(v) $\displaystyle\int \frac{e^x(1 + x)}{\cos^2(xe^x)} dx$ (vi) $\displaystyle\int xe^{x^2} dx$

Solution: (i) $\displaystyle\int \frac{1 + \cos x}{x + \sin x} dx$

Let $x + \sin x = t \;\Rightarrow\; (1 + \cos x) dx = dt$

$\therefore \quad \displaystyle\int \frac{1 + \cos x}{x + \sin x} dx = \int \frac{dt}{t}$

$$= \log t = \log(x + \sin x)$$

(ii) $\displaystyle\int \frac{\sec x}{\log(\sec x + \tan x)} dx$

Let $\log(\sec x + \tan x) = t \;\Rightarrow\; \dfrac{1}{(\sec x + \tan x)}$

$(\sec x \tan x + \sec^2 x) dx = dt$

$\Rightarrow\; \sec x \, dx = dt$

$$\therefore \quad \int \frac{\sec x \, dx}{\log(\sec x + \tan x)} = \int \frac{dt}{t}$$

$$= \log t = \log \log(\sec x + \tan x)$$

(iii) $\displaystyle\int \frac{e^x - \sin x}{e^x + \cos x} dx$

Let $e^x + \cos x = t \Rightarrow (e^x - \sin x) \, dx = dt$

$$\therefore \quad \int \frac{e^x - \sin x}{e^x + \cos x} dx = \int \frac{dt}{t}$$

$$= \log t = \log(e^x + \cos x)$$

(iv) $\displaystyle\int \frac{dx}{\sqrt{1 + \cos 2x}} = \int \frac{dx}{\sqrt{2\cos^2 x}}$

$$= \frac{1}{\sqrt{2}} \int \frac{dx}{\cos x}$$

$$= \frac{1}{\sqrt{2}} \int \sec x \, dx$$

$$= \frac{1}{\sqrt{2}} \log \tan \left(\frac{\pi}{4} + \frac{x}{2} \right)$$

(v) $\displaystyle\int \frac{e^x (1 + x)}{\cos^2 (xe^x)} dx$

Let $xe^x = t \Rightarrow (xe^x + e^x) \, dx = dt$

$$\Rightarrow e^x(x + 1) \, dx = dt$$

$$\therefore \quad \int \frac{e^x (1 + x)}{\cos^2 (xe^x)} dx = \int \frac{dt}{\cos^2 t}$$

$$= \int \sec^2 t \, dt$$

$$= \tan t$$

$$= \tan(xe^x)$$

(vi) $\displaystyle\int xe^{x^2} dx$

Let $x^2 = t \Rightarrow 2x \, dx = dt$

$$\therefore \qquad \int xe^{x^2}dx = \frac{1}{2}\int e^t dt$$

$$= \frac{1}{2}\,e^t = \frac{1}{2}e^{x^2}$$

***Example* 21:** Evaluate

(i) $\quad \displaystyle\int \sqrt{\frac{a+x}{a-x}}\,dx$, $\qquad$ (ii) $\quad \displaystyle\int \sqrt{2-3x}\;dx$

Solution: (i) $\quad \displaystyle\int \sqrt{\frac{a+x}{a-x}}\,dx$

Put $\;x = a\cos\theta \;\Rightarrow\; dx = -a\sin\theta\,d\theta$

$$\therefore \quad \int \sqrt{\frac{a+x}{a-x}}\,dx = \int \sqrt{\frac{a+a\cos\theta}{a-a\cos\theta}}(-a\sin\theta)\,d\theta$$

$$= -a\int \sqrt{\frac{1+\cos\theta}{1-\cos\theta}}\cdot\sin\theta\,d\theta$$

$$= -a\int \sqrt{\frac{2\cos^2\dfrac{\theta}{2}}{2\sin^2\dfrac{\theta}{2}}}\cdot\sin\theta\,d\theta$$

$$= -a\int (\theta + \sin\theta)$$

$$= -a\left[\cos^{-1}\frac{x}{a} + \frac{\sqrt{a^2-x^2}}{a}\right]$$

(ii) $\quad \displaystyle\int \sqrt{2-3x^2}\,dx = \sqrt{3}\int \sqrt{\frac{2}{3}-x^2}\,dx$

$$= \sqrt{3}\int \sqrt{\left(\frac{2}{3}\right)^2 - x^2}\,dx$$

$$\left[\because \sqrt{a^2-x^2} = \frac{x}{2}\sqrt{a^2-x^2} + \frac{1}{2}a^2\sin^{-1}\frac{x}{a}\right]$$

$$= \sqrt{3}\left[\frac{x}{2}\sqrt{\frac{2}{3}-x^2} + \frac{1}{2}\times\frac{2}{3}\sin^{-1}\frac{x}{\sqrt{\frac{2}{3}}}\right]$$

$$\left[\because \sqrt{a^2-x^2} = \frac{x}{2}\sqrt{a^2-x^2} + \frac{1}{2}a^2\sin^{-1}\frac{x}{a}\right]$$

$$= \sqrt{3}\left[\frac{x}{2}\sqrt{\frac{2-3x^2}{3}} + \frac{1}{3}\sin^{-1}x\sqrt{\frac{3}{2}}\right]$$

$$= \frac{x}{2}\sqrt{2-3x^2} + \frac{1}{\sqrt{3}}\sin^{-1}x\frac{\sqrt{3}}{2}$$

Example **22:** Evaluate

(i) $\displaystyle\int x^2\sin 3x\, dx$ (ii) $\displaystyle\int x\cos x\, dx$

(iii) $\displaystyle\int x^2\cos x\, dx$ (iv) $\displaystyle\int xe^x dx$

(v) $\displaystyle\int x\sin dx$

Solution: (i) $\displaystyle\int x^2\sin 3x\, dx = x^2\int\sin 3x\, dx - \int\left[\frac{d}{dx}x^2\int\sin 3x\, dx\right]dx$

$$= -x^2\frac{\cos 3x}{3} + \int 2x\frac{\cos 3x}{3}dx$$

$$= \frac{-x^2\cos 3x}{3} + \frac{2}{3}\left\{x\int\cos 3x\, dx - \int\left[\frac{d}{dx}x\int\cos 3x\, dx\right]dx\right\}$$

$$= \frac{-x^2\cos 3x}{3} + \frac{2}{3}\left\{x\frac{\sin 3x}{3} - \int\frac{\sin 3x}{3}dx\right\}$$

$$= \frac{-x^2\cos 3x}{3} + \frac{2}{9}x\sin 3x + \frac{2}{9}\frac{\cos 3x}{3}$$

$$= \frac{-1}{3}x^2\cos 3x + \frac{2}{9}x\sin 3x + \frac{2}{27}\cos 3x$$

(ii) $\displaystyle\int x\cos x\, dx = x\int\cos x\, dx - \int\left[\frac{d}{dx}x\int\cos x\, dx\right]dx$

$$= x\sin x - \int\sin x\, dx$$

$$= x \sin x + \cos x$$

(iii) $\displaystyle \int x^2 \cos x \, dx = x^2 \int \cos x \, dx - \int \left[\frac{d}{dx} x^2 \int \cos x \, dx \right] dx$

$$= x^2 \sin x - \int 2x \sin x \, dx$$

$$= x^2 \sin x - 2 \left\{ x \int \sin x \, dx - \int \left[\frac{d}{dx} x \int \sin x \, dx \right] dx \right\}$$

$$= x^2 \sin x - 2 \left\{ -x \cos x + \int \cos x \, dx \right\}$$

$$= x^2 \sin x + 2x \cos x - 2 \sin x$$

(iv) $\displaystyle \int x e^x dx = x \int e^x dx - \int \left[\frac{d}{dx} x \int e^x dx \right] dx$

$$= x e^x - \int e^x dx$$

$$= x e^x - e^x$$

$$= e^x (x - 1)$$

(v) $\displaystyle \int x \sin dx = x \int \sin x \, dx - \int \left[\frac{d}{dx} x \int \sin x \, dx \right] dx$

$$= -x \cos x + \int \cos x \, dx$$

$$= -x \cos x + \sin x$$

Example 23: Evaluate

(i) $\displaystyle \int \frac{\cot x}{\log(\sin x)} dx$ (ii) $\displaystyle \int \frac{\sin x \cos x}{a \cos^2 x + b \sin^2 x} dx$

Solution: (i) $\displaystyle \int \frac{\cot x}{\log(\sin x)} dx$

Let $\log (\sin x) = t \implies \cot x \, dx = dt$

$\therefore$ $\displaystyle \int \frac{\cot x \, dx}{\log(\sin x)} = \int \frac{dt}{t}$

$$= \log t$$

$$= \log \log(\sin x)$$

(ii) $\displaystyle \int \frac{\sin x \cos x}{a \cos^2 x + b \sin^2 x} dx$

Let $\qquad a\cos^2 x + b\sin^2 x = t$

$\Rightarrow \qquad (-2a \sin x \cos x + 2b \sin x \cos x)\, dx = dt$

$\Rightarrow \qquad 2(b - a) \sin x \cos x\, dx = dt$

$$\therefore \quad \int \frac{\sin x \cos x}{a\cos^2 x + b\sin^2 x} = \frac{1}{2(b-a)} \int \frac{dt}{t}$$

$$= \frac{1}{2(b-a)} \log t$$

$$= \frac{1}{2(b-a)} \log(a\cos^2 x + b\sin^2 x)$$

***Example* 24:** Evaluate

(i) $\displaystyle\int x \log x\, dx$ \qquad (ii) $\displaystyle\int \tan^{-1} x\, dx$

(iii) $\displaystyle\int \log x\, dx$ \qquad (iv) $\displaystyle\int \log(1 + x^2)\, dx$

***Solution*:** (i) $\displaystyle\int x \log x\, dx = \log x \int x\, dx - \int \left[\frac{d}{dx} \log x \int x\, dx\right] dx$

$$= \log x \cdot \frac{x^2}{2} - \int \frac{1}{x} \cdot \frac{x^2}{2}\, dx$$

$$= \frac{x^2}{2} \log x - \frac{1}{2} \int x\, dx$$

$$= \frac{x^2}{2} \log x - \frac{x^2}{4}$$

(ii) $\displaystyle\int \tan^{-1} x\, dx = \tan^{-1} x \int 1\, dx - \int \left[\frac{d}{dx} \tan^{-1} x \int 1\, dx\right] dx$

$$= \tan^{-1} x \cdot x - \int \frac{1}{1 + x^2} \cdot x\, dx$$

$$= x \tan^{-1} x - \frac{1}{2} \int \frac{2x}{1 + x^2}\, dx$$

$$= x \tan^{-1} x - \frac{1}{2} \log(1 + x^2)$$

(iii) $\displaystyle\int \log x\, dx = \log x \int 1\, dx - \int \left[\frac{d}{dx} \log x \int 1\, dx\right] dx$

$$= x \log x - \int \frac{1}{x} . x \, dx$$

$$= x \log x - x$$

$$= x (\log x - 1)$$

(iv) $\quad \int \log(1 + x^2) \, dx = \log(1 + x^2) \int 1 dx - \int \left[\frac{d}{dx} \log(1 + x^2) \int 1 dx \right] dx$

$$= x \log (1 + x^2) - \int \frac{1}{1 + x^2} . 2x . x \, dx$$

$$= x \log (1 + x^2) - 2 \int \frac{x^2}{x^2 + 1} dx$$

$$= x \log (1 + x^2) - 2 \int \left(1 - \frac{1}{x^2 + 1} \right) dx$$

$$= x \log (1 + x^2) - 2 (x - \tan^{-1} x)$$

$$= x \log(1 + x^2) - 2x + 2 \tan^{-1} x$$

***Example* 25:** Evaluate

    (i) $\quad \int e^{mx} \sin x \cos x \, dx$ (ii) $\quad \int e^{3x} \cos^3 x \, dx$

    (iii) $\quad \int e^x (\sin x + \cos x) \, dx$

***Solution*:**

    (i) $\quad \int e^{mx} \sin x \cos x \, dx$

$$= \frac{1}{2} \int e^{mx} (2 \sin x \cos x) \, dx$$

$$= \frac{1}{2} \int e^{mx} \sin 2x \, dx$$

$$= \frac{1}{2} . \frac{e^{mx}}{\sqrt{m^2 + 4}} \sin \left(2x - \tan^{-1} \frac{2}{m} \right)$$

    (ii) $\quad \int e^{3x} \cos^3 x \, dx$

$$= \int e^{3x} \left(\frac{\cos 3x + 3 \cos x}{4} \right) dx$$

$$(\because \cos 3x = 4 \cos^3 x - 3 \cos x)$$

$$= \frac{1}{4} \int e^{3x} \cos 3x \, dx + \frac{3}{4} \int e^{3x} \cos x \, dx$$

$$= \frac{1}{4} \cdot \frac{e^{3x}}{\sqrt{9+9}}(\cos 3x - \tan^{-1}\frac{3}{3}) + \frac{3}{4} \cdot \frac{e^{3x}}{\sqrt{9+1}}\left(\cos x - \tan^{-1}\frac{1}{3}\right)$$

$$= \frac{e^{3x}}{4\sqrt{18}}(\cos 3x - \tan^{-1}1) + \frac{3e^{3x}}{4\sqrt{10}}\left(\cos x - \tan^{-1}\frac{1}{3}\right)$$

(iii) $\displaystyle\int e^x(\sin x + \cos x)\,dx = \int e^x \sin x\,dx + \int e^x \cos x\,dx$

$$= \int e^x \sin x\,dx + e^x \int \cos x\,dx - \int\left[\frac{d}{dx}e^x \int \cos x\,dx\right]dx$$

$$= \int e^x \sin x\,dx + e^x \sin x - \int e^x \sin x\,dx$$

$$= e^x \sin x$$

Integration by Partial Fractions

***Example* 26:** Evaluate

(i) $\displaystyle\int \frac{x+1}{(x-1)(x+2)}\,dx$ (ii) $\displaystyle\int \frac{3x}{x^2 - x - 2}\,dx$

(iii) $\displaystyle\int \frac{x-1}{(x-3)(x-2)}\,dx$ (iv) $\displaystyle\int \frac{2x+1}{(2x+3)(3x-4)}\,dx$

(v) $\displaystyle\int \frac{2x}{(x-1)(x+3)}\,dx$ (vi) $\displaystyle\int \frac{x}{x^2 + x - 6}\,dx$

***Solution*:** (i) $\displaystyle\int \frac{x+1}{(x-1)(x+2)}\,dx$

Let $\dfrac{x+1}{(x-1)(x+2)} = \dfrac{A}{x-1} + \dfrac{B}{(x+2)}$

$$= \frac{A(x+2) + B(x-1)}{(x-1)(x+2)}$$

$$x + 1 = A(x+2) + B(x-1) \qquad\qquad .....(i)$$

Put $x + 2 = 0 \Rightarrow x = -2$ in equation (2) we get

$$-1 = -3B \Rightarrow B = \frac{1}{3}$$

Put $x - 1 = 0 \Rightarrow x = 1$ in (1)   we get

$$2 = 3A \Rightarrow A = \frac{2}{3}$$

$$2 = 3A \Rightarrow A = \frac{2}{3}$$

$$\therefore \qquad \frac{x+1}{(x-1)(x+2)} = \frac{2}{3}\cdot\frac{1}{x-1} + \frac{1}{3}\cdot\frac{1}{x+2}$$

$$\int \frac{x+1}{(x-1)(x+2)}\, dx = \frac{2}{3}\int \frac{1}{x-1}\, dx + \frac{1}{3}\int \frac{1}{x+2}\, dx$$

$$= \frac{2}{3}\log(x-1) + \frac{1}{3}\log(x+2)$$

(ii) $\quad \displaystyle\int \frac{3x}{x^2-x-2}\, dx = \int \frac{3x}{(x-2)(x+1)}\, dx$

Let $\dfrac{3x}{(x-2)(x+1)} = \dfrac{A}{x-2} + \dfrac{B}{x+1}$

$$= \frac{A(x+1) + B(x-2)}{(x-2)(x+1)}$$

$$3x = A(x+1) + B(x-2) \qquad\qquad\qquad \dots\text{(ii)}$$

Put $x = -1$, and $x = 2$ in (1) we get

$$-3 = -3B \qquad \Rightarrow \qquad B = 1$$

$$6 = 3A \qquad \Rightarrow \qquad A = 2$$

$$\therefore \qquad \frac{3x}{(x-2)(x+1)} = \frac{2}{x-2} + \frac{1}{x+1}$$

$$\int \frac{3x}{(x-2)(x+1)}\, dx = 2\int \frac{1}{x-2}\, dx + \int \frac{1}{x+1}$$

$$= 2\log(x-2) + \log(x+1)$$

(iii) $\quad \displaystyle\int \frac{x-1}{(x-3)(x-2)}\, dx$

Let $\quad \dfrac{x-1}{(x-3)(x-2)} = \dfrac{A}{x-3} + \dfrac{B}{x-2}$

$$= \frac{A(x-2) + B(x-3)}{(x-3)(x-2)}$$

$$x - 1 = A(x-2) + B(x-3) \qquad\qquad\qquad \dots\text{(iii)}$$

Put $x = 2, 3$ in (3) we get

$$1 = -B \qquad \Rightarrow \qquad B = -1$$

$$2 = A \qquad \Rightarrow \qquad A = 2$$

$$\therefore \qquad \frac{x-1}{(x-3)(x-2)} = \frac{2}{x-3} - \frac{1}{x-2}$$

$$\int \frac{x-1}{(x-3)(x-2)}\, dx = 2\int \frac{1}{x-3}\, dx - \int \frac{1}{x-2}\, dx$$

$$= 2 \log (x-3) - \log (x-2)$$

(iv) $\displaystyle \int \frac{2x+1}{(2x+3)(3x-4)}\, dx$

Let $\displaystyle \frac{2x+1}{(2x+3)(3x-4)} = \frac{A}{2x+3} + \frac{B}{3x-4}$

$$= \frac{A(3x-4)+B(2x+3)}{(2x+3)\,(3x-4)}$$

$$2x + 1 = A(3x-4) + B(2x+3) \hspace{3cm} \dots.. \text{(iv)}$$

Put $x = \dfrac{4}{3}, \dfrac{-3}{2}$ in (4), we have

$$B = \frac{11}{17}, A = \frac{4}{17}$$

$\therefore$ $\displaystyle \frac{2x+1}{(2x+3)(3x-4)} = \frac{4}{17}\frac{1}{2x+3} + \frac{11}{17}.\frac{1}{3x-4}$

$$\int \frac{2x+1}{(2x+3)(3x-4)}\, dx = \frac{4}{17}\int \frac{1}{2x+3}\, dx + \frac{11}{17}\int \frac{1}{3x-4}\, dx$$

$$= \frac{4}{17} \log (2x+3) \times \frac{1}{2} + \frac{11}{17} \log (3x-4) \times \frac{1}{3}$$

$$= \frac{4}{34} \log (2x+3) + \frac{11}{51} \log (3x-4)$$

(v) $\displaystyle \int \frac{2x}{(x-1)(x+3)}\, dx$

Let $\displaystyle \frac{2x}{(x-1)(x+3)} = \frac{A}{x-1} + \frac{B}{x+3}$

$$= \frac{A(x+3)+B(x-1)}{(x-1)\,(x+3)}$$

$$2x = A(x+3) + B(x-1) \hspace{3cm} \dots..\text{(v)}$$

Put $x = 1, -3$ in (5), we get

$$A = \frac{1}{2}, \hspace{2cm} B = \frac{3}{2}$$

$$\therefore \quad \frac{2x}{(x-1)(x+3)} = \frac{1}{2(x-1)} + \frac{3}{2(x+3)}$$

$$\int \frac{2x}{(x-1)(x+3)} \, dx = \frac{1}{2} \int \frac{1}{x-1} \, dx + \frac{2}{3} \int \frac{1}{x+3} \, dx$$

$$= \frac{1}{2} \log(x-1) + \frac{2}{3} \log(x+3)$$

(vi) $$\int \frac{x}{x^2+x-6} \, dx = \int \frac{x}{(x+3)(x-2)} \, dx$$

Let $$\frac{x}{(x+3)(x-2)} = \frac{A}{x+3} + \frac{B}{x-2}$$

$$= \frac{A(x-2)+B(x+3)}{(x+3)(x-2)}$$

$$x = A(x-2) + B(x+3) \qquad\qquad ....(\text{vi})$$

Put $\quad x = 2, -3$ in (5) we have

$$A = \frac{3}{5}, \ B = \frac{2}{5}$$

$\therefore \quad$ $$\frac{x}{x^2+x-6} = \frac{3}{5(x+3)} + \frac{2}{5(x-2)}$$

$$\int \frac{x}{x^2+x-6} \, dx = \frac{3}{5} \int \frac{1}{x+3} \, dx + \frac{2}{5} \int \frac{1}{x-2} \, dx$$

$$= \frac{3}{5} \log(x+3) + \frac{2}{5} \log(x-2)$$

***Example* 27:** Evaluate

(i) $$\int \frac{x}{(1+x)(1-x)^2} \, dx$$
(ii) $$\int \frac{x}{(x-1)(x+1)^2} \, dx$$

Solution: (i) $$\int \frac{x}{(1+x)(1-x)^2} \, dx$$

Let $$\frac{x}{(1+x)(1-x)^2} = \frac{A}{1+x} + \frac{B}{1-x} + \frac{C}{(1-x)^2}$$

$$x = A(1-x)^2 + B(1+x)(1-x) + C(1+x) \qquad\qquad ....(\text{i})$$

Put $\quad x = -1, 1$ in (1) we have

When $\quad x = 1, \ c = \dfrac{1}{2}$

$$x = -1, A = \frac{-1}{4}$$

$$x = 0 \quad B = \frac{-1}{4}$$

$$\therefore \quad \frac{x}{(1+x)(1-x)^2} = \frac{-1}{4(1+x)} - \frac{1}{4(1-x)} + \frac{1}{2(1-x)^2}$$

$$\int \frac{x}{(1+x)(1-x)^2} dx = \frac{-1}{4} \int \frac{dx}{1+x} - \frac{1}{4} \int \frac{1}{1-x} dx + \frac{1}{2} \int \frac{1}{(1-x)^2} dx$$

$$= \frac{-1}{4} \log(1+x) + \frac{1}{4} \log(1-x) + \frac{1}{2} \cdot \frac{1}{1-x}$$

$$= \frac{1}{4} [\log(1-x) - \log(1+x)] + \frac{1}{2(1-x)}$$

$$= \frac{1}{4} \log \left(\frac{1-x}{1+x} \right) + \frac{1}{2(1-x)}$$

(ii) $\displaystyle \int \frac{x}{(x-1)(x+1)^2} dx$

Let $\displaystyle \frac{x}{(x-1)(x+1)^2} = \frac{A}{x-1} + \frac{B}{x+1} + \frac{C}{(x+1)^2}$

$$x = A(x+1)^2 + B(x-1)(x+1) + C(x-1) \qquad \qquad .....(ii)$$

Put $x = -1, 1$ in (2) we get

When $x = 1, \quad A = \dfrac{1}{4}$

$$x = -1, \quad C = \frac{1}{2}$$

$$x = 0, \quad B = \frac{-1}{4}$$

$$\therefore \quad \int \frac{x}{(x-1)(x+1)^2} dx = \frac{1}{4} \int \frac{1}{x-1} dx - \frac{1}{4} \int \frac{1}{x+1} dx + \frac{1}{2} \int \frac{1}{(x+1)^2} dx$$

$$= \frac{1}{4} \log(x-1) - \frac{1}{4} \log(x+1) - \frac{1}{2(x+1)}$$

$$= \frac{1}{4} \log \frac{x-1}{x+1} - \frac{1}{2(x+1)}$$

***Example* 28:** Evaluate

(i) $\displaystyle\int \dfrac{x}{(2x+1)(x^2+1)}\,dx$ (ii) $\displaystyle\int \dfrac{dx}{(x^2+1)(x-1)}$ (iii) $\displaystyle\int \dfrac{x+1}{(2x+3)(x^2+1)}\,dx$

Solution:

(i) $\displaystyle\int \dfrac{x}{(2x+1)(x^2+1)}\,dx$

Let $\dfrac{x}{(2x+1)(x^2+1)} = \dfrac{A}{2x+1} + \dfrac{Bx+C}{x^2+1}$

$x = A(x^2+1) + (Bx+C)(2x+1)$

$x = x^2(A+2B) + x(B+2C) + A + C$ $\qquad\qquad\qquad$ ..... (i)

Putting $\qquad x = \dfrac{-1}{2}, \qquad A = \dfrac{-2}{5}$

On comparing coefficients of x^2, x and constants we get

$A + 2B = 0 \;\Rightarrow\; B = \dfrac{-A}{2} = \dfrac{-1}{2}\left(\dfrac{-2}{5}\right) = \dfrac{1}{5}$

$B + 2C = 1$

$A + C = 0 \;\Rightarrow\; C = -A \;\Rightarrow\; C = \dfrac{2}{5}$

$\therefore \quad \displaystyle\int \dfrac{x}{(2x+1)(x^2+1)}\,dx = \dfrac{-2}{5}\int \dfrac{1}{2x+1}\,dx + \int \dfrac{\frac{1}{5}x+\frac{2}{5}}{x^2+1}\,dx$

$\qquad\qquad = \dfrac{-2}{5}\int \dfrac{1}{2x+1}\,dx + \dfrac{1}{5}\int \dfrac{x}{x^2+1}\,dx + \dfrac{2}{5}\int \dfrac{1}{x^2+1}\,dx$

$\qquad\qquad = \dfrac{-2}{5}\log(2x+1)\times\dfrac{1}{2} + \dfrac{1}{10}\log(x^2+1) + \dfrac{2}{5}\tan^{-1}x$

$\qquad\qquad = \dfrac{-1}{5}\log(2x+1) + \dfrac{1}{10}\log(x^2+1) + \dfrac{2}{5}\tan^{-1}x + c$

(ii) $\displaystyle\int \dfrac{dx}{(x^2+1)(x-1)}$

Let $\dfrac{1}{(x^2+1)(x-1)} = \dfrac{Ax+B}{x^2+1} + \dfrac{C}{x-1}$

$1 = (Ax+B)(x-1) + C(x^2+1)$ $\qquad\qquad\qquad$ .....(ii)

Putting $\qquad x = 1, C = \dfrac{1}{2}$

From (2) $\qquad 1 = x^2 (A + C) + x (B - A) + (C - B)$

On comparing the coefficients of different powers of x, we get

$$A + C = 0 \Rightarrow A = - C = \dfrac{-1}{2}$$

$$B - A = 0 \Rightarrow B = A = \dfrac{-1}{2}$$

$$\therefore \qquad \int \dfrac{dx}{(x^2 + 1)(x - 1)} = \dfrac{-1}{2} \int \dfrac{x}{x^2 + 1} dx - \dfrac{1}{2} \int \dfrac{1}{x^2 + 1} dx + \dfrac{1}{2} \int \dfrac{1}{x - 1} dx$$

$$= \dfrac{-1}{4} \log(x^2 + 1) - \dfrac{1}{2} \tan^{-1} x + \dfrac{1}{2} \log x + c$$

(iii) $\quad \displaystyle\int \dfrac{x + 1}{(2x + 3)(x^2 + 1)} dx$

Let $\qquad \dfrac{x + 1}{(2x + 3)(x^2 + 1)} = \dfrac{A}{2x + 3} + \dfrac{Bx + c}{x^2 + 1}$

$$x + 1 = A(x^2 + 1) + (Bx + C)(2x + 3) \qquad\qquad\qquad .....(iii)$$

Putting $\qquad x = \dfrac{-3}{2}$, we get $\qquad A = \dfrac{-2}{13}$

From (3) $\qquad x + 1 = x^2 (A + 2B) + x (3B + 2C) + (A + 3C)$

On comparing the coefficients of different powers of x, we get

$$A + 2B = 0 \quad \Rightarrow B = \dfrac{-A}{2} = \dfrac{-1}{2} \times \left(\dfrac{-2}{13} \right) = \dfrac{1}{13}$$

$$A + 3C = 1 \quad \Rightarrow C = \dfrac{1 - A}{3} = \dfrac{1 + \dfrac{2}{13}}{3} = \dfrac{15}{39}$$

$$\therefore \qquad \int \dfrac{x + 1}{(2x + 3)(x^2 + 1)} dx = \dfrac{-2}{13} \int \dfrac{1}{2x + 3} dx + \dfrac{1}{13} \int \dfrac{x}{x^2 + 1} dx + \dfrac{15}{39} \int \dfrac{1}{x^2 + 1} dx$$

$$= \dfrac{-1}{13} \log(2x + 3) + \dfrac{1}{26} \log(x^2 + 1) + \dfrac{15}{39} \tan^{-1} x + c$$

Definite Integrals

***Example* 29:** Evaluate

(i) $\displaystyle\int_0^1 \frac{(\tan^{-1}x)^2}{1+x^2}\,dx$ (ii) $\displaystyle\int_0^1 \tan^{-1}x\,dx$ (iii) $\displaystyle\int_1^2 \frac{dx}{(x+3)(x+4)}$

***Solution*:**

(i) $\displaystyle\int_0^1 \frac{(\tan^{-1}x)^2}{1+x^2}\,dx$

Put $\tan^{-1}x = t$ $\Rightarrow$ $\dfrac{1}{1+x^2}\,dx = dt$

Also when $x = 0,\ t = 0$

when $x = 1,\ t = \dfrac{\pi}{4}$

$\therefore\qquad \displaystyle\int_0^1 \frac{(\tan^{-1}x)^2}{1+x^2}\,dx = \int_0^{\frac{\pi}{4}} t^2\,dt$

$$= \left[\frac{t^3}{3}\right]_0^{\frac{\pi}{4}}$$

$$= \frac{1}{3}\left[\frac{\pi^3}{64} - 0\right] = \frac{\pi^3}{192}$$

(ii) $\displaystyle\int_0^1 \tan^{-1}x\,dx$

Integrating by parts,

$$= \tan^{-1}x \int_0^1 dx - \int_0^1\left[\frac{d}{dx}\tan^{-1}x \int dx\right]dx$$

$$= \left[x\tan^{-1}x\right]_0^1 - \int_0^1 \frac{x}{1+x^2}\,dx$$

Put $1 + x^2 = t$

$\Rightarrow\qquad 2x\,dx = dt$

when $x = 0,\ t = 1$

when $x = 1,\ t = 2$

$$\therefore \qquad \int\limits_0^1 \tan^{-1} x\, dx = \left[x \tan^{-1} x \right]_0^1 - \frac{1}{2}\int\limits_1^2 \frac{1}{t}\, dt$$

$$= \left(\tan^{-1} 1 - 0 \right) - \frac{1}{2}\left[\log t \right]_1^2$$

$$= \frac{\pi}{4} - \frac{1}{2}\left[\log 2 - \log 1 \right]$$

$$= \frac{\pi}{4} - \frac{1}{2}\log 2 \qquad\qquad (\because \log 1 = 0)$$

(iii) $\displaystyle \int\limits_1^2 \frac{dx}{(x+3)(x+4)}$

Let $\qquad \dfrac{1}{(x+3)(x+4)} = \dfrac{A}{x+3} + \dfrac{B}{x+4}$

$$1 = A(x+4) + B(x+3) \qquad\qquad\qquad .....(i)$$

Putting $\quad x = -3, -4$ in (1), we get

$A = 1, B = -1$

$$\therefore \qquad \int\limits_1^2 \frac{1}{(x+3)(x+4)} = \int\limits_1^2 \frac{1}{x+3}\, dx - \int\limits_1^2 \frac{1}{x+4}\, dx$$

$$= \left[\log(x+3) \right]_1^2 - \left[\log(x+4) \right]_1^2$$

$$= (\log 5 - \log 4) - (\log 6 - \log 5)$$

$$= \log 5 - \log 4 - \log 6 + \log 5$$

$$= 2 \log 5 - \log 4 - \log 6 + \log 5$$

$$= 2 \log 5 - (\log 4 + \log 6)$$

$$= 2 \log 5 - \log (4 \times 6)$$

$$= \log 25 - \log 24 = \log \frac{25}{24}$$

Example 30: Evaluate

(i) $\displaystyle \int\limits_3^5 \frac{\sin(\log x)}{x}\, dx$ \qquad (ii) $\displaystyle \int\limits_0^\pi \frac{dx}{5 + 3\cos x}$ \qquad (iii) $\displaystyle \int\limits_0^\pi \frac{d\theta}{3 + 2\sin\theta + \cos\theta}$

Solution:

(i) $\displaystyle \int\limits_3^5 \frac{\sin(\log x)}{x}\, dx$

Put $\log x = t \implies \dfrac{1}{x}\,dx = dt$

when $x = 3,$ $t = \log 3$

when $x = 5,$ $t = \log 5$

$$\therefore \quad \int_{3}^{5} \frac{\sin(\log x)}{x}\,dx = \int_{\log 3}^{\log 5} \sin t \, dt$$

$$= \Big[-\cos t \Big]_{\log 3}^{\log 5}$$

$$= -\cos(\log 5) + \cos(\log 3)$$

(ii) $\displaystyle \int_{0}^{\pi} \frac{dx}{5 + 3\cos x} = \int_{0}^{\pi} \frac{dx}{5 + 3\dfrac{1 - \tan^2 \dfrac{x}{2}}{1 + \tan^2 \dfrac{x}{2}}}$

$$= \int_{0}^{\pi} \frac{(1 + \tan^2 \frac{x}{2})\,dx}{5(1 + \tan^2 \frac{x}{2}) + 3(1 - \tan^2 \frac{x}{2})}$$

$$= \int_{0}^{\pi} \frac{\sec^2 \frac{x}{2}\,dx}{8 + 2\tan^2 \frac{x}{2}}$$

$$= \frac{1}{2} \int_{0}^{\pi} \frac{\sec^2 \frac{x}{2}\,dx}{4 + \tan^2 \frac{x}{2}}$$

Putting $\tan \dfrac{x}{2} = t \implies \dfrac{1}{2}\sec^2 \dfrac{x}{2}\,dx = dt$

when $x = 0, \ t = 0$

when $x = \pi, \ t = \infty$

$$\therefore \quad \int_{0}^{\pi} \frac{dx}{5 + 3\cos x} = \int_{0}^{\infty} \frac{dt}{4 + t^2}$$

$$= \frac{1}{2}\left[\tan^{-1} \frac{t}{2} \right]_{0}^{\infty}$$

$$= \frac{1}{2}\left[\tan^{-1}\infty - \tan^{-1}0\right]$$

$$= \frac{1}{2}\left[\frac{\pi}{2} - 0\right] = \frac{\pi}{4}$$

(iii) $\displaystyle\int_0^\pi \frac{dx}{3 + 2\sin x + \cos x} = \int_0^\pi \frac{dx}{3 + 2\dfrac{2\tan\dfrac{x}{2}}{1 + \tan^2\dfrac{x}{2}} + \dfrac{1 - \tan^2\dfrac{x}{2}}{1 + \tan^2\dfrac{x}{2}}}$

$$= \int_0^\pi \frac{(1 + \tan^2\frac{x}{2})dx}{3(1 + \tan^2\frac{x}{2}) + 4\tan\frac{x}{2} + 1 - \tan^2\frac{x}{2}}$$

$$= \int_0^\pi \frac{\sec^2\frac{x}{2}\,dx}{4 + 2\tan^2\frac{x}{2} + 4\tan\frac{x}{2}}$$

$$= \frac{1}{2}\int_0^\pi \frac{\sec^2\frac{x}{2}\,dx}{\tan^2\frac{x}{2} + 2\tan\frac{x}{2} + 2}$$

putting $\tan\dfrac{x}{2} = t \Rightarrow \dfrac{1}{2}\sec^2\dfrac{x}{2}\,dx = dt$

when $x = 0,\ t = 0$

when $x = \pi,\ t = \infty$

$\therefore \qquad \displaystyle\int_0^\pi \frac{dx}{3 + 2\sin x + \cos x} = \int_0^\infty \frac{dt}{t^2 + 2t + 2}$

$$= \int_0^\infty \frac{dt}{(t + 1)^2 + 1}$$

$$= \left[\tan^{-1}(t + 1)\right]_0^\infty$$

$$= \tan^{-1}\infty - \tan^{-1}1$$

$$= \frac{\pi}{2} - \frac{\pi}{4} = \frac{\pi}{4}$$

Practice Problems

Integrate the following w.r.t. x.

1. $4x^3 - 4x^{-5}$

2. $\dfrac{x^3 - 1}{x^2}$

3. $5^x + 3e^x + x^3$

4. $5\cos x - 3\sin x + \dfrac{2}{\cos^2 x}$

5. $\displaystyle\int \tan^2 x \, dx$

6. $\displaystyle\int (ax + b)^2 \, dx$

7. $\dfrac{\sin x}{\sin x - \cos x}$

8. $(3x - 4)^3$

9. $\dfrac{1}{1 + x^2} - \dfrac{\cos x}{\sin^2 x}$

10. $\dfrac{1}{\sqrt{1 - x^2}} + \dfrac{1}{1 + x^2}$

11. $\dfrac{1}{\sqrt{5x + 3} + \sqrt{5x + 2}}$

12. $e^{a \log x} + e^{x \log a}$

13. $x^2 \cos x^3$

14. $x^2 e^{x^3}$

15. $\dfrac{e^{\tan^{-1} x}}{1 + x^2}$

16. $\dfrac{x^5}{\sqrt{1 - x^{12}}}$

17. $\dfrac{\cos x}{1 + \sin^2 x}$

18. $\dfrac{\sin x \cos x}{1 + \sin^4 x}$

19. $\dfrac{(1 + \log x)^2}{x}$

20. $\dfrac{1}{x\sqrt{1 - (\log x)^2}}$

21. Evaluate $\displaystyle\int \dfrac{(x + 1)(x + \log x)^3}{2x} \, dx$

22. $\displaystyle\int \sin^5 x \, dx$

23. $\displaystyle\int e^x (e^x + 1)^3 \, dx$

24. $\displaystyle\int \cot x \, \mathrm{cosec}^3 x \, dx$

25. $\displaystyle\int e^x (a + be^x)^n \, dx$

26. $\displaystyle\int \dfrac{(\sin^{-1} x)^3}{\sqrt{1 - x^2}} \, dx$

27. $\displaystyle\int \dfrac{(\tan^{-1} x)^4}{1 + x^2} \, dx$

28. $\displaystyle\int \sin^4 x \cos x \, dx$

29. $\displaystyle\int \dfrac{1}{(\tan^{-1} x)^2 (1 + x^2)} \, dx$

30. $\displaystyle\int (e^x + e^{-x})(e^x - e^{-x}) \, dx$

31. $\displaystyle\int \tan^4 x \, dx$

32. $\displaystyle\int \frac{2x \sin^{-1} x^2}{\sqrt{1-x^4}} \, dx$

33. $\displaystyle\int \frac{1}{\sqrt{1-x^2} \sin^{-1} x} \, dx$

34. $\displaystyle\int \frac{\cot x}{(\log \sin x)^3} \, dx$

35. $\displaystyle\int \frac{x^2 \tan^{-1} x^3}{1+x^6} \, dx$

36. $\displaystyle\int \frac{dx}{\sqrt{25+4x^2}}$

37. $\displaystyle\int \frac{x^2}{x^2-4} \, dx$

38. $\displaystyle\int \sqrt{25-9x^2} \, dx$

39. $\displaystyle\int \frac{dx}{\sqrt{(x+1)^2+4}}$

40. $\displaystyle\int \frac{dx}{x\sqrt{x^2-4}}$

41. $\displaystyle\int \cos x \sqrt{4-\sin^2 x} \, dx$

42. $\displaystyle\int \frac{\sec^2 x \, dx}{\sqrt{\tan^2 x + 4}}$

43. $\displaystyle\int \frac{\sec^2 x \, dx}{\sqrt{\tan^2 x + 4}}$

44. $\displaystyle\int \frac{x \, dx}{a^4 + x^4}$

45. Evaluate $\displaystyle\int x \sin 2x \, dx$

46. $\displaystyle\int x \sin^3 x \, dx$

47. $\displaystyle\int x \tan^2 x \, dx$

48. $\displaystyle\int x \sec^2 x \, dx$

49. $\displaystyle\int x \sin nx \, dx$

50. $\displaystyle\int x^3 e^x \, dx$

51. $\displaystyle\int x^2 e^{3x} \, dx$

52. $\displaystyle\int \tan^{-1} x \, dx$

53. $\displaystyle\int \cos^{-1} x \, dx$

54. $\displaystyle\int x^2 \log x \, dx$

55. $\displaystyle\int x \sin^{-1} x \, dx$

56. $\displaystyle\int x^2 \tan^{-1} x \, dx$

57. $\displaystyle\int x^n \log x \, dx$

58. $\displaystyle\int \frac{\log x}{(1+x)^2} \, dx$

59. $\displaystyle\int (\log x)^2 \, dx$

60. $\displaystyle\int x^2 \sin^{-1} x \, dx$

61. $\displaystyle\int \sec^3 x \, dx$

62. $\displaystyle\int \frac{\sin^{-1} x}{(1-x^2)^{\frac{3}{2}}} \, dx$

63. $\displaystyle\int \frac{x \sin^{-1} x}{\sqrt{1-x^2}} \, dx$

64. $\displaystyle\int \sqrt{x} \tan^{-1} \sqrt{x} \, dx$

65. $\displaystyle \int \tan^{-1}\sqrt{\dfrac{1-x}{1+x}}\,dx$

66. $\displaystyle \int \cos^{-1}\left(\dfrac{1-x^2}{1+x^2}\right)dx$

67. $\displaystyle \int \dfrac{\log(\sec^{-1}x)}{x\sqrt{x^2-1}}\,dx$

68. Evaluate $\displaystyle \int \dfrac{x-1}{(x-2)(x-3)}\,dx$

69. Evaluate $\displaystyle \int \dfrac{x^3+2}{x^3-x}\,dx$

70. Evaluate $\displaystyle \int \dfrac{dx}{x^2-5x+6}$

71. Evaluate $\displaystyle \int \dfrac{2x\,dx}{(x^2+1)(x^2+3)}$

72. Evaluate $\displaystyle \int \dfrac{\cos x\,dx}{(1+\sin x)(2+\sin x)}$

73. Evaluate $\displaystyle \int \dfrac{1-x^2}{x(1-2x)}\,dx$

74. Evaluate $\displaystyle \int \dfrac{x\,dx}{(x-1)^2(x+2)}$

75. Evaluate $\displaystyle \int \dfrac{dx}{1+x+x^2+x^3}$

76. Evaluate $\displaystyle \int \dfrac{dx}{(x-3)(x^2+4)}$

77. Evaluate $\displaystyle \int \dfrac{x\,dx}{(x^2-a^2)(x^2-b^2)}$

78. Evaluate $\displaystyle \int \dfrac{dx}{(x+b)(x^2+a^2)}$

79. $\displaystyle \int \dfrac{x^3\,dx}{x^2-3x+2}$

80. Evaluate $\displaystyle \int \dfrac{dx}{x^3+1}$

81. Evaluate $\displaystyle \int \dfrac{x^2\,dx}{(x+1)(x-2)(x+3)}$

82. Evaluate $\displaystyle \int \dfrac{e^x\,dx}{e^{2x}+4e^x+3}$

83. Evaluate $\displaystyle \int \dfrac{x^2\,dx}{(x-1)^3(x+1)}$

84. Evaluate the following $\displaystyle \int_{0}^{\frac{\pi}{2}} (\sin x + \cos x)\,dx$

85. $\displaystyle \int_{1}^{2} (4x^3-5x^2+6x+9)\,dx$

86. $\displaystyle \int_{0}^{\frac{\pi}{4}} \sec x\,dx$

87. $\displaystyle \int_{0}^{1} \dfrac{dx}{\sqrt{1-x^2}}$

88. $\displaystyle \int_{0}^{1} \sin^{-1}x\,dx$

89. $\displaystyle \int_{0}^{\frac{\pi}{2}} \sin^3 x\,dx$

90. $\displaystyle \int_{1}^{2} \dfrac{dx}{(x+1)(x+2)}$

91. $\displaystyle\int_1^2 \frac{5x^2}{x^2+4x+3}\,dx$

92. $\displaystyle\int_0^{\frac{\pi}{2}} \frac{\sin\theta}{\sqrt{1+\cos\theta}}\,d\theta$

93. $\displaystyle\int_0^1 x\,e^{x^2}\,dx$

94. $\displaystyle\int_0^\pi \frac{dx}{5+4\cos x}$

95. $\displaystyle\int_0^\pi \frac{\sin x\,dx}{1+\cos^2 x}$

96. $\displaystyle\int_0^1 \frac{5x^3}{\sqrt{1-x^8}}\,dx$

97. $\displaystyle\int_0^{\frac{\pi}{2}} \frac{dx}{5+4\sin x}$

98. $\displaystyle\int_0^\infty \frac{x^2\,dx}{(x^2+a^2)(x^2+b^2)}$

99. $\displaystyle\int_0^1 \frac{x\,dx}{\sqrt{1+x^2}}$

100. $\displaystyle\int_0^{\frac{\pi}{3}} \frac{\cos x}{3+4\sin x}\,dx$

Answers

1. $x^4 + x^{-4}$

2. $\dfrac{1}{2}x^2 + \dfrac{1}{x}$

3. $\dfrac{5^x}{\log 5} + 3e^x + \dfrac{x^4}{4}$

4. $5\sin x + 3\cos x + 2\tan x$

5. $\tan x - x$

6. $\dfrac{(ax+b)^3}{3a}$

7. $\dfrac{1}{2}[x + \log(\sin x - \cos x)]$

8. $\dfrac{(3x-4)^4}{12}$

9. $\tan^{-1} x + \operatorname{cosec} x$

10. $\sin^{-1} x + \tan^{-1} x$

11. $\dfrac{2}{15}\left[(5x+3)^{\frac{3}{2}} - (5x+2)^{\frac{3}{2}}\right]$

12. $\dfrac{x^{a+1}}{a+1} + \dfrac{a^x}{\log a}$

13. $\dfrac{1}{3}\sin x^3$

14. $\dfrac{1}{3}e^{x^3}$

15. $e^{\tan^{-1} x}$

16. $\dfrac{1}{6}\sin^{-1}(x^6)$

17. $\tan^{-1}(\sin x)$

18. $\dfrac{1}{2}\tan^{-1}(\sin^2 x)$

19. $\dfrac{1}{3}(\log x + 1)^3$

20. $\sin^{-1}(\log x)$

21. $\dfrac{1}{8}(x + \log x)^4$

22. $-\cos x - \dfrac{\cos^5 x}{5} + \dfrac{2\cos^2 x}{3}$

23. $\dfrac{1}{4}(e^x + 1)^4$

24. $-\dfrac{1}{3}\cos ec^3 x$

25. $\dfrac{(a + be^x)^{n+1}}{b(n+1)}$

26. $\dfrac{1}{4}(\sin^{-1} x)^4$

27. $\dfrac{1}{5}(\tan^{-1} x)^5$

28. $\dfrac{1}{5}(\sin x)^5$

29. $-\dfrac{1}{\tan^{-1} x}$

30. $\dfrac{1}{2}(e^x + e^{-x})^2$

31. $\dfrac{1}{3}\tan^3 x - \tan x + x$

32. $\dfrac{1}{2}(\sin^{-1} x^2)^2$

33. $\log(\sin^{-1} x)$

34. $\dfrac{-1}{2}(\log \sin x)^{-2}$

35. $\dfrac{1}{6}(\tan^{-1} x^3)^2$

36. $\dfrac{1}{2}\log(2x + \sqrt{25 + 4x^2})$

37. $x + \log\dfrac{x - 2}{x + 2}$

38. $\dfrac{x}{2}\sqrt{25 - 9x^2} + \dfrac{25}{6}\sin^{-1}\dfrac{3x}{5}$

39. $\log\left\{(x + 1) + \sqrt{(x+1)^2 + 4}\right\}$

40. $\dfrac{1}{2}\sec^{-1}\dfrac{x}{2}$

41. $\dfrac{1}{2}\sin x\sqrt{4 - \sin^2 x} + 2\sin^{-1}\left(\dfrac{1}{2}\sin x\right)$

42. $\log\left[\tan x + \sqrt{\tan^2 x + 4}\right]$

43. $-\log\left[\cos x + \sqrt{4 + \cos^2 x}\right]$

44. $\dfrac{1}{2a^2}\tan^{-1}\dfrac{x^2}{a^2}$

45. $-\dfrac{1}{2}x\cos 2x + \dfrac{1}{4}\sin 2x$

46. $\dfrac{1}{4}\left[-3x\cos x + 3\sin x + \dfrac{1}{3}x\cos 3x - \dfrac{1}{9}\sin 3x\right]$

47. $x\tan x - \log \sec x - \dfrac{x^2}{2}$

48. $x\tan x + \log \cos x$

49. $\dfrac{1}{n}\left[\dfrac{1}{n}\sin nx - x\cos nx\right]$

50. $e^x[x^3 - 3x^2 + 6x - 6]$

51. $\dfrac{1}{3}x^2 e^{3x} - \dfrac{2}{3^2}xe^{3x} + \dfrac{2}{3^3}e^{3x}$

52. $x\tan^{-1}x - \dfrac{1}{2}\log(1+x^2)$

53. $x\cos^{-1}x - \sqrt{1-x^2}$

54. $\dfrac{1}{3}x^3\log x - \dfrac{x^3}{9}$

55. $\dfrac{1}{2}x^2\sin^{-1}x + \dfrac{1}{4}x\sqrt{1-x^2} - \dfrac{1}{4}\sin^{-1}x$

56. $\dfrac{1}{3}x^3\tan^{-1}x - \dfrac{x^2}{6} + \dfrac{1}{6}\log(x^2+1)$

57. $\dfrac{1}{n+1}x^{n+1}\log x - \dfrac{x^{n+1}}{(n+1)^2}$

58. $\dfrac{-\log x}{x+1} + \log x - \log(x+1)$

59. $x(\log x)^2 - 2x\log x + 2x$

60. $\dfrac{1}{3}x^3\sin^{-1}x - \dfrac{1}{9}(1-x^2)^{\frac{3}{2}} + \dfrac{1}{3}\sqrt{1-x^2}$

61. $\dfrac{1}{2}[\sec x\tan x + \log(\sec x + \tan x]$

62. $\dfrac{x\sin^{-1}x}{\sqrt{1-x^2}} + \dfrac{1}{2}\log(1-x^2) + c$

63. $x - (\sin^{-1}x)\sqrt{1-x^2} + c$

64. $\dfrac{2}{3}x^{\frac{3}{2}}\tan^{-1}\sqrt{x} - \dfrac{x}{3} + \dfrac{1}{3}\log(x+1) + c$

65. $\dfrac{1}{2}\left[x\cos^{-1}x - \sqrt{1-x^2}\right] + c$

66. $2x\tan^{-1}x - 2\log\sqrt{1+x^2} + c$

67. $\sec^{-1}[\log(\sec^{-1}x)-1] + c$

68. $2\log(x-3) - \log(x-2) + c$

69. $x - 2\log x + \dfrac{1}{2}\log(x+1) + \dfrac{3}{2}\log(x-1) + c$

70. $\log\dfrac{x-3}{x-2}$

71. $\dfrac{1}{2}\log\dfrac{x^2+1}{x^2+3}$

72. $\log\left(\dfrac{1+\sin x}{2+\sin x}\right)$

73. $\dfrac{x}{2} + \log x - \dfrac{3}{4}\log(1-2x)$

74. $\dfrac{2}{9}\log\left(\dfrac{x-1}{x+2}\right) - \dfrac{1}{3(x-1)}$

75. $\dfrac{1}{2}\log(1+x) - \dfrac{1}{4}\log(1+x^2) + \dfrac{1}{2}\tan^{-1}x$

76. $\dfrac{1}{13}\left[\log(x-3) - \dfrac{1}{2}\log(x^2+4) - \dfrac{3}{2}\tan^{-1}\dfrac{x}{2}\right]$

77. $\dfrac{1}{2(a^2-b^2)}\log\dfrac{x^2-a^2}{x^2-b^2}$

78. $\dfrac{1}{a^2+b^2}\left[\log\dfrac{x+b}{\sqrt{x^2+a^2}} + \dfrac{b}{a}\tan^{-1}\dfrac{x}{a}\right]$

79. $\dfrac{x^2}{2} + 3x + 8\log(x-2) - \log(x-1)$

80. $\dfrac{1}{3}\log(x+1) - \dfrac{1}{6}\log(x^2-x+1) + \dfrac{1}{\sqrt{3}}\tan^{-1}\dfrac{2x-1}{\sqrt{3}}$

81. $\dfrac{9}{10}\log(x+3)+\dfrac{4}{15}\log(x-2)-\dfrac{1}{6}\log(x+1)$

82. $\dfrac{1}{2}[\log(1+e^x)-\log(3+e^x)]$

83. $\dfrac{1}{8}\log\dfrac{x-1}{x+1}-\dfrac{3}{4(x-1)}-\dfrac{1}{4(x-1)^2}$

84. 2

85. $\dfrac{64}{3}$

86. $\log(1+\sqrt{2})$

87. $\dfrac{\pi}{2}$

88. $\dfrac{\pi}{2}-1$

89. $\dfrac{2}{3}$

90. $\log 9-\log 8$

91. $\dfrac{1}{2}\left[5\log 3+85\log 2-45\log 5\right]+5$

92. $-2(1-\sqrt{2})$

93. $\dfrac{1}{2}(e-1)$

94. $\dfrac{\pi}{3}$

95. $\dfrac{-\pi}{2}$

96. $\dfrac{5\pi}{8}$

97. $\dfrac{2}{3}\tan^{-1}\dfrac{1}{3}$

98. $\dfrac{\pi}{2(a+b)}$

99. $\sqrt{2}-1$

100. $\dfrac{1}{4}\log\left(1+\dfrac{2}{\sqrt{3}}\right)$

Unit – II

Ordinary Derivatives and Applications

- Expansion of Functions
- Maxima & Minima
- Curvature
- Curve Tracing

CHAPTER – 5

Expansion of Functions

5.1 n^{th} Derivatives of Some Standard Functions

(a) $D^n (ax + b)^m = m(m - 1)(m - 2) \ldots (m - n + 1) a^n (ax + b)^{m-n}$

In particular cases

(i) if $m = n$, then $D^n (ax + b)^n = n! . a^n$

(ii) if $m = -1$, then $D^n (ax + b)^{-1} = \dfrac{(-1)^n n! \, a^n}{(ax + b)^{n+1}}$

(b) $D^n \log (ax + b) = \dfrac{(-1)^{n-1} (n - 1)! a^n}{(ax + b)^n}$

(c) $D^n e^{ax} = a^n e^{ax}$

(d) $D^n a^x = a^x (\log_e a)^n$

(e) $D^n \sin (ax + b) - a^n \sin\left(ax + b + \dfrac{n\pi}{2} \right)$

(f) $D^n \cos (ax + b) = a^n \cos\left(ax + b + \dfrac{n\pi}{2} \right)$

(g) $D^n e^{ax + b} = a^n e^{ax + b}$

5.2 Leibnitz's Theorem

If u and v be any two functions of x, then

$$D^n (u.v) = D^n u.v + n_{c_1} D^{n-1} u.Dv + n_{c_2} D^{n-2} u.D^2 v + \ldots + n_{c_r} D^{n-r} u.D^r v + \ldots + u.D^n v$$

(This theorem is used for finding the nth derivative of a product of two functions)

Maclaurin's Theorem: Let f(x) be a function of x which can be expanded in powers of x and this expansion be differentiable term by term in any number of times, then

$$f(x) = f(0) + x f'(0) + \frac{x^2}{2!} f''(0) + \frac{x^3}{3!} f'''(0) + \ldots + \frac{x^n}{n!} f^n (0) + \ldots$$

Proof: Let $f(x) = A_0 + A_1x + A_2 x^2 + A_3 x^3 + A_4 x^4 + ...$ $\qquad$ .....(i)

where $A_0, A_1, A_2, A_3, A_4, ...$ are constants

Now, differentiating (i) w.r.t x, we get

$$f'(x) = A_1 + 2A_2 x + 3 A_3 x^2 + 4 A_4 x^3 + ...$$ $\qquad$ .....(ii)

$$f''(x) = 2 A_2 + 3.2 A_3 x + 4.3 A_4 x^2 +$$ $\qquad$ .....(iii)

$$f'''(x) = 3.2 A_3 + 4.3.2 A_4 x + ...$$ $\qquad$ .....(iv)

Putting $x = 0$ in equation (i), (ii) (iii) and (iv), we have

$$f(0) = A_0, f'(0) = A_1, f''(0) = 2A_2, f'''(0) = 6A_3 \quad , ...$$

substituting these values in equation (i), we have

$$f(x) = f(0) + x f'(0) + \frac{x^2}{2!} f''(0) + \frac{x^3}{3!} f'''(0) + ... + \frac{x^n}{n!} f^n(0) + ...$$

Remark: If $y = f(x)$, then Maclaurin's Theorem takes the form

$$y = (y)_0 + x(y_1)_0 + \frac{x^2}{2!}(y_2)_0 + \frac{x^3}{3!}(y_3)_0 + ... + \frac{x^n}{n!}(y_n)_0 + ...$$

Illustrative Examples

***Example* 1:** Expand $\log(1 + x)$ by Maclaurin's Theorem.

Solution: Let $f(x) = \log(1 + x)$, $\qquad\qquad\qquad$ $f(0) = 0$

$$f'(x) = \frac{1}{1+x}, \qquad\qquad\qquad f'(0) = 1$$

$$f''(x) = -\frac{1}{(1+x)^2}, \qquad\qquad\qquad f''(0) = -1$$

$$f'''(x) = \frac{2}{(1+x)^3}, \qquad\qquad\qquad f'''(0) = 2$$

$$...$$

$$...$$

$$f^n(x) = \frac{(-1)^{n-1}(n-1)!}{(1+x)^n}, \qquad\qquad f^n(0) = (-1)^{n-1}(n-1)!$$

$\therefore$ By Maclaurin's Theorem, we get

$$\log(1 + x) = f(0) + x\ f'(0) + \frac{x^2}{2!}f''(0) + \frac{x^3}{3!}f'''(0) + \dots + \frac{x^n}{n!}f^n(0) + \dots$$

$$= x - \frac{x^2}{2} + \frac{x^3}{3} - \frac{x^4}{4} + \dots + (-1)^{n-1}\frac{x^n}{n} + \dots$$

Example 2: Expand $e^x \sin x$ by Maclaurin's Theorem.   (RGPV Feb 2005)

Solution: Let $f(x) = e^x \sin x$, $f(0) = 0$

$\qquad f'(x) = e^x \cos x + e^x \sin x$, $f'(0) = 1$

'or' $\qquad f'(x) = e^x \cos x + f(x)$

$\qquad f''(x) = e^x \cos x - e^x \sin x + f'(x)$, $f''(0) = 2$

or $\qquad f''(x) = e^x \cos x - f(x) + f'(x)$

$\qquad f'''(x) = e^x \cos x - e^x \sin x - f'(x) + f''(x)$, $f'''(0) = 2$

or $\qquad f'''(x) = e^x \cos x - f(x) - f'(x) + f''(x)$

$\qquad f^{iv}(x) = e^x \cos x - e^x \sin x - f'(x) = f''(x) + f'''(x)$, $f^{iv}(0) = 0$

or $\qquad f^{iv}(x) = e^x \cos x - f(x) - f'(x) = f''(x) + f'''(x)$

$\qquad f^{v}(x) = e^x \cos x - e^x \sin x - f'(x) - f''(x) - f'''(x) + f^{iv}(x)$, $f^{v}(0) = -4$

$\therefore$ By Maclaurin's Theorem

$$f(x) = f(0) + x f'(0) + \frac{x^2}{2!}f''(0) + \frac{x^3}{3!}f'''(0) + \frac{x^4}{4!}f^{iv}(0) + \frac{x^5}{5!}f^{v}(0) + \dots$$

$$e^x . \sin x = x + x^2 + \frac{x^3}{3} - \frac{x^5}{30} \dots..$$

Example 3: Apply Maclaurin's theorem to prove that $\log \sec x = \frac{1}{2}x^2 + \frac{1}{12}x^4 + \frac{1}{45}x^6 + \dots\infty$.

Solution: Let $f(x) = \log \sec x$,

$\qquad f(0) = 0$

$\qquad f'(x) = \dfrac{1}{\sec x}\sec x \tan x = \tan x$, $f'(0) = 0$

$\qquad f''(x) = \sec^2 x$, $f''(0) = 1$

$\qquad f'''(x) = 2\sec^2 x; \tan x$, $f'''(0) = 0$

$$f^{iv}(x) = 4\sec^2 x \tan^2 x + 2\sec^4 x, \qquad\qquad f^{iv}(0) = 2$$

$$f^{v}(x) = 8\sec^2 x \tan^3 x + 16\sec^4 x \tan x, \qquad\qquad f^{v}(0) = 0$$

$$f^{vi}(x) = 16\sec^2 x \tan^4 x + 88\sec^4 x \tan^2 x + 16\sec^2 x, \qquad f^{vi}(0) = 16$$

$\therefore$ By Maclaurin's Theorem, we have

$$f(x) = f(0) + x f'(0) + \frac{x^2}{2!}f''(0) + \frac{x^3}{3!}f'''(0) + \frac{x^4}{4!}f^{iv}(0) + \frac{x^5}{5!}f^{v}(0) + \frac{x^6}{6!}f^{vi}(0) + \dots$$

$$\log\sec x = 0 + x.0 + \frac{x^2}{2}.1 + \frac{x^3}{3.2}.0 + \frac{x^4}{4.3.2}.2 + \frac{x^5}{5.4.3.2}.0 + \frac{x^6}{6.5.4.3.2}16 + \dots$$

$$= \frac{1}{2}x^2 + \frac{1}{12}x^4 + \frac{1}{45}x^6 + \dots\dots$$

***Example* 4:** If $\log\sec\ x = \dfrac{1}{2}x^2 + Ax^4 + Bx^6 + \dots\dots$ then find A and B.   (RGPV Dec. 2004)

***Solution*:** On comparing coefficients of x^4 and x^6 in above example 3, we get $A = \dfrac{1}{12}, B = \dfrac{1}{45}$

***Example* 5:** Find the first five terms in the expansion of log $(1 + \sin x)$ by Maclaurin's theorem.
(RGPV June 2006)

***Solution*:** Let $y = \log(1 + \sin x)$, $(y)_0 = 0$

or $e^y = 1 + \sin x$

Differentiating w.r.t x, we get

$$e^y . y_1 = \cos x, \qquad\qquad (y_1)_0 = 1$$

$$e^y . y_1^2 + e^y . y_2 = -\sin x, \qquad\qquad (y_2)_0 = -1$$

$$e^y . y_1^3 + 2e^y y_1 . y_2 + e^y y_1 y_2 + e^y y_3 = -\cos x, \qquad (y_3)_0 = 1$$

$$e^y . y_1^4 + 3e^y y_1^2 y_2 + 3e^y y_1^2 y_2 + 3e^y y_2^2 + 3e^y y_1 y_3 + e^y . y_1 y_3 + e^y . y_4 = \sin x,$$

$$(y_4)_0 = -2$$

or $e^y . y_1^4 + 6e^y y_1^2 y_2 + 3e^y y_2^2 + 4e^y y_1 y_3 + e^y y_4 = \sin x$

$$e^y . y_1^5 + 10e^y y_1^3 y_2 + 10e^y y_1^2 y_3 + 15e^y y_1 y_2^2 + 5e^y y_1 y_4 + 10e^y y_2 y_3 + e^y y_5 = \cos x \quad (y_5)_0 = 5$$

$\therefore$ By Maclaurin's Theorem, we get

$$y = (y)_0 + x(y_1)_0 + \frac{x^2}{2!}(y_2)_0 + \frac{x^3}{3!}(y_3) + \frac{x^4}{4!}(y_4)_0 + \frac{x^5}{5!}(y_5)_0 + \dots.$$

$$\log(1+\sin x) = x - \frac{x^2}{2} + \frac{x^3}{6} - \frac{x^4}{12} + \frac{x^5}{24} + ...$$

***Example* 6:** Expand $e^{\sin x}$ by Maclaurin's theorem upto the terms containing x^4.

(RGPV June 2007, June 2009)

***Solution*:** Let $y = y = e^{\sin x}$, $(y)_0 = 1$

Differentiating w.r.t x, we get

$$y_1 = e^{\sin x} . \cos x = y \cos x, \qquad (y_1)_0 = 1$$

$$y_2 = y_1 \cos x - y \sin x, \qquad (y_2)_0 = 1$$

$$y_3 = y_2 \cos x - 2y_1 \sin x - y\cos x, \qquad (y_3)_0 = 0$$

$$y_4 = y_3 \cos x - 3y_2 \sin x - 3y_1 \cos x + y \sin x, \qquad (y_4)_0 = -3$$

$$y_5 = y_4 \cos x - 4y_3 \sin x - 6y_2 \cos x + 4y_1 \sin x + y \cos x \qquad (y_5)_0 = -8$$

$\therefore$ By Maclaurin's Theorem

$$y = (y)_0 + x(y_1)_0 + \frac{x^2}{2!}(y_2)_0 + \frac{x^3}{3!}(y_3)_0 + \frac{x^4}{4!}(y_4)_0 + \frac{x_5}{5!}(y_5)_0 + ...$$

$$e^{\sin x} = 1 + x + \frac{x^2}{2} - \frac{x^4}{8} - \frac{x^5}{15} + ...$$

***Example* 7:** Using Maclaurin's series to prove that:

$$\log(1 + e^x) = \log 2 + \frac{x}{2} + \frac{x^2}{8} - \frac{x^4}{192} + ...$$

(RGPV June 2005, Jan 2008)

***Solution*:** Let $y = \log (1 + e^x)$ $(y)_0 = \log 2$

Differentiating w.r.t. x, we get

$$y_1 = \frac{1}{1+e^x}.e^x = 1 - \frac{1}{1+e^x}, \qquad (y_1)_0 = \frac{1}{2}$$

$$y_2 = \frac{e^x}{\left(1+e^x\right)^2} = \frac{e^x}{\left(1+e^x\right)}.\frac{1}{\left(1+e^x\right)}$$

or $y_2 = y_1(1 - y_1), \qquad (y_2)_0 = \frac{1}{4}$

$$y_3 = y_2(1 - y_1) + y_1(-y_2),$$

$$= y_2 - 2y_1 y_2, \qquad (y_3)_0 = 0$$

$$y_4 = y_3 - 2y_3 y_1 - 2 y_2^2, \qquad (y_4)_0 = -\frac{1}{8}$$

$\therefore$ By Maclaurin's Theorem, we get

$$y = (y)_0 + x(y_1)_0 + \frac{x^2}{2!}(y_2)_0 + \frac{x^3}{3!}(y_3)_0 + \frac{x^4}{4!}(y_4)_0 + \dots$$

$$\log(1 + e^x) = \log 2 + \frac{x}{2} + \frac{x^2}{8} - \frac{x^4}{192} + \dots$$

Example 8: Expand $e^{x \cos x}$ by Maclaurin's Theorem.

Solution: Let $y = e^{x \cos x}$, $\hspace{3cm}$ $(y)_0 = 1$

$\hspace{1cm}$ Then $y_1 = e^{x \cos x}(\cos x - x \sin x) = y(\cos x - x \sin x)$, $\hspace{1cm}$ $(y_1)_0 = 1$

$\hspace{2cm}$ $y_2 = y_1(\cos x - x \sin x) + y(-2 \sin x - x \cos x)$, $\hspace{1cm}$ $(y_2)_0 = 1$

$\hspace{2cm}$ $y_3 = y_2(\cos x - x \sin x) + y_1(-\sin x - \sin x - x \cos x) +$

$\hspace{3cm}$ $y_1(-2 \sin x - x \cos x) + y(-2 \cos x - \cos x + x \sin x)$

$\hspace{3cm}$ $= y_2(\cos x - x \sin x) + y_1(-4 \sin x - 2x \cos x) +$

$\hspace{3cm}$ $y(-3 \cos x + x \sin x)$, $\hspace{3cm}$ $(y_3)_0 = -2$

$\hspace{2cm}$ $y_4 = y_3(\cos x - x \sin x) + y_2(-2 \sin x - x \cos x) +$

$\hspace{3cm}$ $y_2(-4 \sin x - 2x \cos x) + y_1(-6 \cos x + 2 x \sin x) +$

$\hspace{3cm}$ $+ y_1(-3 \cos x + x \sin x) + y(4 \sin x + x \cos x)$

$\hspace{3cm}$ $= y_3(\cos x - x \sin x) + y_2(-6 \sin x - 3x \cos x)$

$\hspace{3cm}$ $y_1(-9 \cos x + 3x \sin x) + y(4 \sin x + x \cos x)$, $\hspace{1cm}$ $(y_4)_0 = -11$

$\hspace{1cm}$ $\therefore$ By Maclaurin's Theorem, we get

$$e^{x \cos x} = (y)_0 + x(y_1)_0 + \frac{x^2}{2!}(y_2)_0 + \frac{x^3}{3!}(y_3)_0 + \frac{x^4}{4!}(y_4)_0 + \dots$$

$$= 1 + x + \frac{x^2}{2!} - \frac{2x^3}{3!} - \frac{11x^4}{4!} + \dots$$

Example 9: Show that $\log \cos h\, x = \frac{1}{2}x^2 - \frac{1}{12}x^4 + \frac{1}{45}x^6 \dots$

Solution: Let $y = \log \cos h\, x$ $\hspace{3cm}$ $(y)_0 = 0$

$\therefore \hspace{2cm}$ $y_1 = \dfrac{1}{\cosh x} \sinh x = \tanh x$, $\hspace{2cm}$ $(y_1)_0 = 0$

$\hspace{2cm}$ $y_2 = \sec h^2 x$, $\hspace{3.5cm}$ $(y_2)_0 = 1$

$\hspace{2cm}$ $y_3 = -2 \sec h^2 x \tanh x = -2y_2 y_1$, $\hspace{1.5cm}$ $(y_3)_0 = 0$

$\hspace{2cm}$ $y_4 = \left(2y_3 y_1 + 2y_2^2\right)$, $\hspace{3cm}$ $(y_4)_0 = -2$

$\hspace{2cm}$ $y_5 = -(2y_4 y_1 + 6 y_3 y_2)$, $\hspace{2.5cm}$ $(y_5)_0 = 0$

$\hspace{2cm}$ $y_6 = -(2y_5 y_1 + 8y_4 y_2 + 6 y_3^2)$, $\hspace{2cm}$ $(y_6)_0 = 16$

$\therefore$ By Maclaurin's Theorem, we get

$$\log \cosh x = (y)_0 + x(y_1)_0 + \frac{x^2}{2!}(y_2)_0 + \frac{x^3}{3!}(y_3)_0 + \frac{x^4}{4!}(y_4)_0 + \frac{x^5}{5!}(y_5)_0 + \frac{x^6}{6!}(y_6)_0 + \ldots$$

$$= \frac{x^2}{2} - \frac{x^4}{12} + \frac{x^6}{45} + \ldots\ldots$$

***Example* 10:** Prove that $\sin^{-1}x = x + \frac{1}{2}\cdot\frac{x^3}{3} + \frac{1.3}{2.4}\frac{x^5}{5} + \frac{1.3.5}{2.4.6}\cdot\frac{x^7}{7} + \ldots .$

***Solution*:** Let $y = \sin^{-1} x$, $\qquad\qquad\qquad (y)_0 = 0$

Differentiating w.r.t x, we get

$$y_1 = \frac{1}{\sqrt{1-x^2}}$$

'or' $y_1^2\left(1-x^2\right)=1$, $\qquad\qquad\qquad (y_1)_0 = 1$

$$2y_1 y_2 (1-x^2) + y_1^2 (-2x) = 0$$

'or' $y_2(1 - x_2) - xy_1 = 0$, $\qquad\qquad\qquad (y_2)_0 = 0$

Differentiating n times by Leibnitz's theorem, we get

$$y_{n+2}(1-x^2) + n\cdot y_{n+1}(-2x) + \frac{n(n-1)}{2!}y_n(-2) - y_{n+1}.x - n\,y_n = 0$$

'or' $(1-x^2)\,y_{n+2} - (2n+1)\,x\,y_{n+1} - n^2\,y_n = 0$

Putting $x = 0$, we get $(y_{n+2})_0 = n^2(y_n)_0$

Putting $n = 1, 2, 3,\ldots..$, we get

$\qquad (y_3)_0 = 1,\ (y_4)_0 = 0,\ (y_5)_0 = 3^2,\ (y_6)_0 = 0,\ (y_7)_0 = 5^2\cdot 3^2$

$\therefore$ By Maclaurin's Theorem, we get

$$\sin^{-1}x = (y)_0 + x(y_1)_0 + \frac{x^2}{2!}(y_2)_0 + \ldots.$$

$$= 0 + x + \frac{x^2}{2!}(0) + \frac{x^3}{3!}(1) + \frac{x^4}{4!}(0) + \frac{x^5}{5!}(3^2) + \frac{x^6}{6!}(0) + \frac{x^7}{7!}\cdot(5^2.3^2)$$

$$= x + \frac{1}{2}\frac{x^3}{3} + \frac{1.3}{2.4}\frac{x^5}{5} + \frac{1.3.5}{2.4.6}\frac{x^7}{7} + \ldots.$$

***Example* 11:** Expand $e^{a\sin^{-1}x}$ by Maclaurin's theorem and find the general term. Hence show that

$$e^{\theta} = 1 + \sin\theta + \frac{1}{2!}\sin^2\theta + \frac{2}{3!}\sin^3\theta + ...$$

(RGPV Dec 2001, June 2002, June 2008, April 2009)

***Solution*:** Let $y = e^{a\sin^{-1}x}$ .....(i)

Differentiating w.r.t x, we get

$$y_1 = e^{a\sin^{-1}x} \frac{a}{\sqrt{1-x^2}}$$

'or' $y_1^2\left(1-x^2\right) = a^2y^2$ .....(ii)

Differentiating again w.r.t x, we get

$$y_1^2(-2x) + 2y_1 y_2 (1-x^2) = a^2 \, 2y \, y_1$$

'or' $(1-x^2) y_2 - xy_1 - a^2y = 0$ .....(iii)

Differentiating (iii) n times by Leibnitz's Theorem, we get

$$\left(1-x^2\right)y_{n+2} + ny_{n+1}(-2x) + \frac{n(n-1)}{2!} \, y_n(-2) - xy_{n+1} - ny_n - a^2y_n = 0$$

'or; $(1-x^2) y_{n+2} - (2n+1) x \, y_{n+1} - (n^2 + a^2) \, y_n = 0$ .....(iv)

Putting $x = 0$ in equation (i), (ii), (iii) and (iv), we get

$(y)_0 = 1$, $(y_1)_0 = a$, $(y_2)_0 = a^2$ and

$(y_{n+2})_0 = (n^2 + a^2) (y_n)_0$ .....(v)

Putting $n = 1, 2, 3, ...$ in equation (v), we have

$(y_3)_0 = (1^2 + a^2)a$, $(y_4)_0 = (2^2 + a^2)a^2$ ....

$\therefore$ By Maclaurin's Theorem, we get

$$e^{a\sin^{-1}x} = (y)_0 + x(y_1)_0 + \frac{x^2}{2!} \, (y_2)_0 + \frac{x^3}{3!}(y_3)_0 + \frac{x^4}{4!}(y_4)_0 + ...$$

$$= 1 + ax + a^2 \frac{x^2}{2!} + (1^2 + a^2) \, a \, \frac{x^3}{3!} + (2^2 + a^2) \, a^2 \, \frac{x^4}{4!} + ...$$

Also putting $x = \sin\theta$, $a = 1$, we get

$$e^\theta = 1 + \sin\theta + \frac{1}{2!}\sin^2\theta + \frac{2}{3!}\sin^3\theta + \ldots\ldots$$

***Example* 12:** Expand $\sin(m\,\sin^{-1}x)$ by Maclaurin's Theorem in terms of x^5. Hence or otherwise expand $\sin m\theta$ in powers of $\sin\theta$.

***Solution*:** Let $y = \sin(m\,\sin^{-1}x)$ $\qquad\qquad\qquad\qquad\qquad$ .....(i)

Differentiating w.r.t x, we get

$$y_1 = \cos(m\,\sin^{-1}x)\;\frac{m}{\sqrt{1-x^2}}$$

or $\qquad y_1^2(1-x^2) = m^2\cos^2(m\,\sin^{-1}x)$

or $\qquad y_1^2(1-x^2) = m^2(1-y^2)$ $\qquad\qquad\qquad\qquad$ .....(ii)

Again differentiating w.r.t x, we get

$$2y_1\,y_2(1-x^2) + y_1^2(-2x) = m^2(-2y\,y_1)$$

or $\qquad y_2(1-x^2) - xy_1 + m^2y = 0$ $\qquad\qquad\qquad$ .....(iii)

Differentiating equation (iii) n times by Leibnitz's Theorem, we get

$$(1-x^2)y_{n+2} + ny_{n+1}(-2x) + \frac{n(n-1)}{2!}\,y_n(-2) - xy_{n+1} - ny_n + m^2\,y_n = 0$$

or $\qquad (1-x^2)\,y_{n+2} - (2n+1)\,x\,y_{n+1} - (n^2 - m^2)\,y_n = 0$ $\qquad$ .....(iv)

Putting $x = 0$ in equation (i), (ii), (iii) and (iv) , we have

$$(y)_0 = 0,\ (y_1)_0 = m,\ (y_2)_0 = 0,\ \text{and}$$

$$(y_{n+2})_0 = (n^2 - m^2)\,(y_n)_0$$

$$(y_3)_0 = (1^2 - m^2)m,\ (y_4)_0 = 0,\ (y_5)_0 = (3^2 - m^2)\,(1^2 - m^2)m$$

$\therefore$ By Maclaurin's Theorem, we get

$$\sin(m\,\sin^{-1}x) = (y)_0 + x(y_1)_0 + \frac{x^2}{2!}\,(y_2)_0 +$$

$$\frac{x^3}{3!}\,(y_3)_0 + \frac{x^4}{4!}\,(y_4)_0 + \frac{x^5}{5!}\,(y_5)_0 + \ldots..$$

$$= 0 + xm + \frac{x^2}{2!}\,(0) + \frac{x^3}{3!}\,m(1^2 - m^2) +$$

$$\frac{x^4}{4!}(0) + \frac{x^5}{5!}(3^2 - m^2)(1^2 - m^2)m + \dotsb$$

$$= mx + m(1^2 - m^2)\ \frac{x^3}{3!} + m(1^2 - m^2)(3^2 - m^2)\frac{x^5}{5!} + \dotsb$$

Putting $x = \sin\theta$, we get

$$\sin m\theta = m\sin\theta + m(1^2 - m^2)\frac{\sin^3\theta}{3!} + m(1^2 - m^2)\left(3^2 - m^2\right)\frac{\sin^5\theta}{5!} + \dotsc$$

***Example* 13:** Expand $\left\{x + \sqrt{1+x^2}\right\}^m$ in ascending powers of x and find the general term.

***Solution*:** Let $y = \left\{x + \sqrt{1+x^2}\right\}^m$ .....(i)

$$\therefore \qquad y_1 = m\left\{x + \sqrt{1+x^2}\right\}^{m-1} \cdot \left[1 + \frac{x}{\sqrt{1+x^2}}\right]$$

$$= \frac{m\left[x + \sqrt{1+x^2}\right]^m}{\sqrt{1+x^2}} = \frac{my}{\sqrt{1+x^2}}$$

or $\qquad y_1^2\,(1 + x^2) = m^2\,y^2$ .....(ii)

Differentiating again, we get

$$(1 + x^2)\,y_2 + xy_1 - m^2 y = 0 \qquad\qquad .....(iii)$$

Differentiating equation (iii) n times by Leibnitz's Theorem, we get

$$(1 + x^2)\,y_{n+2} + n\,y_{n+1}\,2x + \frac{n(n-1)}{2!}\,y_n.2 +$$

$$xy_{n+1} + ny_n - m^2\,y_n = 0$$

'or' $\qquad (1 + x^2)\,y_{n+2} + (2n + 1)\,xy_{n+1} + (n^2 - m^2)\,y_n = 0 \qquad .....(iv)$

Putting $x = 0$ in equation (i), (ii), (iii) and (iv), we get

$$(y)_0 = 1,\ (y_1)_0 = m,\ (y_2)_0 = m^2 \text{ and,}$$

$$(y_{n+2})_0 = (m^2 - n^2)\,(y_n)_0 \qquad\qquad .....(v)$$

Putting $\ n = 1, 2, 3 \dotsc$ in equation (v), we have

$$(y_3)_0 = (m^2 - 1^2)m,\ (y_4)_0 = (m^2 - 2^2)\,m^2$$

$\therefore$ By Maclaurin's Theorem, we get

$$\left[x + \sqrt{1+x^2}\right]^m = 1 + mx + \frac{m^2 x^2}{2!} + \frac{m(m^2 - 1^2)}{3!} x^3 + \frac{m^2\left(m^2 - 2^2\right)}{4!} x^4 + \dots$$

For the general term:

When n is even then by equation (v), we get

$$(y_n)_0 = [m^2 - (n-2)^2]\,[m^2 - (n-4)^2]\,\dots$$
$$(m^2 - 4^2)\,(m^2 - 2^2)\,m^2 \qquad\qquad\qquad\qquad \dots(vi)$$

When n is odd, then

$$(y_n)_0 = [m^2 - (n-2)^2]\,[m^2 - (n-4)^2]\,\dots$$
$$(m^2 - 3^2)\,(m^2 - 1^2)\,m\dots\dots\dots \qquad\qquad\qquad\qquad \dots(vii)$$

Then general term $= \dfrac{x^n}{n!}(y_n)_0$, where $(y_n)_0$ is given by equation (vi) and (vii).

5.3   Taylor's Theorem

Let f(x + h) be a function of h, which can be expanded in ascending powers of h and let the expansion be differentiable any number of times, then

$$f(x+h) = f(x) + hf'(x) + \frac{h^2}{2!}f''(x) + \dots + \frac{h^n}{n!}f^n(x) + \dots$$

Proof: Let $f(x+h) = A_0 + A_1 h + A_2 h^2 + A_3 h^2 + A_3 h^3 + A_4 h^4 +$ $\dots(i)$

where $A_0, A_1, A_2 \dots$ are constants.

By successive differentiation of equation (i) with respect to h, we get

$$f'(x+h) = A_1 + 2A_2 h + 3A_3 h^2 + 4A_4 h^3 + \dots\dots \qquad\qquad \dots(ii)$$

$$f''(x+h) = 2A_2 + 3.2A_3 h + 4.3A_4 h^2 + \dots\dots \qquad\qquad \dots(iii)$$

$$f'''(x+h) = 3.2A_3 + 4.3.2A_4 h + \dots\dots \qquad\qquad \dots(iv)$$

etc

Putting h = 0 in equation (i), (ii), (iii) and (iv), we get

$$f(x) = A_0, \;\; f'(x) = A_1, f''(x) = 2A_2, f'''(x) = 3.2A_3\,,\dots$$

Putting these values of $A_0, A_1, A_2, A_3 \dots$ in equation (i), we have

$$f(x+h) = f(x) + hf'(x) + \frac{h^2}{2!}f''(x) + \frac{h^3}{3!}f'''(x) + \dots + \frac{h^n}{n!}f^n(x) + \dots$$

Remark:

(i)   If we take x = 0, h = x then we get the Maclaurin's Theorem

(ii) Putting x = a, we have

$$f(a+h) = f(a) + hf'(a) + \frac{h^2}{2!}f''(a) + \dots \frac{h^n}{n!}f^n(a) + \dots$$

(iii) Putting x = h, h = a, we have

$$f(a+h) = f(h) + af'(h) + \frac{a^2}{2!}f''(h) + \dots \frac{a^n}{n!}f^n(h) + \dots$$

(iv) Putting h = x –a, we have

$$f(x) = f(a) + (x-a)f'(a) + \frac{(x-a)^2}{2!}f''(a) + \dots + \frac{(x-a)^n}{n!}f^n(a) + \dots$$

***Example* 14:** Show that $\log(x+h) = \log h + \dfrac{x}{h} - \dfrac{x^2}{2h^2} + \dfrac{x^3}{3h^3} + \dots$ (RGPV Jan 2007)

Solution: Let f(x + h) = log (x + h)

∴   f(h) = log h

$$f'(h) = \frac{1}{h}, \ f''(h) = \frac{-1}{h^2}, \ f'''(h) = \frac{2}{h^3}\dots$$

∴ By Taylor's Theorem, we have

$$f(x+h) = f(h) + xf'(h) + \frac{x^2}{2!}f''(h) + \frac{x^3}{3!}f'''(h) + \dots$$

$$\log(x+h) = \log h + \frac{x}{h} - \frac{x^2}{2h^2} + \frac{x^3}{3h^3}\dots$$

Example 15: Expand log sin (x + h) in powers of h by Taylor's Theorem.

Solution: Let f(x + h) = log sin(x + h)

∴   f(x) = log sin x

$$f'(x) = \frac{1}{\sin x}\cos x = \cot x$$

$$f''(x) = -\cos ec^2\, x$$

$$f'''(x) = 2\cos ec^2 x \cot x \ \text{etc}$$

$\therefore$ By Taylor's Theorem, we get

$$f(x+h) = f(x) + hf'(x) + \frac{h^2}{2!}f''(x) + \frac{h^3}{3!}f'''(x) + ...$$

$$\log \sin(x+h) = \log \sin x + h \cot x$$

$$- \frac{h^2}{2}\cos ec^2 x + \frac{h^3}{3}\cos ec^2 x \cot x + ...$$

***Example* 16:** Expand $\sin^{-1}(x+h)$ in powers of x as far as term in x^3.

***Solution*:** Let $f(x+h) = \sin^{-1}(x+h)$

$\therefore \qquad f(h) = \sin^{-1} h$

$$f'(h) = \frac{1}{\sqrt{1-h^2}}, \ f''(h) = \frac{h}{\left(1-h^2\right)^{3/2}}, \ f'''(h) = \frac{1+2h^2}{\left(1-h^2\right)^{5/2}}$$

$\therefore$ By Taylor's Theorem, we have

$$f(x+h) = f(h) + xf'(h) + \frac{x^2}{2!}f''(h) + \frac{x^3}{3!}f'''(h) + ...$$

$$\sin^{-1}(x+h) = \sin^{-1} h + \frac{x}{\sqrt{1-h^2}} + \frac{x^2 h}{2\left(1-h^2\right)^{3/2}} + \frac{x^3\left(1+2h^2\right)}{6\left(1-h^2\right)^{5/2}} + ...$$

***Example* 17:** Use Taylor's Theorem to prove that

$$\tan^{-1}(x+h) = \tan^{-1} x + h \sin z \frac{\sin z}{1} - (h \sin z)^2 \frac{\sin 2z}{2} + (h \sin z)^3 \frac{\sin 3z}{3} + ...$$

where $z = \cot^{-1} x$. $\hspace{3cm}$ (RGPV June 2001)

***Solution*:** Let $f(x+h) = \tan^{-1}(x+h)$

$\therefore \qquad f(x) = \tan^{-1} x$

$$f'(x) = \frac{1}{1+x^2} = \frac{1}{1+\cot^2 z}, \hspace{2cm} (\because z = \cot^{-1} x)$$

$$= \sin^2 z$$

$$f''(x) = 2 \sin z \cos z \frac{dz}{dx}$$

$$= \sin 2z \left(-\frac{1}{1+x^2} \right) \qquad \left(\because \frac{dz}{dx} = \frac{-1}{1+x^2} \right)$$

$$= -\sin 2z \cdot \sin^2 z$$

$$f'''(x) = -\left(\sin 2z.2\sin z \cos z + 2\cos 2z \sin^2 z \right) \frac{dz}{dx}$$

$$= -2\sin^2 z \left(2\cos^2 z + \cos 2z \right) \left(-\frac{1}{1+x^2} \right)$$

$$= 2\sin^2 z \left(2 - 2\sin^2 z + 1 - 2\sin^2 z \right) \left(\frac{1}{1+\cot^2 z} \right)$$

$$= 2\sin^3 z(3\ \sin z - 4\ \sin^3 z)$$

$$= 2\ \sin^3 z\ \sin^3 z$$

$\therefore$ By Taylor's Theorem, we get

$$f(x+h) = f(x) + hf'(x) + \frac{h^2}{2!}f''(x) + \frac{h^3}{3!}f'''(x) + ...$$

$$\tan^{-1}(x+h) = \tan^{-1}x + h\sin z \frac{\sin z}{1} - (h\sin z)^2 \frac{\sin 2z}{2} + (h\sin z)^3 \frac{\sin 3z}{3} + ...$$

Expand **18:** Expand sin x in powers of $\left(x - \dfrac{\pi}{2} \right)$.   (RGPV Dec 2005)

***Solution*:** By Taylor's Theorem, we have

$$f(x) = f(a) + (x-a)f'(a) + \frac{(x-a)^2}{2!}f''(a) + \frac{(x-a)^3}{3!}f'''(a) + ... \qquad .....(i)$$

Now $f(x) = \sin x$ and $a = \dfrac{\pi}{2}$, $f\left(\dfrac{\pi}{2} \right) = 1$

$$\therefore\ f'(x) = \cos x, \qquad\qquad f'\left(\frac{\pi}{2} \right) = 0,$$

$$f''(x) = -\sin x, \qquad\qquad f''\left(\frac{\pi}{2} \right) = -1$$

$$f'''(x) = -\cos x, \qquad\qquad f'''\left(\frac{\pi}{2} \right) = 0$$

$$f^{iv}(x) = \sin x, \qquad\qquad f^{iv}\left(\frac{\pi}{2}\right) = 1$$

Putting these values in equation (i), we have

$$\sin x = 1 - \frac{1}{2!}\left(x - \frac{\pi}{2}\right)^2 + \frac{1}{4!}\left(x - \frac{\pi}{2}\right)^4 \cdots\cdots\cdots$$

***Example* 19:** Find the Taylor's series expansion of the function about the point $\frac{\pi}{3}$, If $f(x) = \log \cos x$.   (RGPV Dec 2003)

***Solution*:** By Taylor's theorem, we have

$$f(x) = f(a) + (x-a)f'(a) + \frac{(x-a)^2}{2!}f''(a) + \frac{(x-a)^3}{3!}f'''(a) + \cdots \qquad .....(i)$$

Now, Let

$$f(x) = \log \cos x, \text{ and } a = \frac{\pi}{3}, \qquad f\left(\frac{\pi}{3}\right) = \log\left(\frac{1}{2}\right)$$

$$f'(x) = \frac{-\sin x}{\cos x} = -\tan x, \qquad f'\left(\frac{\pi}{3}\right) = -\sqrt{3}$$

$$f''(x) = -\sec^2 x \qquad\qquad f''\left(\frac{\pi}{3}\right) = -4$$

$$f'''(x) = -2\sec^2 x \tan x \qquad f'''\left(\frac{\pi}{3}\right) = -8\sqrt{3} \quad \text{and so on}$$

Putting these values in equation (i), we have

$$\log \cos x = \log\left(\frac{1}{2}\right) - \sqrt{3}\left(x - \frac{\pi}{3}\right) - \frac{4\left(x - \frac{\pi}{3}\right)^2}{2!} - 8\sqrt{3}\frac{\left(x - \frac{\pi}{3}\right)^3}{3!} + \cdots$$

***Example* 20:** Expand the polynomial $2x^3 + 7x^2 + x - 1$ in powers of $(x - 2)$.

***Solution*:** By Taylor's theorem, we have

$$f(x) = f(a) + (x-a)f'(a) + \frac{(x-a)^2}{2!}f''(a) + \frac{(x-a)^3}{3!}f'''(a) + \qquad .....(i)$$

Here $f(x) = 2x^3 + 7x^2 + x - 1$ and $a = 2$, $f(2) = 45$

$$f'(x) = 6x^2 + 14x + 1, \qquad\qquad f'(2) = 53$$

$$f''(x) = 12x + 14 \qquad\qquad f''(2) = 38$$

$$f'''(x) = 12 \qquad\qquad\qquad f'''(2) = 12$$

$$f^{iv}(x) = 0 \qquad\qquad\qquad f^{iv}(2) = 0$$

Putting these values in equation (i), we have

$$f(x) = 45 + 53(x-2) + \frac{(x-2)^2}{2!} 38 + \frac{(x-2)^3}{3!}.12 + 0$$

$$= 45 + 53(x-2) + 19(x-2)^2 + 2(x-2)^3$$

***Example* 21:** Expand $\tan\left(x + \frac{\pi}{4}\right)$ as far as the term x^4 and evaluate $\tan 46.5°$ up to four significant digits. $\hfill$ (RGPV Jan 2006)

***Solution*:** By Taylor's series, we have

$$f(x+h) = f(h) + xf'(h) + \frac{x^2}{2!}f''(h) + \frac{x^3}{3!}f'''(h) + \qquad\qquad .....(i)$$

Now, let $f(x) = \tan x,$ $\qquad\qquad\qquad f\left(\frac{\pi}{4}\right) = 1$

$$f'(x) = \sec^2 x, \qquad\qquad\qquad f'\left(\frac{\pi}{4}\right) = 2$$

$$f''(x) = 2\sec^2 x \tan x, \qquad\qquad f''\left(\frac{\pi}{4}\right) = 4$$

$$f'''(x) = 2 + 8\tan^2 x + 6\tan^4 x, \qquad f'''\left(\frac{\pi}{4}\right) = 16 \text{ and so on}$$

Putting these values in equation (i) and h = $\frac{\pi}{4}$, we have

$$f\left(x + \frac{\pi}{4}\right) = f\left(\frac{\pi}{4}\right) + xf'\left(\frac{\pi}{4}\right) + \frac{x^2}{2!}f''\left(\frac{\pi}{4}\right) + \frac{x^3}{3!}f'''\left(\frac{\pi}{4}\right) + ...$$

$$\tan\left(x + \frac{\pi}{4}\right) = 1 + 2x + \frac{x^2}{2!}.4 + \frac{x^3}{3!}.16 + ...$$

$$= 1 + 2x + 2x^2 + \frac{8}{3}x^3 + ... \qquad\qquad ....(ii)$$

Putting x = $1.5° = 1.5 \times \frac{\pi}{180} = 0.02617$ in eqn (ii), we get

$$\tan(1.5° + 45°) = 1 + 2(0.02617) + 2(0.02617)^2 + \frac{8}{3}(0.02617)^3 + ...$$

$$\tan 46.5° = 1.0537 \text{ (approx.)}$$

***Example* 22:** Expand $\tan^{-1}x$ in powers of $\left(x - \dfrac{\pi}{4}\right)$.

***Solution*:** By Taylor's Theorem, we have

$$f(x) = f\left(\frac{\pi}{4}\right) + \left(x - \frac{\pi}{4}\right)f'\left(\frac{\pi}{4}\right) + \frac{\left(x - \frac{\pi}{4}\right)^2}{2!}f''\left(\frac{\pi}{4}\right) + \frac{\left(x - \frac{\pi}{4}\right)^3}{3!}f'''\left(\frac{\pi}{4}\right) + \quad \dots\text{(i)}$$

Now, let $f(x) = \tan^{-1}x,$ $\qquad f\left(\dfrac{\pi}{4}\right) = \tan^{-1}\left(\dfrac{\pi}{4}\right)$

$$f'(x) = \frac{1}{1+x^2}, \qquad\qquad f'\left(\frac{\pi}{4}\right) = \frac{1}{1+\dfrac{\pi^2}{16}}$$

$$f''(x) = \frac{-2x}{\left(1+x^2\right)^2}, \qquad\qquad f''\left(\frac{\pi}{4}\right) = \frac{-\pi}{2\left(1+\dfrac{\pi^2}{16}\right)^2} \quad \text{and so on}$$

Putting these values in equation (i), we have

$$\tan^{-1}x = \tan^{-1}\frac{\pi}{4} + \left(x - \frac{\pi}{4}\right)\frac{1}{1+\dfrac{\pi^2}{16}} - \frac{\left(x - \frac{\pi}{4}\right)^2}{2!}\frac{\pi}{2\left(1+\dfrac{\pi^2}{16}\right)^2} + \dots$$

***Example* 23:** Evaluate the approximate value of $\log_e(1.1)$ correct to four decimal places, by using an appropriate Taylor's expansion.

or

Expand $\log_e x$ in powers of $(x - 1)$ and hence evaluate $\log(1.1)$ correct to four decimal places.

(RGPV Dec 2002, June 2003)

***Solution*:** By Taylor's Theorem, we have

$$f(x+h) = f(a) + hf'(a) + \frac{h^2}{2!}f''(a) + \frac{h^3}{3!}f'''(a) + \dots$$

Here $a = 1, h = x - 1$ then

$$f(x) = f(1) + (x-1)f'(1) + \frac{(x-1)^2}{2!}f''(1) + \frac{(x-1)^3}{3!}f'''(1) + \dots \qquad \dots(i)$$

Now, let $f(x) = \log x$, $\qquad\qquad\qquad f(1) = 0$

$$f'(x) = \frac{1}{x}, \qquad\qquad\qquad\qquad f'(1) = 1$$

$$f''(x) = \frac{-1}{x^2}, \qquad\qquad\qquad\qquad f''(1) = -1$$

$$f'''(x) = \frac{-2}{x^3}, \qquad\qquad\qquad\qquad f'''(1) = 2$$

$$f^{iv}(x) = \frac{-6}{x^4}, \qquad\qquad\qquad\qquad f^{iv}(1) = -6$$

Putting these values in equation (i), we have

$$\log x = 0 + (x-1) \cdot 1 + \frac{(x-1)^2}{2}(-1) + \frac{(x-1)^3}{3.2}(2) + \frac{(x-1)^4}{4.3.2}(-6) + \dots$$

$$= (x-1) - \frac{1}{2}(x-1)^2 + \frac{1}{3}(x-1)^3 - \frac{1}{4}(x-1)^4 + \dots$$

Putting $x = 1.1$, we have

$$\log 1.1 = (1.1 - 1) - \frac{1}{2}(1.1-1)^2 + \frac{1}{3}(1.1-1)^3 - \frac{1}{4}(1.1-1)^4 + \dots$$

$$= 0.1 - 0.005 + 0.00033 - 0.000025 + \dots$$

$$= 0.0953$$

Practice Problems

1. Show that $e^x \cos x = 1 + x - \frac{2x^3}{3!} - \frac{2^2 x^4}{4!} - \frac{2^2 x^5}{5!} + \dots$

2. Find the first five terms in the expansion of $\log(1 + \tan x)$ by Maclaurin's theorem.

$$\left[\textbf{Ans. } x - \frac{x^2}{2} + \frac{2x^3}{3} + \dots\right]$$

3. Expand $e^x \cdot \sec x$ by Maclaurin's Theorem.

$$\left[\textbf{Ans. } 1 + x + \frac{2x^2}{2!} + \frac{4x^3}{3!} + \dots\right]$$

4. Apply Maclaurin's Theorem to obtain the terms up to x^4 in the expansion of $\log(1 + \sin^2 x)$.

$$\left[\textbf{Ans. } x^2 - \frac{5}{6}x^4 + ... \right]$$

5. Expand $\cos x \sinh x$ to the fifth powers of x.

$$\left[\textbf{Ans. } x - \frac{x^3}{3} - \frac{x^5}{30} + ... \right]$$

6. Expand $\tan^{-1}x$ by Maclaurin's Theorem.

$$\left[\textbf{Ans. } x - \frac{x^3}{3} + \frac{x^5}{5} + ... \right]$$

7. Expand $(\sin^{-1} x)^2$ in ascending powers of x.

$$\left[\textbf{Ans. } \frac{2x^2}{2!} + 2.2^2 \frac{x^4}{4!} + 2.2^2.4^2 \frac{x^6}{6!} + ... \right]$$

8. Expand $e^{a\cos^{-1}x}$ by Maclaurin's Theorem.

$$\left[\textbf{Ans. } e^{a\frac{\pi}{2}} - axe^{a\frac{\pi}{2}} + \frac{x^2}{2!}a^2 e^{a\frac{\pi}{2}} + ... \right]$$

9. If $\log y = \tan^{-1} x$, then find the coefficient of x^5 in the expansion of y by Maclaurin's theorem.

10. If $y = e^{m\tan^{-1}x}$, then prove that $y = 1 + mx + \frac{m^2 x^2}{2!} + \frac{m(m^2 - 2)}{3!}x^3 + ...$

11. Expand $\log\left\{x + \sqrt{1 + x^2}\right\}$ in ascending powers of x and find the general term.

$$\left[\textbf{Ans. } x - \frac{x^3}{3!} + \frac{x^5}{5!}\left(3^2.1^2\right) - \frac{x^7}{7!}\left(5^2.3^2.1^2\right) + ... \right]$$

12. Use Maclaurin's Theorem to prove that $\sin\left(e^x - 1\right) = x + \frac{1}{2}x^2 - \frac{5}{24}x^4 + ...$

13. Expand $\sin^{-1} (x + h)$ in powers of x as far as term in x^3.

$$\left[\textbf{Ans. } \sin^{-1}h + \frac{x}{\sqrt{1 - h^2}} + \frac{x^2 h}{2\left(1 - h^2\right)^{\frac{3}{2}}} + \frac{x^3\left(1 + 2h^2\right)}{6\left(1 - h^2\right)^{5/2}} + ... \right]$$

14. Expand $\log \sin x$ in powers of $(x - 2)$.

$$[\textbf{Ans. } \log\sin 2 + (x-2)\cot 2 - \frac{(x-2)^2}{2!}\cos ec_2^2 + ...]$$

15.   Expand $\sin\left(x + \dfrac{\pi}{4}\right)$ in power of x.

$$\left[\textbf{Ans. } \frac{1}{\sqrt{2}}\left(1 + x - \frac{x^2}{2!} - \frac{x^2}{2!} - \frac{x^3}{3!} + \frac{x^4}{4!} + ...\right).\right]$$

16.   Expand $2 + x^2 - 3x^5 + 7x^6$ in powers of $(x-1)$.

(RGPV Dec 2001)

$$[\textbf{Ans. } 7 + 29(x-1) + 76(x-1)^2 + 110(x-1)^3 + 40(x-1)^4 + 39(x-1)^5 + 7(x-1)^6]$$

17.   Expand $\tan\theta$ in powers of $\left(\theta - \dfrac{\pi}{4}\right)$.

(RGPV June 2004)

$$\left[\textbf{Ans. } 1 + 2\left(\theta - \frac{\pi}{4}\right) + 2\left(\theta - \frac{\pi}{4}\right)^2 + ...\right]$$

CHAPTER 6

Maxima and Minima

6.1   Maxima and Minima of a Function of One Variable

Definition: A function $f(x)$ is said to have a maximum for a value a of x if $f(a)$ is greater than any other value that the function can have in the small neighbourhood of $x = a$.

Similarly a function $f(x)$ is said to have a minimum for a value a of x if $f(a)$ is less than any other value that the function can have in the small neighbourhood of $x = a$.

Working method: If a function $y = f(x)$ be given. Find $\dfrac{dy}{dx}$ and equate it to zero, solve this equation for the values x. Let the values of x are $a_1, a_2, a_3, \ldots.$

Find $\dfrac{d^2y}{dx^2}$ and hence find $\left(\dfrac{d^2y}{dx^2}\right)_{x=a_1}$, $\left(\dfrac{d^2y}{dx^2}\right)_{x=a_2}$, $\ldots$

If $\left(\dfrac{d^2y}{dx^2}\right)_{x=a_1}$ is positive, we have a minimum at $x = a_1$.

If $\left(\dfrac{d^2y}{dx^2}\right)_{x=a_1}$ is negative, we have a maximum at $x = a_1$.

If $\left(\dfrac{d^2y}{dx^2}\right)_{x=a_1} = 0$, find $\dfrac{d^3y}{dx^3}$ and then $\left(\dfrac{d^3y}{dx^3}\right)_{x=a_1}$.

If $\left(\dfrac{d^3y}{dx^3}\right)_{x=a_1} \neq 0$, then there is neither maxima nor minima at $x = a_1$.

If $\left(\dfrac{d^3y}{dx^3}\right)_{x=a_1} = 0$, find $\left(\dfrac{d^4y}{dx^4}\right)$ and then $\left(\dfrac{d^4y}{dx^4}\right)_{x=a_1}$

If $\left(\dfrac{d^4y}{dx^4}\right)_{x=a_1}$ is negative, then $f(x)$ is maximum at $x = a_1$.

If $\left(\dfrac{d^4y}{dx^4}\right)_{x=a_1}$ is positive, then f(x) is minimum at $x = a_1$.

If $\left(\dfrac{d^4y}{dx^4}\right)_{x=a_1} = 0$, then find $\left(\dfrac{d^5y}{dx^5}\right)_{x=a_1}$ and so on.

***Example* 1:** Show that $x^3 - 3x^2 + 3x + 7$ has neither a maximum nor a minimum at $x = 1$.

Solution: Let $f(x) = x^3 - 3x^2 + 3x + 7$

$\therefore \qquad f'(x) = 3x^2 - 6x + 3$

$\qquad\qquad f''(x) = 6x - 6$

$\qquad\qquad f'''(x) = 6$

Equating $f'(0)$ to zero, we get

$\qquad\qquad 3x^2 - 6x + 3 = 0$

or $\qquad 3(x - 1)^2 = 0$

or $\qquad x = 1$

at $x = 1$, $\;f''(x) = 6 - 6 = 0$

and $\qquad f'''(x) = 6 \neq 0$

Hence, the function has neither maximum nor minimum at $x = 1$.

***Example* 2:** Find the largest and smallest values of $x^3 - 18x^2 + 96x$ in the interval $[0, 9]$

(RGTV Jan 2006)

Solution: Let $f(x) = x^3 - 18x^2 + 96x$

$\therefore \qquad f'(x) = 3x^2 - 36x + 96$

and $\qquad f''(x) = 6x - 36$

For maxima and minimum $f'(x) = 0$

or $\qquad 3x^2 - 36x + 96 = 0$

or $\qquad x = 4$ and $x = 8$

at $\qquad x = 4$, $f''(4) = 6 \times 4 - 36 = -12 < 0$

$\therefore \qquad f(x)$ has maximum at $x = 4$

$\therefore \qquad f(4) = (4)^3 - 18\,(4)^2 + 96(4) = 160$

at $\qquad x = 8$, $f''(8) = 6 \times 8 - 36 = 12 > 0$

$\therefore$ f(x) has minimum at x = 8

$\therefore$ $f(8) = (8)^3 - 18(8)^2 + 96(8) = 128$

Also, for the given interval [0, 9] i.e.,

at   x = 0, f(0) = 96

at $x = 9, f(9) = (9)^3 - 18\ (9)^2 + 96(9) = 135$

Hence maximum value is 160 at x = 4 and minimum value is 96 at x = 0

***Example* 3:** Find the minimum and maximum values of the function $f(x) = x^2 - 4x + 5$ in the interval $1 \leq x \leq 4$.   (RGPV Dec 2005)

***Solution*:**

Let $f(x) = x^2 - 4x + 5$

$\therefore$ $f'(x) = 2x - 4$

and $f''(x) = 2$

For maxima and minima, $f'(x) = 0$

or $2x - 4 = 0$

or $x = 2$

at $x = 2,\ f''(x) = 2 > 0$

$\therefore$ f(x) has minimum value at x = 2

$\therefore$ $f(2) = (2)^2 - 4(2) + 5 = 1$

Also, for the given interval $1 \leq x \leq 4$, i.e.,

at   x = 1, $f(1) = (1)^2 - 4(1) + 5 = 2$

at   x = 4, $f(4) = (4)^2 - 4(4) + 5 = 5$

Hence, minimum value is 1 at x = 2.

***Example* 4:** Investigate for maximum and minimum values, the function (sin x + cos 2x).

(RGPV Dec. 2004)

***Solution*:** Let f(x) = sin x + cos 2x

$\therefore$ $f'(x) = \cos x - 2 \sin 2x$

and $f''(x) = - \sin x - 4 \cos 2x$

For maxima and minima, $f'(x) = 0$

or $\cos x - 2 \sin 2x = 0$

or $\cos x - 4 \sin x \cos x = 0$

or $\cos x \, (1 - 4 \sin x) = 0$

or $\cos x = 0$ and $1 - 4 \sin x = 0$

or $x = \pi/2$ and $\sin x = 1/4$

at $x = \pi/2$, $f'\left(\dfrac{\pi}{2}\right) = -\sin \dfrac{\pi}{2} - 4\cos\pi$

$\qquad = -1 + 4 = 3 > 0$

$\therefore$ $f(x)$ is minimum at $x = \dfrac{\pi}{2}$ and the minimum value is

$$f\left(\dfrac{\pi}{2}\right) = + \sin \dfrac{\pi}{2} + \cos \pi$$

$$= 1 - 1 = 0$$

at $\sin x = \dfrac{1}{4}$, $f''(x) = -\sin x - 4 \, (1 - 2 \sin^2 x)$

$$= -\dfrac{1}{4} - 4\left(1 - 2 \times \dfrac{1}{16}\right) = \dfrac{-15}{4} < 0$$

$\therefore$ $f(x)$ is maximum at $\sin x = \dfrac{1}{4}$ and the maximum value is $f(x) = \sin x + 1 - 2 \sin^2 x$

$$= \dfrac{1}{4} + 1 - 2 \times \dfrac{1}{16}$$

$$= \dfrac{9}{8}$$

Example 5: Find the maximum and minimum values of $x + \sin 2x$ for $0 < x < 2\pi$.

Solution: Let $f(x) = x + \sin 2x$

$\therefore$ $f'(x) = 1 + 2 \cos 2x$

and $f''(x) = -4 \sin 2x$

For maxima and minima, $f'(x) = 0$

or $1 + 2 \cos 2x = 0$

or $\cos 2x = \dfrac{-1}{2}$

or $2x = \dfrac{2\pi}{3}$ or $2x = \dfrac{4\pi}{3}$

$\Rightarrow$ $x = \dfrac{\pi}{3}$ or $x = \dfrac{2\pi}{3}$

at $\quad x = \dfrac{\pi}{3}$, $f''(x) = -4 \sin \dfrac{2\pi}{3} < 0$

$\therefore \quad$ f(x) is maximum at $x = \dfrac{\pi}{3}$, and

Maximum value f(x) = x + sin 2x

$$= \dfrac{\pi}{3} + \sin \dfrac{2\pi}{3}$$

$$= \dfrac{\pi}{3} + \dfrac{\sqrt{3}}{2}$$

At $x = \dfrac{2\pi}{3}$, $f''(x) = -4 \sin \dfrac{4\pi}{3}$

$$= -4 \left(\dfrac{\sqrt{3}}{2} \right) > 0$$

$\therefore$ f(x) is minimum at $x = \dfrac{2\pi}{3}$ and

Minimum value is f(x) = x + sin 2x

$$= \dfrac{2\pi}{3} + \sin \dfrac{4\pi}{3}$$

$$= \dfrac{2\pi}{3} - \dfrac{\sqrt{3}}{2}$$

***Example* 6:** Show that sin x (1 + cos x) is a maximum when $x = \dfrac{\pi}{3}$.   (RGPV 2008)

***Solution*:** Let $f(x) = \sin x (1 + \cos x) = \sin x + \dfrac{1}{2} \sin x$

$\therefore \qquad f'(x) = \cos x + \cos 2x$

and $\quad f''(x) = -\sin x - 2 \sin 2x$

For maxima and minima, $f'(x) = 0$

or $\quad \cos x + \cos 2x = 0$

or $\quad \cos 2x = -\cos x$

or $\quad \cos 2x = \cos (\pi - x)$

or $\quad 2x = \pi - x$

or $x = \dfrac{\pi}{3}$

at $x = \dfrac{\pi}{3}$, $f''(x) = -\sin x - 2\sin 2x$

$$= -\sin\dfrac{\pi}{3} - 2 \quad \sin\dfrac{2\pi}{3} < 0$$

$\therefore$ $f(x)$ is maximum at $x = \dfrac{\pi}{3}$ and the maximum value is

$$f(x) = \sin\dfrac{\pi}{3}\left(1 + \cos\dfrac{\pi}{3}\right)$$

$$= \dfrac{\sqrt{3}}{2}\left(1 + \dfrac{1}{2}\right) = \dfrac{3\sqrt{3}}{4}$$

***Example* 7:** Find the volume of the largest possible right circular cylinder that can be inscribed in a sphere of radius a.   (RGPV June 2007)

***Solution*:** Let x be the radius of base of cylinder and height be 2h; therefore volume of cylinder be $V = \pi x^2 (2h)$

$$V = 2\pi x^2 \sqrt{a^2 - x^2}$$

(In Δ OAB, $h = \sqrt{a^2 - x^2}$)

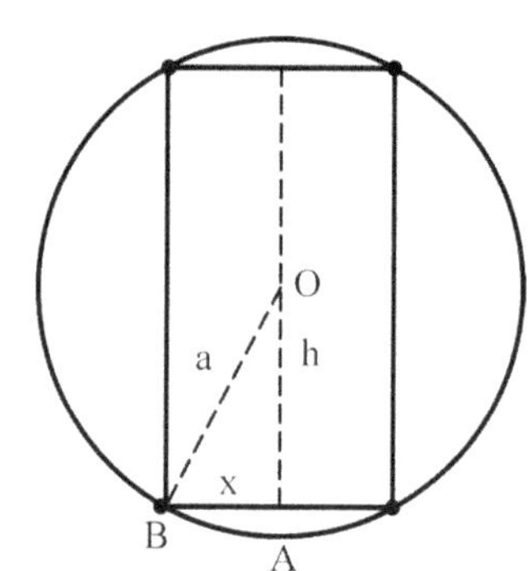

$\therefore$ $\dfrac{dv}{dx} = 2\pi x^2 . \dfrac{1}{2}\left(\dfrac{-2x}{\sqrt{a^2 - x^2}}\right) + 4\pi x \sqrt{a^2 - x^2}$

$$= \dfrac{-2\pi x^3}{\sqrt{a^2 - x^2}} + 4\pi x \sqrt{a^2 - x^2}$$

For maxima $\dfrac{dv}{dx} = 0$

or $\dfrac{-2\pi x^3}{\sqrt{a^2 - x^2}} + 4\pi x \sqrt{a^2 - x^2} = 0$

or $x = a\sqrt{\dfrac{2}{3}}$

since $h = \sqrt{a^2 - x^2}$

$$= \sqrt{a^2 - \dfrac{2a^2}{3}} = \dfrac{a}{\sqrt{3}}$$

$\therefore$ maximum volume of the cylinder be

$$V = \pi x^2 . 2h$$

$$= \pi \left(\frac{2a^2}{3} \right) \frac{2a}{\sqrt{3}} = \frac{4\pi a^3}{3\sqrt{3}}$$

***Example* 8:** Show that the maximum rectangle that can be inseribed in a circle is a square.

(RGPV 2000)

***Solution*:** Let ABCD be the rectangle inscribed in the circle with centre O and radius a. Let AB = 2x and OB = a, then OM = $\sqrt{a^2 - x^2}$ and hence BC = $2\sqrt{a^2 - x^2}$

$\therefore$ The area of the rectangle be

$$A = 2x \, . \, 2\sqrt{a^2 - x^2}$$

$$= 4x\sqrt{a^2 - x^2}$$

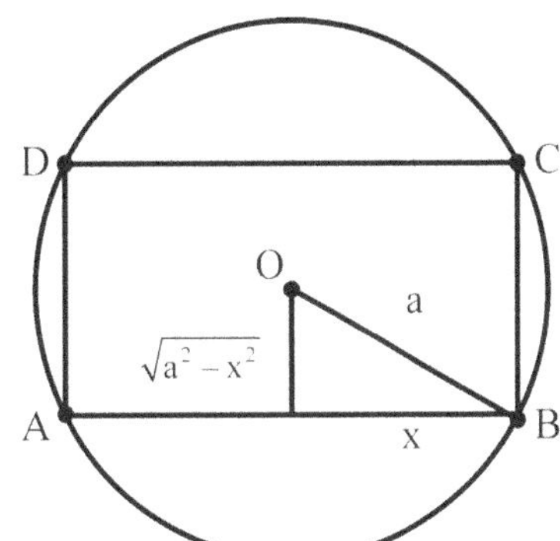

$$\therefore \quad \frac{dA}{dx} = 4\left[x\left(\frac{-2x}{2\sqrt{a^2 - x^2}} \right) + \sqrt{a^2 - x^2} \right] = \frac{4\left(a^2 - 2x^2\right)}{\sqrt{a^2 - x^2}}$$

and $$\frac{d^2A}{dx^2} = -4\left[\frac{4x\left(a^2 - x^2\right) - x\left(a^2 - x^2\right)}{\left(a^2 - x^2\right)^{3/2}} \right]$$

For maxima,

$$\frac{dA}{dx} = 0$$

or $$\frac{4\left(a^2 - 2x^2\right)}{\sqrt{a^2 - x^2}} = 0$$

or $a^2 - 2x^2 = 0 \quad \Rightarrow \quad x = \dfrac{a}{\sqrt{2}}$

at $x = \dfrac{a}{\sqrt{2}}, \dfrac{d^2A}{dx^2} = \dfrac{16\left(\dfrac{a}{\sqrt{2}} \right)\left(a^2 - \dfrac{a^2}{2} \right)}{\left(a^2 - \dfrac{a^2}{2} \right)^{3/2}} < 0$

$\therefore$ A is maximum at $x = \dfrac{a}{\sqrt{2}}$

$$\therefore \quad AB = 2x = \frac{2a}{\sqrt{2}} = a\sqrt{2}$$

$$\text{and} \quad BC = 2\sqrt{a^2 - x^2} = a\sqrt{2}$$

$$\text{i.e.,} \quad AB = BC \text{ at } x = \frac{a}{\sqrt{2}}$$

Hence the maximum rectangle inscribed in a circle is a square.

***Example* 9:** Show that the right circular cylinder of given surface and maximum volume is such that its height is equal to the diameter of the base.

***Solution*:** Let r be the radius of the base and h be the height of the cylinder.

The whole surface of the cylinder be

$$S = 2\pi rh + 2\pi r^2 \qquad \qquad \dots\dots(i)$$

Also, volume of the cylinder be

$$V = \pi r^2 h \qquad \qquad \dots\dots(ii)$$

Eliminating h from (i) and (ii), we get

$$V = \pi r^2 \left(\frac{S - 2\pi r^2}{2\pi r} \right) = \frac{1}{2}\left(Sr - 2\pi r^3 \right)$$

$$\therefore \quad \frac{dv}{dr} = \frac{1}{2}\left(S - 6\pi r^2 \right)$$

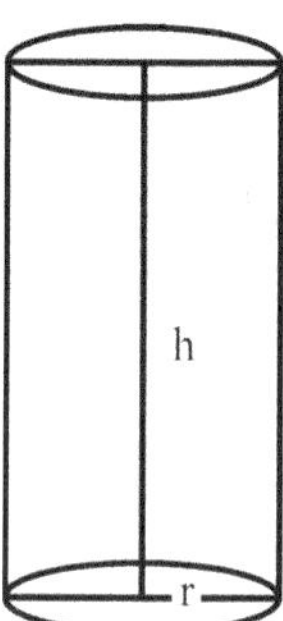

$$\text{and} \quad \frac{d^2v}{dr^2} = -6\pi r$$

For maxima,

$$\frac{dv}{dr} = 0$$

$$\text{or} \quad \frac{1}{2}\left(S - 6\pi r^2 \right) = 0 \Rightarrow r = \sqrt{\frac{S}{6\pi}}$$

$$\text{at} \quad r = \sqrt{\frac{S}{6\pi}}, \frac{d^2v}{dr^2} = -6\pi\sqrt{\frac{S}{6\pi}} < 0$$

Hence for maximum volume v at $r = \sqrt{\frac{S}{6\pi}}$

$$\text{i.e.,} \quad S = 6\pi r^2$$

and from (i) $6\pi r^2 = 2\pi rh + 2\pi r^2$

$$\text{or} \quad 2\pi rh = 4\pi r^2$$

or $h = 2r$

i.e.,   height is equal to diameter of the base of the cylinder

***Example* 10:** The shape of a hole bored by a drill is a cone surmounted by cylinder. If the cylinder be of height h and radius r and the semi vertical angle of the cone be α, where $\tan \alpha = h/r$, show that for a total height H of

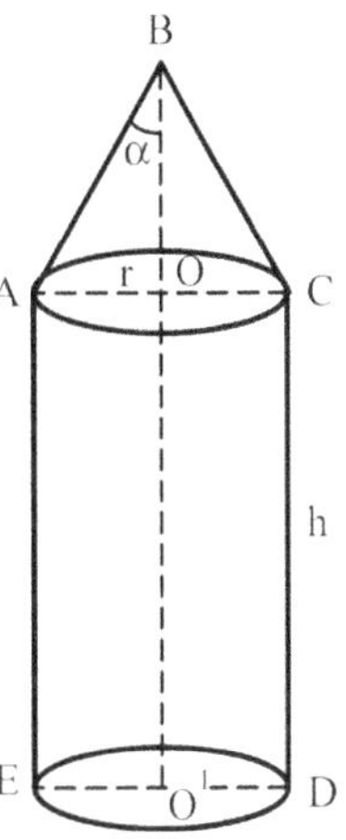

the hole, the volume removed is maximum if $h = \dfrac{H}{6}\left(\sqrt{7}+1\right)$

(RGPV June 2003)

***Solution*:** Let ACDE be the cylinder of height h and radius r and ABC be the cone of radius r. Let H be the total height of the shape. Then

$$H = BO + OO' \qquad\qquad .....(i)$$

In $\Delta ABO,$ $\tan \alpha = \dfrac{AO}{OB} = \dfrac{r}{OB}$

$\therefore$ $OB = r \cot \alpha$

$\therefore$ from (i), $H = r \cot \alpha + h$

$$= r\left(r/h\right) + h \qquad (\text{given } \tan \alpha = \dfrac{h}{r})$$

$$r^2 = (H - h)h \qquad\qquad .....(ii)$$

Let V be the volume removed, then

$$V = \frac{1}{3}\pi\, OA^2 . BO + \pi r^2 h$$

$$= \frac{1}{3}\pi r^2 . r \cot \alpha + \pi r^2 h$$

$$= \frac{1}{3}\pi r^3 . \frac{r}{h} + \pi r^2 h$$

$$= \pi r^2 \left(\frac{r^2}{3h} + h\right)$$

$$= \pi (H - h) h \left(\frac{H - h}{3} + h\right), \quad \text{from (ii)}$$

$$= \frac{1}{3}\pi (H - h)(H + 2h).h$$

$$V = \frac{1}{3}\pi (H^2 h + Hh^2 - 2h^3)$$

$$\therefore \qquad \frac{dv}{dh} = \frac{1}{3}\pi\left(H^2 + 2Hh - 6h^2\right)$$

$$\text{and} \qquad \frac{d^2v}{dh^2} = \frac{1}{3}\pi\left(2H - 12h\right)$$

For maxima,

$$\frac{dv}{dh} = 0$$

$$\Rightarrow \qquad \frac{1}{3}\pi\left(H^2 + 2Hh - 6h^2\right) = 0$$

$$\Rightarrow \qquad 6h^2 - 2Hh - H^2 = 0$$

$$\Rightarrow \qquad h = \frac{2H \pm \sqrt{4H^2 + 24H^2}}{12}$$

$$\Rightarrow \qquad h = \frac{H}{6}\left(1 + \sqrt{7}\right)$$

Also, for h, $\dfrac{d^2v}{dh^2} = \text{negative}$

i.e.,   v is maximum

Hence the volume removed is maximum when

$$h = \frac{H}{6}\left(1 + \sqrt{7}\right)$$

Hence proved.

6.2   Maxima and Minima of a Function of Two Variable

Definition: A function f(x, y) of two independent variables x and y is said to be maximum at (a, b) if f(a, b) > f(a + h, b +k) for small values of h and k, positive or negative.

A function f(x, y) of two independent variables x and y is said to be minimum at (a, b) if f(a, b) < f(a + h, b + k) for small values of h and k, positive or negative.

Working Rule: Let f(x, y) be a given function, then

(i)   Find $\dfrac{\partial f}{\partial x}$ and $\dfrac{\partial f}{\partial y}$

(ii)   Solve the equations $\dfrac{\partial f}{\partial x} = 0$ and $\dfrac{\partial f}{\partial y} = 0$ for solving x and y.

(iii) Find $r = \dfrac{\partial^2 f}{\partial x^2}$, $s = \dfrac{\partial^2 f}{\partial x \partial y}$, $t = \dfrac{\partial^2 f}{\partial y^2}$ at the roots obtained in (ii) and calculate $rt - s^2$.

(iv) (a) If $rt - s^2 > 0$ and $r < 0$, then f is maximum.

   (b) If $rt - s^2 > 0$ and $r > 0$, then f is minimum

   (c) If $rt - s^2 < 0$, then f is neither maximum nor minimum

   (d) If $rt - s^2 = 0$, then the case is doubtful and further investigation is required.

***Example* 11:** Discuss the maxima and minima of the function $x^3 + y^3 - 3axy$.

(RGPV Dec 2003, June 2004, Feb 2005)

***Solution*:** Let $f = x^3 + y^3 - 3axy$

$\therefore \quad \dfrac{\partial f}{\partial x} = 3x^2 - 3ay$ and $\dfrac{\partial f}{\partial y} = 3y^2 - 3ax$

For maxima and minima

$$\dfrac{\partial f}{\partial x} = 0 \Rightarrow 3x^2 - 3ay = 0$$

and $\dfrac{\partial f}{\partial y} = 0 \Rightarrow 3y^2 - 3ax = 0$

By solving these two equations, we get $x = y = a$

Now $r = \dfrac{\partial^2 f}{\partial x^2} = 6x, \, S = \dfrac{\partial^2 f}{\partial x \partial y} = -3a, \, t = \dfrac{\partial^2 f}{\partial y^2} = 6y$

at the point (a, a), we have

$$r = 6a, \, S = -3a, \, t = 6a$$

and $rt - S^2 = (6a)(6a) - (-3a)^2 = 27\,a^2 > 0$

since $rt - S^2 > 0$ and $r > 0$ or $r < 0$ according as $a > 0$ or $a < 0$

we have a maximum or minimum according as $a < 0$ or $a > 0$ at $x = y = a$

***Example* 12:** Discuss the maximum or minimum value of $x^3 y^2 (1 - x - y)$.

(RGPV Dec 2002, Jan 2007)

***Solution*:** Let $f = x^3 y^2 (1 - x - y)$

$\therefore \quad \dfrac{\partial f}{\partial x} = 3x^2 y^2 (1 - x - y) - x^3 y^2$

and $\dfrac{\partial f}{\partial y} = 2x^3 y (1 - x - y) - x^3 y^2$

for maxima and minima, we have

$$\frac{\partial f}{\partial x} = 3x^2y^2\left(1-x-y\right) - x^3y^2 = 0 \qquad\qquad .....(i)$$

$$\frac{\partial f}{\partial y} = 2x^3y\left(1-x-y\right) - x^3y^2 = 0 \qquad\qquad .....(ii)$$

By solving equation (i) and (ii), we get $x = \dfrac{1}{2}, y = \dfrac{1}{3}$

Now $r = \dfrac{\partial^2 f}{\partial x^2} = 6xy^2 - 12x^2y^2 - 6xy^3$

$t = \dfrac{\partial^2 f}{\partial y^2} = 2x^3 - 2x^4 - 6x^3y$

$s = \dfrac{\partial^2 f}{\partial x \partial y} = 6x^2y - 8x^3y - 9x^2y^2$

at $\left(\dfrac{1}{2}, \dfrac{1}{3}\right)$, $r = \dfrac{-1}{9}, t = \dfrac{-1}{8}, s = \dfrac{-1}{12}$

and $rt - s^2 = \left(-\dfrac{1}{9}\right)\left(\dfrac{-1}{8}\right) - \left(\dfrac{-1}{12}\right)^2 = \dfrac{1}{72}$

since $rt - s^2 > 0$ and $r < 0$. Hence f has a maximum at $\left(\dfrac{1}{2}, \dfrac{1}{3}\right)$

$\therefore$ The maximum value of $f = \left(\dfrac{1}{2}\right)^3 \left(\dfrac{1}{3}\right)^2 \left(1 - \dfrac{1}{2} - \dfrac{1}{3}\right) = \dfrac{1}{432}$

***Example* 13:** Find the maxima and minima of the function $\sin x + \sin y + \sin (x + y)$.

***Solution*:** Let $f = \sin x + \sin y + \sin(x + y)$

$\therefore \quad \dfrac{\partial f}{\partial x} = \cos x + \cos(x + y)$

and $\quad \dfrac{\partial f}{\partial y} = \cos y + \cos(x + y)$

for maxima and minima

$$\frac{\partial f}{\partial x} = \cos x + \cos(x + y) = 0 \qquad\qquad .....(i)$$

$$\frac{\partial f}{\partial y} = \cos y + \cos(x + y) = 0 \qquad\qquad .....(ii)$$

By solving (i) and (ii), we get

$$\cos x = \cos y \;\Rightarrow\; x = y$$

Also from (i) $\cos x + \cos (x + x) = 0$

$\Rightarrow \qquad \cos x + \cos 2x = 0$

$\Rightarrow \qquad 2 \cos^2 x - 1 + \cos x = 0$

$\Rightarrow \qquad 2 \cos^2 x + 2 \cos x - \cos x - 1 = 0$

$\Rightarrow \qquad (2 \cos x - 1) (\cos x + 1) = 0$

$\Rightarrow \qquad 2 \cos x - 1 = 0$ and $\cos x + 1 = 0$

$\Rightarrow \qquad \cos x = \dfrac{1}{2} \qquad$ and $\qquad \cos x = -1$

$\Rightarrow \qquad x = \dfrac{\pi}{3} \qquad$ and $\qquad x = \pi$

Now, $r = \dfrac{\partial^2 f}{\partial x^2} = -\sin x - \sin(x + y)$

$t = \dfrac{\partial^2 f}{\partial y^2} = -\sin y - \sin(x + y)$

$s = \dfrac{\partial^2 f}{\partial x \partial y} = -\sin (x + y)$

at $\left(\dfrac{\pi}{3}, \dfrac{\pi}{3}\right), r = -\sin \dfrac{\pi}{3} - \sin \dfrac{2\pi}{3} = -\sqrt{3}$

$t = -\sin \dfrac{\pi}{3} - \sin \dfrac{2\pi}{3} = -\sqrt{3}$

$s = -\sin \dfrac{2\pi}{3} = \dfrac{\sqrt{3}}{2}$

and $rt - s^2 = \left(-\sqrt{3}\right)\left(-\sqrt{3}\right) - \left(\dfrac{-\sqrt{3}}{2}\right)^2 = \dfrac{9}{4}$

since $rt - s^2 > 0$ and $r < 0$, therefore f has a maxima at $\left(\dfrac{\pi}{3}, \dfrac{\pi}{3}\right)$.

at $(\pi, \pi), r = -\sin \pi - \sin 2\pi = 0, t = 0, s = 0$ and $rt - s^2 = 0$

This case is doubtful and hence we shall leave it

***Example* 14:** Find the maximum and minimum value of the function sin x sin y sin (x + y).

Solution: Let $f = \sin x \sin y \sin(x + y)$

For maxima or minima

$$\frac{\partial f}{\partial x} = \cos x \sin y \sin (x + y) + \sin x \sin y \cos (x + y)$$

$$= \sin y \sin (2x + y) = 0 \qquad\qquad \text{.....(i)}$$

$$\frac{\partial f}{\partial y} = \sin x \sin (x + 2y) = 0 \qquad\qquad \text{.....(ii)}$$

From (i) $\sin y = 0$ or $\sin (2x + y) = 0 \Rightarrow y = 0$ or

$$2x + y = \pi \text{ or } 2\pi \qquad\qquad \text{.....(iii)}$$

From (ii) $\sin x = 0$ or $\sin(x + 2y) = 0 \Rightarrow x = 0$ or

$$x + 2y = \pi \text{ or } 2\pi \qquad\qquad \text{.....(iv)}$$

By solving (iii) and (iv), we get $\left(\dfrac{\pi}{3}, \dfrac{\pi}{3}\right)$ and $\left(\dfrac{2\pi}{3}, \dfrac{2\pi}{3}\right)$

Now, $r = \dfrac{\partial^2 f}{\partial x^2} = 2 \sin y \cos (2x + y)$

$$t = \frac{\partial^2 f}{\partial y^2} = 2 \sin x \cos (x + 2y)$$

$$s = \frac{\partial^2 f}{\partial x \partial y} = \sin(2x + 2y)$$

at $\left(\dfrac{\pi}{3}, \dfrac{\pi}{3}\right)$, $r = 2 \cdot \sin \dfrac{\pi}{3} \cos \pi = 2\left(\dfrac{\sqrt{3}}{2}\right)(-1) = -\sqrt{3}$

$$t = -\sqrt{3}, \qquad s = \frac{-\sqrt{3}}{2}$$

and $rt - s^2 = \left(-\sqrt{3}\right)\left(-\sqrt{3}\right) - \left(\dfrac{-\sqrt{3}}{2}\right)^2 = \dfrac{9}{4}$

since $rt - s^2 > 0$ and $r < 0$,

Hence f is maximum at $\left(\dfrac{\pi}{3}, \dfrac{\pi}{3}\right)$

and maximum value is $\sin \dfrac{\pi}{3} \sin \dfrac{\pi}{3} \sin \dfrac{2\pi}{3} = \dfrac{3\sqrt{3}}{8}$.

at $\left(\dfrac{2\pi}{3}, \dfrac{2\pi}{3}\right)$, $r = \sqrt{3}$, $t = \sqrt{3}$, $s = \dfrac{\sqrt{3}}{2}$

and $rt - s^2 = \dfrac{9}{4}$

since $rt - s^2 > 0$ and $r > 0$, Hence f is minimum at $\left(\dfrac{2\pi}{3}, \dfrac{2\pi}{3}\right)$

and minimum value $= \sin \dfrac{2\pi}{3} \; \sin \dfrac{2\pi}{3} \; \sin \dfrac{4\pi}{3} = \dfrac{-3\sqrt{3}}{8}$

6.3 Lagrange's Method of Undetermined Multipliers

Consider the function $u = f(x, y, z)$ .....(1)

with the condition $\phi(x, y, z) = 0$ .....(2)

Consider the Lagrange's function

$$F(x, y, z) = f(x, y, z) + \lambda \, \phi(x, y, z)$$

$\therefore \quad dF = \dfrac{\partial f}{\partial x}dx + \dfrac{\partial f}{\partial y}dy + \dfrac{\partial f}{\partial z}dz + \lambda\left[\dfrac{\partial \phi}{\partial x}dx + \dfrac{\partial \phi}{\partial y}dy + \dfrac{\partial \phi}{\partial z}dz\right]$

$\quad = \left(\dfrac{\partial f}{\partial x} + \lambda \dfrac{\partial \phi}{\partial x}\right)dx + \left(\dfrac{\partial f}{\partial y} + \lambda \dfrac{\partial \phi}{\partial y}\right)dy + \left(\dfrac{\partial f}{\partial z} + \lambda \dfrac{\partial \phi}{\partial z}\right)dz$

Now, $dF = 0$ implies that

$$\dfrac{\partial f}{\partial x} + \lambda \dfrac{\partial \phi}{\partial x} = 0, \; \dfrac{\partial f}{\partial y} + \lambda \dfrac{\partial \phi}{\partial y} = 0, \; \dfrac{\partial f}{\partial z} + \lambda \dfrac{\partial \phi}{\partial z} = 0 \qquad \text{.....(3)}$$

By solving these equations, we get four unknowns x, y, z and λ

Where u may be stationary.

The Lagrange's method gives us only a stationary value.

***Example* 15:** Prove that the rectangular solid of maximum volume which can be inscribed in a given sphere is a cube.   (RGPV June 2002)

***Solution*:** Let the length, breadth and height of a rectangular solid be x, y and z respectively. If V be the volume of the solid, then

$$V = xyz \qquad \text{.....(i)}$$

Since each diagonal of the solid passes through the centre of the sphere, therefore

Each diagonal = Diameter of sphere d (say)

Let the equation of sphere be $x^2 + y^2 + z^2 = d^2$

or $\quad z = \sqrt{d^2 - x^2 - y^2}$

$\therefore$ from equation (i), we have

$$V = xy\sqrt{d^2 - x^2 - y^2}$$

$$\Rightarrow \quad V^2 = x^2 y^2 (d^2 - x^2 - y^2) = f \text{ (say)}$$

$$\therefore \quad \frac{\partial f}{\partial x} = 2xy^2 \left(d^2 - 2x^2 - y^2\right)$$

and $\quad \dfrac{\partial f}{\partial y} = 2x^2 y \left(d^2 - x^2 - 2y^2\right)$

and $\quad \dfrac{\partial^2 f}{\partial x^2} = r = 2d^2 y^2 - 12x^2 y^2 - 2y^4$

$$\frac{\partial^2 f}{\partial y^2} = t = 2d^2 x^2 - 2x^4 - 12x^2 y^2$$

$$\frac{\partial^2 f}{\partial x \partial y} = s = 4d^2 xy - 8x^3 y - 8xy^3$$

Now, for maxima or minima.

$$\frac{\partial f}{\partial x} = 0 = 2xy^2 \left(d^2 - 2x^2 - y^2\right) \qquad\qquad .....(ii)$$

and $\quad \dfrac{\partial f}{\partial y} = 0 = 2x^2 y \left(d^2 - x^2 - 2y^2\right) \qquad\qquad .....(iii)$

By solving (ii) and (iii), we get $x = y = \dfrac{d}{\sqrt{3}}$

and $\quad z = \dfrac{d}{\sqrt{3}}$

at $\quad x = y = \dfrac{d}{\sqrt{3}},$

$$r = 2d^2 \left(\frac{d^2}{3}\right) - 12\left(\frac{d^2}{3}\right)\left(\frac{d^2}{3}\right) - 2\left(\frac{d^4}{9}\right)$$

$$= -\frac{8d^4}{9} < 0$$

$$t = 2d^2 \left(\frac{d^2}{3}\right) - 2\frac{d^4}{9} - 12\left(\frac{d^2}{3}\right)\left(\frac{d^2}{3}\right)$$

$$= -\frac{8d^4}{9}$$

$$s = 4d^2\left(\frac{d^2}{3}\right) - 8\left(\frac{d^3}{3\sqrt{3}}\right)\left(\frac{d}{\sqrt{3}}\right) - 8\left(\frac{d}{\sqrt{3}}\right)\left(\frac{d^3}{3\sqrt{3}}\right) = \frac{-4d^4}{9}$$

$$\therefore \quad rt - s^2 = \left(-\frac{8d^4}{9}\right)\left(\frac{-8d^4}{9}\right) - \left(-\frac{4d^4}{9}\right)^2 = \frac{16d^8}{27} > 0$$

Here, $rt - s^2 > 0$ and $r < 0$

$\therefore$ f is maximum at $x = y = z$

i.e., when the rectangular solid is a cube, volume of rectangular solid is maximum at $x = y = z$.

Hence proved.

***Example* 16:** Find the maxima and minima of $u = x^2 + y^2 + z^2$ where $ax^2 + by^2 + cz^2 = 1$.

***Solution*:** Let $u = x^2 + y^2 + z^2$ .....(i)

and $ax^2 + by^2 + cz^2 = 1$ .....(ii)

For maxima or minima

From (i) $du = 2x\,dx + 2y\,dy + 2z\,dz = 0$

or $x\,dx + y\,dy + z\,dz = 0$ .....(iii)

from (ii) $2ax\,dx + 2by\,dy + 2cz\,dz = 0$ .....(iv)

or $ax\,dx + by\,dy + cz\,dz = 0$

multiplying (iv) by λ and adding to (iii) and then equating the coefficients of dx, dy and dz to zero, we have

$x + a\lambda x = 0$ .....(v)

$y + b\lambda y = 0$ .....(vi)

$z + c\lambda z = 0$ .....(vii)

Multiply (v), (vi) and (vii) by x, y, z and adding we get

$(x^2 + y^2 + z^2) + \lambda(ax^2 + by^2 + cz^2) = 0$

$u + \lambda = 0 \quad \Rightarrow \lambda = -u$

substituting the value of λ in (v), (vi) and (vii), we get

$1 - au = 0, \quad\quad 1 - bu = 0, \quad 1 - cu = 0$

or $\dfrac{1}{a} - u = 0, \quad\quad \dfrac{1}{b} - u = 0, \quad \dfrac{1}{c} - u = 0$

Hence the maximum or minimum value of u is

$$\left(\frac{1}{a} - u\right)\left(\frac{1}{b} - u\right)\left(\frac{1}{c} - u\right) = 0$$

Example **17:** If x, y z are the angles of a triangle, show that $u = \cos x \cos y \cos z$ is maximum when the triangle is equilateral.

Solution: Let $u = \cos x \cos y \cos z$ .....(i)

with the condition $\phi = x + y + z = \pi$ .....(ii)

consider the Lagrange's method

$$F(x, y, z) = \cos x \cos y \cos z + \lambda(x + y + z - \pi)$$

$$\therefore \ dF = (-\sin x \cos y \cos z) \, dx + (-\cos x \sin y \cos z) \, dy$$

$$+ (-\cos x \cos y \sin z)dz + \lambda(dx + dy + dz)$$

$$= (-\sin x \cos y \cos z + 1) \, dx + (-\cos x \sin y \cos z + \lambda) \, dy$$

$$+ (-\cos x \cos y \sin z + \lambda) \, dz$$

since $dF = 0 \ \Rightarrow \ -\sin x \cos y \cos z + \lambda = 0$

$$-\cos x \sin y \cos z + \lambda = 0$$

$$-\cos x \cos y \sin z + \lambda = 0 \qquad\qquad .....(iii)$$

From equation (iii) $\sin x \cos y \cos z = \cos x \sin y \cos z = \cos x \cos y \sin z = \lambda$

Dividing by $\cos x \cos y \cos z$, we get $\tan x = \tan y = \tan z$

Hence $x = y = z$

Which proves that the triangle is equilateral

Practice Problems

1. Investigate the maxima of the ordinates of the curve $y = (x - 2)^6 (x - 3)^5$.

   [**Ans.** $x = 2, 3, \dfrac{28}{11}$, (i) at $x = 2$, maxima, (ii) at $x = 3$, neither maximum nor minimum,

   (iii) at $x = \dfrac{28}{11}$, minima]

2. Prove that $\dfrac{x}{1 + x \tan x}$ is a maximum when $x = \cos x$.

3. Show that $x^3 - 3x^2 + 3x + 7$ has neither a maximum nor minimum at $x = 1$.

4. Find the maximum and minimum values of $a \sec x + b \, \mathrm{cosec}\, x$, $0 < a < b$.

$$\textbf{[Ans.}\ \tan x = \left(\frac{b}{a}\right)^{\frac{1}{3}}, \text{minimum},\ \left(a^{2/3} + b^{2/3}\right)^{3/2}\ \textbf{]}.$$

5.   Show that the triangle of maximum area that can be inscribed in a given circle is an equilateral triangle.

6.   Show that the Semi-vertical angle of the cone of maximum volume and of given slant height is $\tan^{-1}\sqrt{2}$.

7.   Prove that a conical tent of a given capacity will require the least amount of canvas when the height is $\sqrt{2}$ times the radius of the base.

8.   Find the surface of the right circular cylinder or greatest surface which can be inscribed in a sphere of radius a.

$$\textbf{[Ans.}\ \pi\, a^2 \left(1 + \sqrt{5}\right)\textbf{]}$$

9.   Find the maximum and minimum value of $2x^3 - 15\,x^2 + 36\,x + 10$.

$$\textbf{[Ans.}\ 38, 37\textbf{]}$$

10.   Find the largest and smallest values of $3x^4 - 2x^3 - 6x^2 + 6x + 1$ in the interval $(0, 2)$

$$\textbf{[Ans.}\ 21\ \text{and}\ 1\textbf{]}$$

11.   Discuss the maximum or minimum values of the function $u = x^3 - 4xy + 2y^2$.

$$\textbf{[Ans.}\ \left(\frac{4}{3}, \frac{4}{3}\right)\ \text{max value} = \frac{-32}{37}\,\textbf{]}$$

12.   Discuss the maxima and minima of the function $ax^3 y^2 - x^4 y^2 - x^3 y^3$.

$$\textbf{[Ans.}\ \left(\frac{a}{2}, \frac{a}{3}\right),\ \text{maximum}\textbf{]}$$

13.   If $f = 2x^4 + y^4 - 2x^2 - 2y^2$, then show that f has maximum at $(0, 0)$ and a minimum at $\left(\frac{1}{\sqrt{2}}, 1\right)$.

14.   Find maximum and minimum values of the function $f = xy(a - x - y)$.

$$\textbf{[Ans.}\ \left(\frac{a}{3}, \frac{a}{3}\right),\ \text{max value} = \frac{a^3}{27}\,\textbf{]}$$

15.   If x, y, z are angles of a triangle, then find the maximum value of sin x sin y sin z.

$$\textbf{[Ans.}\ x = y = z = \frac{\pi}{3},\ \text{u is maximum}\textbf{]}$$

16.   Find the minimum value of $x^2 + y^2 + z^2$ when $ax + by + cz = p$.

$$\textbf{[Ans.}\ \frac{p^2}{\left(a^2 + b^2 + c^2\right)}\,\textbf{]}$$

17.   In a plane triangle, find the maximum value of u = cos A cos B cos C.

$$[\textbf{Ans. } A = B = C = \frac{\pi}{3}, \ u_{max} = \frac{1}{8}]$$

18.   Find the maxima and minima of $y = \sin x + \frac{1}{2} \cos 2x$.

$$[\textbf{Ans. } x = \frac{\pi}{2}, \frac{\pi}{6}, \ (i) \text{ at } x = \frac{\pi}{2}, y \text{ is min (ii) at } x = \frac{\pi}{6}, y = \max]$$

19.   Show that $\dfrac{\log x}{x}$ is maximum at x = e.

20.   Verify whether (−1, 0) is a stationary value of $z = x^3 + 3xy^2 - 3x$.

Curvature

7.1 Definition

Let P and Q be two points on a given curve and let the arc be measured along the curve from the fixed point A. Let arc AP = s and arc AQ = s + δs. Let the tangents PT and QT′ to the curve at P and Q make angle ψ and ψ + δψ with x-axis, then

(a) The curvature of the curve at the point P is denoted by K(Kappa) i.e., curvature

$$(k) = \frac{d\psi}{ds}.$$

(b) The length CP is called the radius of curvature and is usually denoted by ρ(rho). In other words the reciprocal of curvature is called the radius of curvature $= \dfrac{ds}{d\rho}.$

(c) C is defined as the centre of curvature.

(d) A circle with centre C and the radius CP is called the circle of curvature at the point P.

(e) A chord of the circle drawn through P is called the chord of curvature.

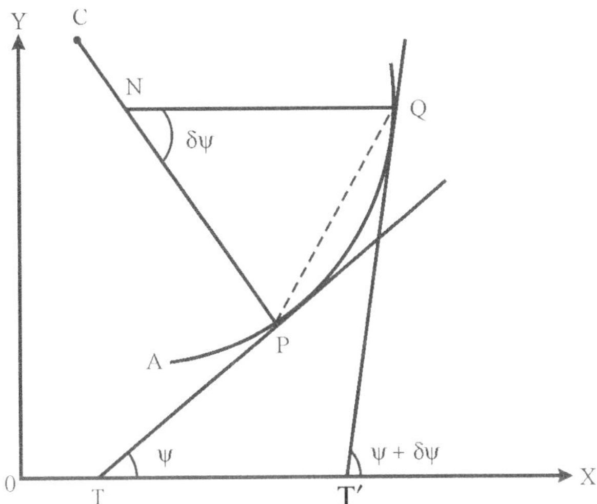

7.2   Intrinsic Formula for the Radius of Curvature

Let P and Q be any two points of a given curve (see above figure) and let the arc be measured along the curve from the fixed point A. Let arc AP = s and arc AQ = s + δs.

Therefore arc PQ = δs.

Let the tangent PT and QT′ to the curve at P and Q make angle ψ and $\psi + \delta\psi$ with x-axis. Since PN and QN are perpendicular to PT and QT′. Then $\angle PNQ = \delta\psi$.

$$\text{In } \Delta PNQ, \quad \frac{PN}{\sin \angle PQN} = \frac{\angle PQ}{\sin \angle PNQ} = \frac{\delta s}{\sin \delta\psi}$$

$$\text{or} \quad PN = \frac{\delta s}{\sin \delta\psi} \sin \angle PQN$$

$$= \frac{\delta s}{\sin \delta\psi} \cdot \frac{\delta\psi}{\delta\psi} \sin \angle PQN$$

$$\therefore \text{ Radius of curvature } \rho = CP = \lim_{Q \to P} PN$$

Substituting the values of PN and $\delta\psi \to O$ as $Q \to P$

Therefore,

$$\rho = \left(\lim_{\delta\psi \to 0} \frac{\delta\psi}{\sin \delta\psi} \right) \left(\lim_{\delta\psi \to 0} \frac{\delta s}{\delta\psi} \right) \left(\lim_{Q \to P} \sin \angle PQN \right)$$

$$= 1. \frac{ds}{d\psi} . \sin 90°,$$

(since as $Q \to P$ then $\delta\psi \to 0$ and $\dfrac{\delta s}{\delta\psi} \to \dfrac{ds}{d\psi}$ and $\angle PQN = 90°$)

$$\text{Hence} \quad \rho = \frac{ds}{d\psi}$$

7.3   Cartesian Formula for Radius of Curvature

Let the given curve be y = f(x)

Then, we know that

$$\frac{dy}{dx} = \tan \psi \qquad \qquad .....(1)$$

Differentiating eqn (i) w.r.t. x, we get

$$\frac{d^2y}{dx^2} = \sec^2\psi \cdot \frac{d\psi}{dx}$$

$$= \left(1 + \tan^2\psi\right)\frac{d\psi}{ds} \cdot \frac{ds}{dx}$$

$$= \left[1 + \left(\frac{dy}{dx}\right)^2\right]\cdot\frac{1}{\rho}\sqrt{1 + \left(\frac{dy}{dx}\right)^2} \qquad \left(\because \frac{ds}{dx} = \sqrt{1 + \left(\frac{dy}{dx}\right)^2}\right)$$

$$\therefore \ \rho = \frac{\left[1 + \left(\frac{dy}{dx}\right)^2\right]^{3/2}}{\dfrac{d^2y}{dx^2}}$$

7.4 Parametric Formula for Radius of Curvature

Let the given curve be $x = f(t)$ and $y = \phi(t)$, where t is the parameter.

Then $\dfrac{dy}{dx} = \dfrac{\dfrac{dy}{dt}}{\dfrac{dx}{dt}} = \dfrac{y'}{x'}$

Also $\dfrac{d^2y}{dx^2} = \dfrac{d}{dt}\left(\dfrac{y'}{x'}\right)\cdot\dfrac{dt}{dx}$

$$= \frac{x'.y'' - y'x''}{\left(x'\right)^2}\cdot\frac{1}{x'}$$

$$= \frac{x'y'' - y'x''}{\left(x'\right)^3}$$

$$\therefore \qquad \rho = \frac{\left[1 + \left(\frac{dy}{dx}\right)^2\right]^{3/2}}{\dfrac{d^2y}{dx^2}}$$

$$= \frac{\left[1 + \left(\frac{y'}{x'}\right)^2\right]^{3/2}}{\dfrac{\left(x'y'' - y'x''\right)}{\left(x'\right)^3}}$$

$$\Rightarrow \quad \rho = \frac{\left[\left(x'\right)^2 + \left(y'\right)^2\right]^{3/2}}{x'y'' - y'x''}$$

Illustrative Examples

***Example* 1:** Find the radius of Curvature at (s, ψ) on the tractrix $s = c \log \sec \psi$.

***Solution*:** Let the given curve be $s = c \log \sec \psi$

∴ Radius of curvature

$$\rho = \frac{ds}{d\psi}$$

$$= \frac{c}{\sec \psi} . \sec \psi \tan \psi$$

$$= c \tan \psi$$

***Example* 2:** Find the radius of curvature at the point (x, y) on parabola $y^2 = 4ax$.

***Solution*:** Let the given curve be $y^2 = 4ax$ .....(i)

Differentiating w.r.t x, we get $2y\dfrac{dy}{dx} = 4a$ or

$$\frac{dy}{dx} = \frac{2a}{y} \qquad\qquad .....(ii)$$

Again differentiating w.r.t x, we get

$$\frac{d^2y}{dx^2} = \frac{-2a}{y^2}\frac{dy}{dx} = \frac{-4a^2}{y^3}$$

∴ Radius of curvature

$$\rho = \frac{\left[1 + \left(\dfrac{dy}{dx}\right)^2\right]^{3/2}}{\dfrac{d^2y}{dx^2}}$$

$$= \frac{\left(1 + \dfrac{4a^2}{y^2}\right)^{3/2}}{-\dfrac{4a^2}{y^3}}$$

$$= \frac{-\left(y^2 + 4a^2\right)^{3/2}}{4a^2}$$

$$= \frac{\left(4ax + 4a^2\right)^{3/2}}{4a^2},$$

neglecting the negative sign and putting y^2 from (i)

$$= \frac{\left(4a\right)^{3/2}\left(x + a\right)^{3/2}}{4a^2}$$

$$= \frac{2}{\sqrt{a}}\left(x + a\right)^{3/2}$$

***Example* 3:** Find the radius of curvature at (x, y) on the curve $y = c \cos h \left(\dfrac{x}{c}\right)$.

Solution: The given curve be $y = c \cos h \left(\dfrac{x}{c}\right)$

Differentiating w.r.t x, we get

$$\frac{dy}{dx} = \sinh\left(\frac{x}{c}\right)$$

and $\quad \dfrac{d^2y}{dx^2} = \dfrac{1}{c}\cosh\left(\dfrac{x}{c}\right)$

$\therefore$ Radius of curvature

$$\rho = \frac{\left[1+\left(\dfrac{dy}{dx}\right)^2\right]^{3/2}}{d^2y\Big/dx^2}$$

$$= \frac{\left[1+\sinh^2\left(\dfrac{x}{c}\right)\right]^{3/2}}{\left(\dfrac{1}{c}\right)\cos h\left(x\Big/c\right)}$$

$$= \frac{c\left[\cos h^2\left(\dfrac{x}{c}\right)\right]^{\frac{3}{2}}}{\cos h\left(x\Big/c\right)} = c \cos h^2\left(\dfrac{x}{c}\right)$$

***Example* 4:** In the cycloid $x = a(\theta + \sin \theta)$, $y = a(1 - \cos \theta)$, then prove that radius of curvature $\rho = 4a \cos \dfrac{\theta}{2}$.

(RGPV Dec 2001 June 2006)

***Solution*:** Let $x = a(\theta + \sin \theta)$ and $y = a(1 - \cos \theta)$

$$\therefore \qquad x' = \frac{dx}{dt} = a(1 + \cos\theta), \ y' = \frac{dy}{dt} = a \sin\theta$$

$$\text{and} \quad x'' = \frac{d^2x}{dt^2} = -a\sin\theta, \quad y'' = a\cos\theta$$

$\therefore$ Radius of curvature

$$\rho = \frac{\left[(x')^2 + (y')^2\right]^{3/2}}{x'\,y'' - y'x''}$$

$$= \frac{\left[a^2(1+\cos\theta)^2 + a^2\sin^2\theta\right]^{3/2}}{a^2(1+\cos\theta)\cos\theta + a^2\sin^2\theta}$$

$$= \frac{a\left(1 + \cos^2\theta + 2\cos\theta + \sin^2\theta\right)^{3/2}}{\cos\theta + \cos^2\theta + \sin^2\theta}$$

$$= \frac{a\left[2(1+\cos\theta)\right]^{3/2}}{1+\cos\theta}$$

$$= 2^{3/2}a(1+\cos\theta)^{1/2}$$

$$= 2^{3/2}a\left(2\cos^2\frac{\theta}{2}\right)^{1/2} = 4a\cos\frac{\theta}{2}$$

***Example* 5:** Show that the radius of curvature at $(a\cos^3\theta, b\sin^3\theta)$ on the curve $x^{2/3} + y^{2/3} = a^{2/3}$ is $3a \sin\theta \cos\theta$.

(RGPV Dec 2002, Dec 2005)

***Solution*:** The given curve be $x^{2/3} + y^{2/3} = a^{2/3}$

.....(i)

Differentiating w.r.t x, we get

$$\frac{2}{3}x^{-1/3} + \frac{2}{3}y^{\frac{-1}{3}}\frac{dy}{dx} = 0$$

$$\text{or} \quad \frac{dy}{dx} = -\left(\frac{y}{x}\right)^{1/3}$$

Differentiating again, we get

$$\frac{d^2y}{dx^2} = -\left[\frac{x^{1/3}.\frac{1}{3}y^{\frac{-2}{3}}\frac{dy}{dx} - y^{1/3}\frac{1}{3}x^{\frac{-2}{3}}}{x^{2/3}}\right]$$

$$= -\left[\frac{-x^{1/3}.y^{\frac{-2}{3}}\left(y/x\right)^{1/3} - y^{\frac{1}{3}}x^{\frac{-2}{3}}}{3x^{2/3}}\right]$$

$$= \frac{1}{3}\frac{y^{\frac{-1}{3}} + y^{\frac{1}{3}}x^{\frac{-2}{3}}}{x^{2/3}} = \frac{1}{3}\frac{x^{2/3} + y^{2/3}}{x^{4/3}.y^{1/3}}$$

$$= \frac{1}{3}.\frac{a^{2/3}}{x^{4/3}.y^{1/3}}, \qquad \text{from (i)}$$

$\therefore$ Radius of curvature

$$\rho = \frac{\left[1+\left(\dfrac{dy}{dx}\right)^2\right]^{3/2}}{\dfrac{d^2y}{dx^2}}$$

$$= \frac{\left[1+\left(\dfrac{y}{x}\right)^{2/3}\right]^{3/2}}{\dfrac{1}{3}.a^{2/3}\Big/x^{4/3}.y^{1/3}}$$

$$= \frac{3\left(x^{2/3}+y^{2/3}\right)^{3/2}.x^{4/3}\,y^{1/3}}{a^{2/3}}$$

$$= \frac{3a\,x^{1/3}\,y^{1/3}}{a^{2/3}}$$

$$\rho = 3a^{1/3}\,x^{1/3}\,y^{1/3}$$

Now, putting $x = \cos^3\theta$, $y = a\sin^3\theta$, we have

$$\rho = 3a^{1/3}\left(a\cos^3\theta\right)^{1/3}\left(a\sin^3\theta\right)^{1/3}$$

$$= 3\,a\cos\theta\sin\theta$$

Hence proved

***Example* 6:** Find the radius of curvature at the point 't' of the following curve

$$x = 3a\cos t - a\cos 3t,\quad y = 3a\sin t - a\sin 3t. \hspace{2cm} \text{(RGPV June 2004)}$$

***Solution*:** The given curve be

$$x = 3a\cos t - a\cos 3t, \hspace{2cm} y = 3a\sin t - a\sin 3t$$

Differentiating, we get

$$x' = \frac{dx}{dt} = -3a\sin t + 3a\sin 3t, \hspace{1cm} y' = \frac{dy}{dt} = 3a\cos t - 3a\cos 3t$$

Differentiating again, we get

$$x'' = -3a\cos t + 9a\cos 3t, \hspace{1.5cm} y'' = -3a\sin t + 9a\sin 3t$$

∴   Radius of curvature

$$\rho = \frac{\left[\left(x'\right)^2 + \left(y'\right)^2\right]^{3/2}}{x'y'' - y'x''}$$

$$= \frac{\left[\left(-3a\sin t + 3a\sin 3t\right)^2 + \left(3a\cos t - 3a\cos 3t\right)^2\right]^{3/2}}{\left(-3a\sin t + 3a\sin 3t\right)\left(-3a\sin t + 9a\sin 3t\right) - \left(3a\cos t - 3a\cos 3t\right) - 3a\cos t + 9a\cos 3t\right)}$$

$$= \frac{9a^2\left[\sin^2 t + \sin^2 3t - 2\sin t\sin 3t + \cos^2 t + \cos^2 3t - \cos t\cos 3t\right]^{3/2}}{\left[\begin{array}{l}9a^2\sin^2 t - 27a^2\sin t\sin 3t - 9a^2\sin t\sin 3t + 27a^2\sin^2 3t + \\ 9a^2\cos^2 t - 27a^2\cos t\cos 3t - 9a^2\cos t\cos 3t + 27a^2\cos^2 3t\end{array}\right]}$$

$$= \frac{\left[9a^2\left(2 - 2\cos 2t\right)\right]^{3/2}}{36a^2\left(1 - \cos 2t\right)}$$

$$= \frac{\left[18a^2\left(1 - \cos 2t\right)\right]^{3/2}}{36a^2\left(1 - \cos 2t\right)}$$

$$= \frac{\left[36a^2\sin^2 t\right]^{3/2}}{36a^2 \cdot 2\sin^2 t} = 3a\sin t.$$

***Example* 7:** Find the radius of curvature of $\sqrt{x} + \sqrt{y} = \sqrt{a}$ at $\left(\dfrac{1}{4}, \dfrac{1}{4}\right)$.   (RGPV June 2008)

***Solution*:** The given curve be $\sqrt{x} + \sqrt{y} = \sqrt{a}$

Differentiating w.r.t, x, we get

$$\frac{1}{2}x^{\frac{-1}{2}} + \frac{1}{2}y^{\frac{-1}{2}}\frac{dy}{dx} = 0$$

or $\quad \dfrac{dy}{dx} = -y^{1/2}\,x^{-1/2}$

Differentiating again, we have

$$\frac{d^2y}{dx^2} = -\frac{1}{2}y^{-1/2}\frac{dy}{dx}\cdot x^{-\frac{1}{2}} + y^{\frac{1}{2}}\frac{1}{2}x^{\frac{-3}{2}}$$

$$= \frac{1}{2}y^{\frac{-1}{2}}y^{\frac{1}{2}}x^{-\frac{1}{2}}x^{\frac{-1}{2}} + \frac{1}{2}x^{\frac{-3}{2}}y^{\frac{1}{2}}$$

$$= \frac{1}{2x} + \frac{y^{\frac{1}{2}}}{2x^{\frac{3}{2}}}$$

$$= \frac{x^{\frac{1}{2}} + y^{\frac{1}{2}}}{2x^{\frac{3}{2}}} = \frac{\sqrt{a}}{2x^{\frac{3}{2}}}$$

$\therefore$ Radius of curvature

$$\rho = \frac{\left[1 + \left(dy/dx\right)^2\right]^{3/2}}{\dfrac{d^2y}{dx^2}}$$

$$= \frac{\left(1 + y/x\right)^{3/2}}{\sqrt{a}\Big/2x^{3/2}}$$

$$\rho = \frac{2(x+y)^{\frac{3}{2}}}{\sqrt{a}}$$

Now, at the point $\left(\dfrac{1}{4}, \dfrac{1}{4}\right)$, we have

$$\rho = \frac{1}{\sqrt{2a}}$$

***Example* 8:** Find the radius of curvature of $y = 4 \sin x - \sin 2x$ at $x = \dfrac{\pi}{2}$.

***Solution*:** The given curve be $y = 4 \sin x - \sin 2x$

Differentiating w.r.t x, we get

$$\frac{dy}{dx} = 4 \cos x - 2 \cos 2x$$

and

$$\frac{d^2y}{dx^2} = -4\sin x + 4\sin 2x$$

at

$$x = \frac{\pi}{2}, \quad \frac{dy}{dx} = -4\cos\frac{\pi}{2} - 2\cos 2 \times \frac{\pi}{2} = 2$$

and

$$\frac{d^2y}{dx^2} = -4\sin\frac{\pi}{2} + 4\sin 2 \times \frac{\pi}{2} = -4$$

$\therefore$ Radius of curvature

$$\rho = \frac{\left[1 + \left(\dfrac{dy}{dx}\right)^2\right]^{\frac{3}{2}}}{\dfrac{d^2y}{dx^2}}$$

$$= \frac{(1+4)^{3/2}}{-4}$$

$$= \frac{5\sqrt{5}}{4}$$

Neglecting negative sign

***Example* 9:** In the ellipse $\dfrac{x^2}{a^2} + \dfrac{y^2}{b^2} = 1$, show that the radius of curvature at an end of the major

axis is equal to the semi latus rectum of the ellipse.   (RGPV Dec 2000)

***Solution*:** The equation of ellipse be $\dfrac{x^2}{a^2} + \dfrac{y^2}{b^2} = 1$

Differentiating w.r.t x, we get

$$\frac{2x}{a^2} + \frac{2y}{b^2}\frac{dy}{dx} = 0$$

or

$$\frac{dy}{dx} = \frac{-b^2x}{a^2y}$$

and $\quad \dfrac{d^2y}{dx^2} = \dfrac{-b^2}{a^2}\left[\dfrac{y.1 - x\frac{dy}{dx}}{y^2}\right]$

$$= \dfrac{-b^2}{a^2 y^2}\left[y - x\left(\dfrac{-b^2 x}{a^2 y}\right)\right]$$

$$= \dfrac{-b^2}{a^2 y^2}\left[\dfrac{a^2 y^2 + b^2 x^2}{a^2 y}\right] = -\dfrac{b^4}{a^2 y^3}\left(\dfrac{x^2}{a^2} + \dfrac{y^2}{b^2}\right)$$

$$= -\dfrac{b^4}{a^2 y^3}$$

$\therefore$ Radius of curvature

$$\rho = \dfrac{\left[1 + \left(\frac{dy}{dx}\right)^2\right]^{\frac{3}{2}}}{d^2y \Big/ dx^2}$$

$$= \dfrac{\left[1 + \left(-b^2 x \Big/ a^2 y\right)^2\right]^{\frac{3}{2}}}{b^4 \Big/ a^2 y^3}, \text{ neglecting negative sign}$$

$$\rho = \dfrac{\left(a^4 y^2 + b^4 x^2\right)^{3/2}}{a^4 b^4}$$

Now, the coordinate of one end of major axis are (a, 0). Hence

$$\rho = \dfrac{\left(a^4 \times 0 + b^4 a^2\right)^{3/2}}{a^4 b^4} = \dfrac{b^6 a^3}{a^4 b^4} = \dfrac{b^2}{a}$$

$$= \text{ semi latus rectum of the ellipse}$$

***Example* 10:** Prove that, for the ellipse $\dfrac{x^2}{a^2} + \dfrac{y^2}{b^2} = 1$, $\rho = \dfrac{a^2 b^2}{p^3}$, p being the length of perpendicular form the centre upon the tangent at (x, y) .

(RGPV Dec 2001)

***Solution*:** Proceeding as above Example 9

$$\rho = \dfrac{\left(b^4 x^2 + a^4 y^2\right)^{3/2}}{a^4 b^4} \qquad\qquad \text{.....(i)}$$

Also, the equation of tangent to the ellipse at (x, y) is

$$\frac{X.x}{a^2} + \frac{Y.y}{b^2} = 1$$

The length of perpendicular from the centre (0, 0) of the ellipse to this tangent be

$$P = \frac{1}{\sqrt{\left(\frac{x}{a^2}\right)^2 + \left(\frac{y}{b^2}\right)^2}}$$

$$P = \frac{a^2 b^2}{\sqrt{b^4 x^2 + a^4 y^2}}$$

or $$P^3 = \frac{a^6 b^6}{\left(b^4 x^2 + a^4 y^2\right)^{3/2}}$$

or $$\left(b^4 x^2 + a^4 y^2\right)^{3/2} = \frac{a^6 b^6}{p^3}$$

$\therefore$ We have

$$\rho = \frac{\frac{a^6 b^6}{p^3}}{a^4 b^4}$$

or $$\rho = \frac{a^2 b^2}{p^3}$$

Hence proved

***Example* 11:** Prove that, the radius of curvature at any point on the hyper cycloid $x^{2/3} + y^{2/3} = a^{2/3}$ is three times the perpendicular from the origin to the tangent at that point.

(RGPV Dec 2002, June 2005)

***Solution*:** Proceeding as above Example 5, we have

$$\rho = 3a \sin \theta \cos \theta \qquad\qquad \text{.....(i)}$$

Let the parametric equations of the cuve be

$$x = a \cos^3 \theta, \, y = b \sin^3 \theta$$

Then $$x' = \frac{dx}{d\theta} = -3a \cos^2 \theta \sin \theta \quad \text{and} \quad y' = \frac{dy}{d\theta} = - 3b \sin^2 \theta \cos \theta$$

$\therefore$ $$\frac{dy}{dx} = \frac{dy/d\theta}{dx/d\theta} = \frac{-\sin \theta}{\cos \theta}$$

$\therefore$ Equation of tangent at the point 'θ' is

$$y - y_1 = \left(\frac{dy}{dx}\right)(x - x_1)$$

$$y - b \sin^3 \theta = \frac{-\sin \theta}{\cos \theta} (x - a \cos^3 \theta)$$

'or' $y \cos \theta - b \sin^3 \theta \cos \theta = -x \sin \theta + a \cos^3 \theta \sin \theta$

'or' $x \sin \theta + y \cos \theta = a \sin \theta \cos \theta (\sin^2 \theta + \cos^2 \theta)$.

'or' $x \sin \theta + y \cos \theta = a \sin \theta \cos \theta$

Now, perpendicular length P from the origin (0, 0) to the tangent be

$$P = \frac{a \sin \theta \cos \theta}{\sqrt{\sin^2 \theta + \cos^2 \theta}}, \quad \text{neglecting negative sign}$$

$$P = a \sin \theta \cos \theta$$

Hence (i) becomes

$$\rho = 3p$$

Hence proved.

***Example* 12:** If ρ_1, ρ_2 be the radii of curvature at the extremities of two conjugate diameters of an ellipse, prove that $\left[(\rho_1)^{2/3} + (\rho_2)^{2/3}\right](ab)^{2/3} = a^2 + b^2$.   (RGPV Dec 2003)

***Solution*:** Proceeding as above example 9, we have

$$\rho = \frac{\left(a^4 y^2 + b^4 x^2\right)^{3/2}}{a^4 b^4} \qquad \qquad .....(i)$$

Since the extremities of a pair of conjugate diameters are (a cos ϕ, b sin ϕ) and (−a sin ϕ, b cos ϕ)

Hence $$\rho_1 = \frac{\left(a^4 b^2 \sin^2 \phi + b^4 a^2 \cos^2 \phi\right)^{3/2}}{a^4 b^4} = \frac{\left(a^2 \sin^2 \phi + b^2 \cos^2 \phi\right)^{3/2}}{ab}$$

and $$\rho_2 = \frac{\left(a^4 b^2 \cos^2 \phi + b^4 a^2 \sin^2 \phi\right)^{3/2}}{a^4 b^4} = \frac{\left(a^2 \cos^2 \phi + b^2 \sin^2 \phi\right)^{3/2}}{ab}$$

$\therefore$ $$(\rho_1)^{2/3} + (\rho_2)^{2/3} = \frac{a^2 + b^2}{(ab)^{2/3}}$$

or $$\left[(\rho_1)^{2/3} + (\rho_2)^{2/3}\right](ab)^{2/3} = a^2 + b^2$$

Hence proved.

***Example* 13:** Find the radius of curvature at the highest point of the arc of the cycloid

$x = a(\theta - \sin \theta), y = a(1 - \cos 0)$ (RGPV June 2006)

***Solution*:** The given equations are

$$x = a (\theta - \sin \theta), y = a(1 - \cos \theta)$$

Differentiating, we get $x' = a(1 - \cos\theta), y' = a\sin\theta$

and $\quad x'' = a\sin\theta, y'' = a\cos\theta$

$\therefore$ Radius of curvature

$$\rho = \frac{\left[(x')^2 + (y')^2\right]^{3/2}}{x'y'' - y'x''}$$

$$= \frac{\left[a^2(1-\cos\theta)^2 + a^2\sin^2\theta\right]^{3/2}}{a(1-\cos\theta)a\cos\theta - a\sin\theta\, a\sin\theta}$$

$$= \frac{a^3\left[1 + \cos^2\theta - 2\cos\theta + \sin^2\theta\right]^{3/2}}{a^2\left[\cos\theta - \cos^2\theta - \sin^2\theta\right]}$$

$$= \frac{a\left[2(1-\cos\theta)\right]^{3/2}}{-(1-\cos\theta)}$$

$$\rho = 4a\,\sin\frac{\theta}{2}, \text{ neglecting negative sign}$$

But for the given cycloid the highest point of the arc is $\theta = \pi$

Then $\rho = 4a\,\sin\frac{\pi}{2}$

$\qquad = 4a$

***Example* 14:** Find the radius of curvature of the curve $y = e^x$. At the point where it crosses the y-axis.

***Solution*:** The given curve be $y = e^x$

Differentiating w.r.t x, we get

$$\frac{dy}{dx} = e^x$$

and $\quad \dfrac{d^2y}{dx^2} = e^x$

$\therefore$ Radius of curvature

$$\rho = \frac{\left[1+\left(\dfrac{dy}{dx}\right)^2\right]^{3/2}}{\dfrac{d^2y}{dx^2}}$$

$$= \frac{\left(1+e^{2x}\right)^{3/2}}{e^x}$$

Now the curve crosses y-axis at the point where $x = 0$

$$\therefore \quad \rho = \frac{\left(1+1\right)^{3/2}}{1} = 2^{3/2} = 2\sqrt{2}$$

7.5 Pedal Formula for Radius of Curvature

Let $P(r,\ \theta)$ be a point on the given curve. Let the tangent at P makes angle ψ with x-axis.

Then $\angle TOP = \phi$ and

$$\psi = \theta + \phi \qquad\qquad .....(1)$$

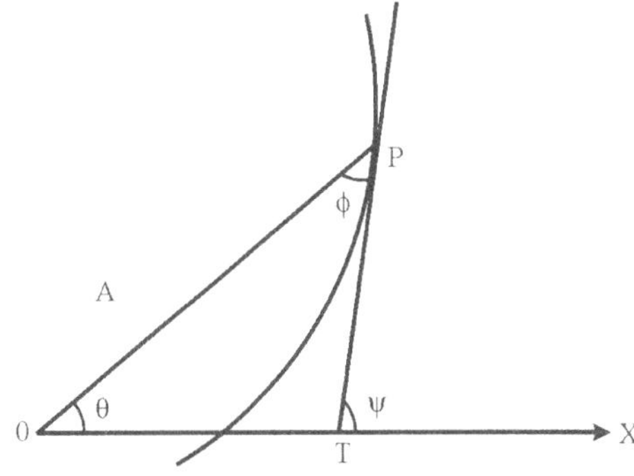

Differentiating w.r.t s, we get

$$\frac{d\psi}{ds} = \frac{d\theta}{ds} + \frac{d\phi}{ds}$$

or $\quad \dfrac{1}{\rho} = \dfrac{1}{r}.r\dfrac{d}{ds} \quad \dfrac{d\phi}{dr}.\dfrac{dr}{ds}$

$$= \frac{1}{r}\sin\phi + \frac{d\phi}{dr}\cos\phi \qquad\qquad \left(\text{since } r\frac{d\theta}{ds}\sin\phi,\ \frac{dr}{ds} = \cos\phi\right)$$

$$= \frac{1}{r}\left[\sin\phi + r\cos\phi\frac{d\phi}{dr}\right]$$

$$= \frac{1}{r}\frac{d}{dr}\left(r\sin\phi\right)$$

$$= \frac{1}{r}\frac{dP}{dr} \qquad\qquad (\because P = r\sin\phi)$$

$$\therefore \quad \rho = r\frac{dr}{dP}$$

7.6 Polar Formula for Radius of Curvature

We know that

$$\frac{1}{P^2} = \frac{1}{r^2} + \frac{1}{r^4}\left(\frac{dr}{d\theta}\right)^2$$

Differentiating w.r.t r, we get

$$\frac{-2}{P^3}\frac{dP}{dr} = \frac{-2}{r^3} - \frac{4}{r^5}\left(\frac{dr}{d\theta}\right)^2 + \frac{1}{r^4}\cdot 2\frac{dr}{d\theta}\frac{d}{dr}\left(\frac{dr}{d\theta}\right)$$

$$= \frac{-2}{r^3} - \frac{4}{r^5}\left(\frac{dr}{d\theta}\right)^2 + \frac{2}{r^4}\cdot\frac{dr}{d\theta}\frac{d}{d\theta}\left(\frac{dr}{d\theta}\right)\frac{d\theta}{dr}$$

$$= \frac{-2}{r^3} - \frac{4}{r^5}\left(\frac{dr}{d\theta}\right)^2 + \frac{2}{r^4}\frac{d^2r}{d\theta^2}$$

or $\qquad \dfrac{1}{P^3}\dfrac{dP}{dr} = \dfrac{1}{r^3} + \dfrac{2}{r^5}\left(\dfrac{dr}{d\theta}\right)^2 - \dfrac{1}{r^4}\dfrac{d^2r}{d\theta^2}$

Now $\quad \rho = \dfrac{dr}{dP}$

$$= \frac{r\cdot\left(1\big/P^3\right)}{\dfrac{1}{r^3} + \dfrac{2}{r^5}\left(\dfrac{dr}{d\theta}\right)^2 - \dfrac{1}{r^4}\dfrac{d^2r}{d\theta^2}}$$

$$= \frac{r^6\left[\dfrac{1}{r^2} + \dfrac{1}{r^4}\left(\dfrac{dr}{d\theta}\right)^2\right]^{3/2}}{r^2 + 2\left(\dfrac{dr}{d\theta}\right)^2 - r\dfrac{d^2r}{d\theta^2}}$$

$$\therefore \quad \rho = \frac{\left[r^2 + \left(\dfrac{dr}{d\theta}\right)^2\right]^{3/2}}{r^2 + 2\left(\dfrac{dr}{d\theta}\right)^2 - r\dfrac{d^2r}{d\theta^2}}$$

Illustrative Examples

***Example* 15:** Find the radius of curvature at (P, r) on the curve $P^2 = ar$.

Solution: The given curve be $P^2 = ar$

Differentiating w.r.t r, we get

$$2P\frac{dP}{dr} = a$$

or $\quad \dfrac{dr}{dP} = \dfrac{2P}{a}$

∴ Radius of curvature

$$\rho = r\frac{dr}{dP}$$

$$= \frac{2\,Pr}{a}$$

$$= \frac{2P}{a}\cdot\frac{P^2}{a} = \frac{2P^3}{a^2}$$

***Example* 16:** Show that the radius of curvature at any point on.

The cardioid $r = a(1 - \cos\theta)$ is $\dfrac{2}{3}\sqrt{2ar}$.

Solution: The given curve be $r = a(1 - \cos\theta)$

Differentiating w.r.t θ, we get

$$\frac{dr}{d\theta} = a\sin\theta \quad\text{and}\quad \frac{d^2r}{d\theta^2} = a\cos\theta$$

∴ Radius of curvature

$$\rho = \frac{\left[r^2 + \left(dr\big/d\theta\right)^2\right]^{3/2}}{r^2 + 2\left(dr\big/d\theta\right)^2 - r\,d^2r\big/d\theta^2}$$

$$= \frac{\left[a^2(1-\cos\theta)^2 + a^2\sin^2\theta\right]^{3/2}}{a^2(1-\cos\theta)^2 + 2a^2\sin^2\theta - a(1-\cos\theta)a\cos\theta}$$

$$= \frac{a^3\left[1 + \cos^2\theta - 2\cos\theta + \sin^2\theta\right]^{3/2}}{a^2\left[1 + \cos^2\theta - 2\cos\theta + 2\sin^2\theta - \cos\theta + \cos^2\theta\right]}$$

$$= \frac{a\left[2(1-\cos\theta)\right]^{3/2}}{3(1-\cos\theta)}$$

$$= \frac{2\sqrt{2}}{3}a(1-\cos\theta)^{1/2}$$

$$= \frac{2\sqrt{2}}{3}a\left(r/a\right)^{1/2} = \frac{2}{3}\sqrt{2ar}$$

Hence proved.

***Example* 17:** Find the radius of curvature at (r, θ) on the curve $r^2 = a^2 \cos 2\theta$.

***Solution*:** The given curve be $r^2 = a^2 \cos 2\theta$ .....(i)

Taking logarithms on both sides, we get

$$2 \log r = 2 \log a + \log \cos 2\theta$$

Differentiating w.r.t θ, we have

$$\frac{2}{r}\frac{dr}{d\theta} = \frac{-2\sin 2\theta}{\cos 2\theta}$$

or $r\dfrac{d\theta}{dr} = -\cot 2\theta = \tan\left(\pi/2 + 2\theta\right)$

or $\tan\phi = \tan\left(\pi/2 + 2\theta\right)$

or $\phi = \pi/2 + 2\theta$

$\therefore$ From $P = r \sin\phi$, we get $P = r \sin\left(\pi/2 + 2\theta\right) = r \cos 2\theta$

or $P = r\left(r^2/a^2\right),$ from (i)

$$= r^3/a^2$$

or $r^3 = a^2 P$

Differentiating w.r.t r, we get

$$3r^2 = a^2 \frac{dp}{dr}$$

'or' $\dfrac{dr}{dp} = \dfrac{a^2}{3r^2}$

$\therefore$ Radius of curvature $\rho = r\dfrac{dr}{dP}$

$$= r. \ a^2 \Big/ 3r^2 \ = \ \dfrac{a^2}{3r}$$

***Example* 18:** Find the radius of curvature at (r, θ) on the curve $r^n = a^n \cos n\theta$.

***Solution*:** The given curve be $r^n = a^n \cos n\theta$ .....(i)

Taking log on both sides, we get

$$n \log r = n \log a + \log \cos n\theta$$

Differentiating w.r.t θ, we get

$$\dfrac{n}{r}\dfrac{dr}{d\theta} = \dfrac{-n\sin n\theta}{\cos n\theta}$$

or $r\dfrac{d\theta}{dr} = -\cot n\theta = \tan\left(\dfrac{\pi}{2} + n\theta\right)$

or $\tan\phi = \tan\left(\dfrac{\pi}{2} + n\theta\right) \ \Rightarrow \ \phi = \dfrac{\pi}{2} + n\theta$

$\therefore$ From $P = r \sin\phi$, we get

$$P = r \sin\left(\dfrac{\pi}{2} + n\theta\right) = r \cos n\theta$$

or $P = r\left(r^n \Big/ a^n\right)$

$\Rightarrow$ $P a^n = r^{n+1}$

Differentiating, we get $a^n \dfrac{dP}{dr} = (n+1) r^n$

$\therefore$ Radius of curvature

$$\rho = r\dfrac{dr}{dP}$$

$$= \dfrac{r a^n}{(n+1)r^n} = \dfrac{a^n}{(n+1)r^{n-1}}$$

***Example* 19:** Show that at any point on the equi angular spiral $r = a\, e^{\theta \cot \alpha}$, $\rho = r\, \text{cosec}\ \alpha$, and that it subtends a right angle at the pole.

***Solution*:** The given curve be $r = a\, e^{\theta \cot \alpha}$ .....(i)

Differentiating w.r.t θ, we get

$$\frac{dr}{d\theta} = ae^{\theta \cot \alpha} \cot \alpha$$

$$= r \cot \alpha$$

and $\quad \dfrac{d^2 r}{d\theta^2} = \cot \alpha \dfrac{dr}{d\theta} = r \cot^2 \alpha$

$\therefore$ Radius of curvature

$$\rho = \frac{\left[r^2 + \left(\dfrac{dr}{d\theta} \right)^2 \right]^{3/2}}{r^2 + 2\left(\dfrac{dr}{d\theta} \right)^2 - r\dfrac{d^2 r}{d\theta^2}}$$

$$= \frac{\left[r^2 + r^2 \cot^2 \alpha \right]^{3/2}}{r^2 + 2r^2 \cot^2 \alpha - r^2 \cot^2 \alpha}$$

$$= \frac{r^3 \left[1 + \cot^2 \alpha \right]^{3/2}}{r^2 \left(1 + \cot^2 \alpha \right)}$$

$$= r(\text{cosec}^2\ \alpha)^{1/2} = r\ \text{cosec}\ \alpha \qquad\qquad .....(ii)$$

Now, let $P(r, \theta)$ be any point on the curve and let PT and PC are tangent and normal to the curve at P. Let C be the centre of curvature, then

$$\angle OPT = \phi = \alpha$$

and $\quad \angle OPC = 90^\circ - \alpha$

Let $\quad \angle POC = \beta$

Then in $\triangle OPC$, we have $\angle OCP = 90^\circ + \alpha - \beta$

and $\quad \dfrac{OP}{\sin \angle OCP} = \dfrac{PC}{\sin \beta}$

or $\quad \dfrac{r}{\sin\left(90^\circ + \alpha - \beta\right)} = \dfrac{\rho}{\sin \beta}$

or $\quad \dfrac{r}{\cos(\alpha - \beta)} = \dfrac{r\cos ec\ \alpha}{\sin \beta}$, from (ii)

or $\quad \sin \alpha \sin \beta = \cos (\alpha - \beta)$

$$= \cos \alpha \cos \beta + \sin \alpha \sin \beta$$

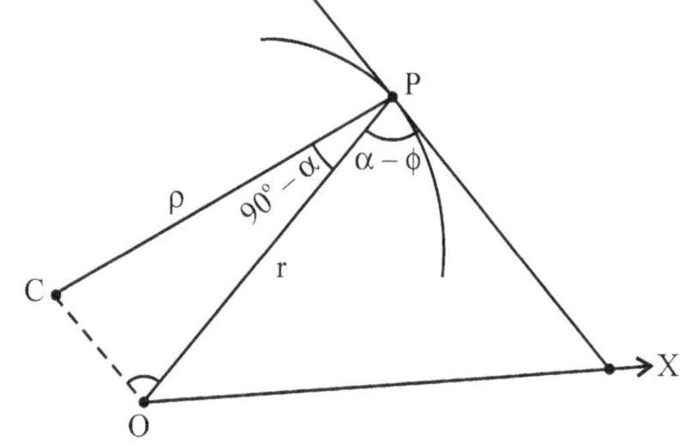

or $\cos \alpha \cos \beta = 0$

or $\cos \beta = 0$ $(\because \cos \alpha \neq 0)$

or $\beta = \pi/2$

7.7 Curvature at Origin

The following methods are used to find the radius of curvature at the origin :

(i)   ***Substitution method***: we know that the radius of curvature at any point (x, y) is

$$\rho = \frac{\left[1+\left(\dfrac{dy}{dx}\right)^2\right]^{3/2}}{\dfrac{d^2y}{dx^2}}$$

$\therefore$ Radius of curvature at the origin can be found by putting x = 0, y = 0.

(ii)   ***Expansion method***: Let the equation of the curve be y = f(x) and y can be expanded in powers of x, then we can write

$$y = px + \frac{qx^2}{2!} + \text{.......}$$

then we find the curve passes through the origin and

$$\left(\frac{dy}{dx}\right)_{x=0,y=0} = p \quad \text{and} \quad \left(\frac{d^2y}{dx^2}\right)_{x=0,y=0} = q$$

$\therefore$ Radius of curvature

$$\rho = \frac{\left[1+\left(\dfrac{dy}{dx}\right)^2\right]^{3/2}}{\dfrac{d^2y}{dx^2}}$$

$$\text{or } \rho(\text{at the origin}) = \frac{\left(1+p^2\right)^{3/2}}{q} \qquad \text{.....(1)}$$

(iii)   ***Newtonian method***

  (a)   ***If tangent at origin is x-axis***: If a curve passes through (0, 0) and tangent at x-axis, then x = 0, y = 0 and $\dfrac{dy}{dx} = 0$ at the origin. Therefore, the expansion of y by Maclaurin's theorem, we have

$$y = 0 + 0.x + \frac{q}{2!} x^2 + \text{........}$$

Dividing each term by x^2 and taking limit as $x \to 0$, we have

$$\lim_{x \to 0}\left(\frac{2y}{x^2}\right) = q \qquad\qquad .....(2)$$

$\therefore$ from equation (ii), we have

$$\rho = \frac{\left(1 + P^2\right)^{3/2}}{q} = \frac{1}{q} \quad \text{as } P = 0 \text{ at the origin}$$

$\therefore \qquad\qquad \rho(\text{at the origin}) = \lim_{x \to 0}\left(\frac{x^2}{2y}\right), \qquad\qquad \text{from (2)}$

(b)   *If tangent at origin is y-axis*: Proceeding as above, we have

$$\rho(\text{at the origin}) = \lim_{x \to 0}\left(\frac{y^2}{2x}\right)$$

(iv)   *Curvature at pole*: If initial line is tangent to the curve at the pole (r, θ) and if $r = 0$ and $\theta \to 0$

Then $\qquad\qquad \rho = \lim_{x \to 0}\left(\frac{x^2}{2y}\right)$

$$= \lim_{\theta \to 0}\left(\frac{r^2 \cos^2 \theta}{2r \sin \theta}\right)$$

$$(\because x = r\cos\theta,\ y = r\sin\theta)$$

$$= \lim_{\theta \to 0}\left[\frac{r}{2\theta}\cdot\frac{\theta}{\sin\theta}\cdot\cos^2\theta\right]$$

or $\qquad \rho(\text{at the pole})$

$$= \lim_{\theta \to 0}\left(\frac{r}{2\theta}\right) \qquad \left(\because \lim_{\theta \to 0}\frac{\theta}{\sin\theta} = 1,\ \lim_{\theta \to 0}\cos^2\theta = 1\right)$$

But if $r = 0$ and $\theta = \frac{\pi}{2}$, then x-axis is the tangent to the curve at the pole

Hence $\qquad\qquad \rho = \lim_{x \to 0}\left(\frac{y^2}{2x}\right)$

$$= \lim_{\theta \to 0}\left(\frac{r^2 \sin^2 \theta}{2r \cos \theta}\right)$$

$$= \lim_{\theta \to 0}\left(\frac{r\theta^2}{2}\cdot\left(\frac{\sin\theta}{\theta}\right)^2\cdot\frac{1}{\cos\theta}\right)$$

$\rho(\text{at the pole}) \qquad = \lim_{\theta \to 0}\left(\frac{r\theta^2}{2}\right)$

***Example* 20:** Find the radius of curvature at the origin of the curve $y = x^4 - 4x^3 - 18x^2$.

***Solution*:** The given curve be

$$y = x^4 - 4x^3 - 18x^2$$

Differentiating, we get

$$\frac{dy}{dx} = 4x^3 - 12x^2 - 36x \text{ and } \frac{d^2y}{dx^2} = 12x^2 - 24x - 36$$

$\therefore$ At the origin i.e., at $x = 0$, $y = 0$, we have

$$\frac{dy}{dx} = 0, \text{ and } \frac{d^2y}{dx^2} = -36$$

ρ(at the origin)

$$= \frac{\left[1 + \left(\frac{dy}{dx}\right)^2\right]^{3/2}}{\frac{d^2y}{dx^2}}$$

$$= \frac{-1}{36} = \frac{1}{36} \quad \text{(neglecting negative sign)}$$

***Example* 21:** Find the radii of curvature of the curve $a(y^2 - x^2) = x^3$ at the origin.

***Solution*:** The given curve be $a(y^2 - x^2) = x^3$

or $$y^2 = \frac{x^3}{a} + x^2$$

or $$y = \pm\, x\sqrt{1 + \frac{x}{a}}$$

$$= \pm\, x\left(1 + \frac{1}{2}\cdot\frac{x}{a} + ...\right) \qquad \text{(by using Binomial expansion)}$$

or $$y = \pm\left(x + \frac{1}{2}\cdot\frac{x^2}{a} + ...\right)$$

which is of the form $y = px + \dfrac{qx^2}{2!} + ...$

and (at the origin) $= \dfrac{\left(1 + p^2\right)^{3/2}}{2}$

$$= \frac{(1+1)^{3/2}}{2/a} = a\sqrt{2}$$

Equating the coefficients of x and x^2, we get

$$p = 1, q = \frac{1}{a} \qquad \text{or} \qquad p = -1, q = -\frac{1}{a}$$

Now ρ(at the origin)

$$= \frac{\left(1+p^2\right)^{3/2}}{q}$$

$$= \frac{(1+1)^{3/2}}{1/a} = 2a\sqrt{2}$$

and ρ(at the origin)

$$= \frac{(1+1)^{3/2}}{-1/a} = -2a\sqrt{2}$$

***Example* 22:** Show that the radii of curvature of the curve

$$y^2 = x^2(a + x)/(a - x) \text{ at the origin are } \pm a\sqrt{2} \ .$$

***Solution*:** Let the curve be $y^2 = \dfrac{x^2\left(a + x\right)}{a - x}$

$$\text{or} \qquad y = \pm x \sqrt{\frac{a+x}{a-x}}$$

$$= \pm x \left(1+\frac{x}{a}\right)^{1/2}\left(1-x/a\right)^{-1/2}$$

$$= \pm x\left(1+\frac{x}{2a}+.....\right)\left(1+x/2a+....\right)$$

$$= \pm x \left(1+\frac{x}{a}+.....\right)$$

$$= \pm \left(x+\frac{x^2}{a}+.....\right)$$

which is the form $y = px + \dfrac{qx^2}{2!} + ...$

$\therefore$ $p = 1, q = \dfrac{2}{a}$ and $p = -1, q = \dfrac{-2}{a}$

$\therefore$ ρ(at the origin) $= \dfrac{(1+1)^{3/2}}{-\dfrac{2}{a}} = -a\sqrt{2}$

Hence ρ(at the origin) $= \pm\ a\sqrt{2}$

***Example* 23:** Find the radius of curvature at the origin for the curve $x^3 + y^3 - 2x^2 + 6y = 0$

***Solution*:** The curve passes through $(0, 0)$ and the tangent with x-axis is $y = 0$. (obtained by equating the lowest degree term in the equation to zero)

Now, the given curve is $x^3 + y^3 - 2x^2 + 6y = 0$

Dividing each term by 2y, we get

$$x\left(\dfrac{x^2}{2y}\right) + \dfrac{1}{2}y^2 - 2\left(\dfrac{x^2}{2y}\right) + 3 = 0$$

taking limit as $x \to 0$ $y \to 0$ we get

$$\lim_{x \to 0}\left[x\left(\dfrac{x^2}{2y}\right) + \dfrac{1}{2}y^2 - 2\left(\dfrac{x^2}{2y}\right) + 3\right] = 0$$

or $0 + 0 - 2\rho + 3 = 0$ $\qquad\qquad\left(\because \lim_{x \to 0} \dfrac{x^2}{2y} = \rho\right)$

or $\rho = \dfrac{3}{2}$

***Example* 24:** Find the radius of curvature at the origin for the cycloid

$$x = a(\theta + \sin\theta),\ y = a(1 - \cos\theta)$$

***Solution*:**

Let the curve $x = a(\theta + \sin\theta),\ y = a(1 - \cos\theta)$

Differentiating,

$$\dfrac{dx}{d\theta} = a(1 + \cos\theta),\ \dfrac{dy}{d\theta} = a\sin\theta$$

$\therefore$

$$\dfrac{dy}{dx} = \dfrac{\dfrac{dy}{d\theta}}{\dfrac{dx}{d\theta}} = \dfrac{a\sin\theta}{a(1 + \cos\theta)} = \tan\dfrac{\theta}{2}$$

Now, at the origin $x = 0, y = 0$, hence $\theta = 0$

$\therefore$

$$\dfrac{dy}{dx} = 0,$$

Hence x-axis is the tangent at the origin

$$\therefore \qquad \rho(\text{at the origin}) = \lim_{x \to 0}\left(\frac{x^2}{2y}\right)$$

$$= \lim_{\theta \to 0} \frac{a^2(\theta + \sin\theta)^2}{2a(1 - \cos\theta)}, \quad \text{limit is of the form } \frac{0}{0}$$

$$= \lim_{\theta \to 0} \frac{a}{2} \frac{2(\theta + \sin\theta)(1 + \cos\theta)}{\sin\theta}, \left(\frac{0}{0}\right) \text{ form}$$

$$= \lim_{\theta \to 0} a \frac{(\theta + \sin\theta)(-\sin\theta) + (1 + \cos\theta)^2}{\cos\theta} = 4a$$

***Example* 25:** Find the radius of curvature of the curve r = a sin n θ at the pole.

***Solution*:** Let the curve be r = a sin nθ

Differentiating,

$$\frac{dr}{d\theta} = an\cos n\theta$$

$\therefore$ At the pole r = 0, θ = 0, we have

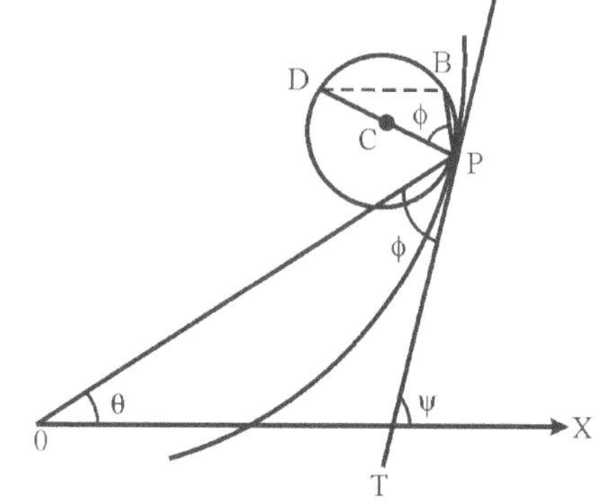

$$\rho(\text{at the pole}) = \lim_{\theta \to 0}\left(\frac{r}{2\theta}\right)$$

$$= \lim_{\theta \to 0}\left(\frac{a\sin n\theta}{2\theta}\right), \qquad \text{limit is of the form } (0/0)$$

$$= \lim_{\theta \to 0}\left(\frac{an\cos n\theta}{2}\right)$$

$$= \frac{an}{2}$$

7.8 Chord of Curvature through the Origin

Let P be any point on the curve and C be the centre of curvature at P. Let PT be the tangent to the curve at point P, then $\angle OPT = \phi$.

$\therefore \quad \angle PAB = \phi \quad \text{and} \quad \angle BPA = 90^\circ - \phi$

$\therefore$ In $\triangle APB$, we get, the chord of curvature through the origin O = BP

$$= PA \cos(90^\circ - \phi)$$

$$= 2\rho \sin\phi$$

$\because \ PA = 2\rho$

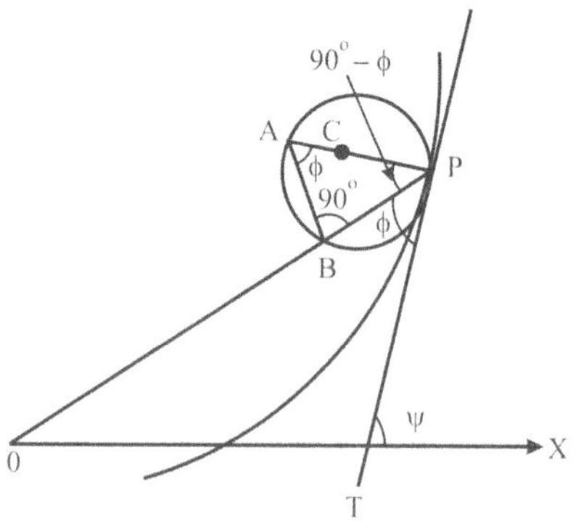

Chord of Curvature Parallel to Coordinate axes

Let P be any point on the curve and C be the centre of curvature. Let PT be the tangent to the curve at P making an angle ψ with x-axis. PA and PB are parallel to x and y axes then

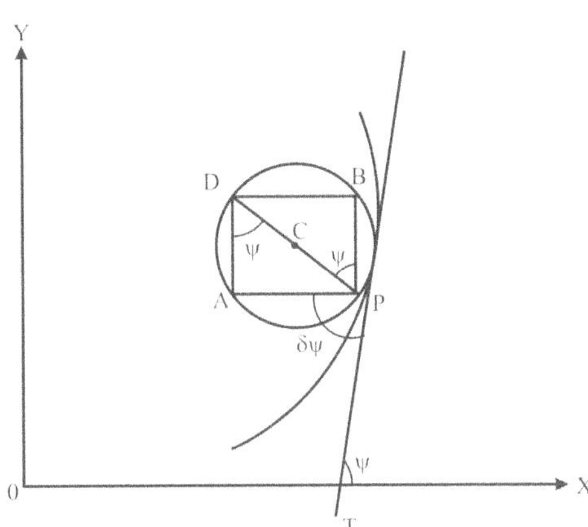

$$\angle APT = \angle BPD = \angle ADP = \psi .$$

$\therefore$ Chord of curvature parallel to x-axis = PA

$$= PD \sin \psi$$

$$= 2\rho \sin \psi$$

and chord of curvature parallel to y-axis = PB

$$= 2 \rho \cos \psi$$

Chord of Curvature Perpendicular to Radius Vector

Let $P(r, \theta)$ be any point on the curve and C be the centre of curvature. Let PT be the tangent to the curve at P.

then $\angle OPT = \phi$

$\therefore$ Chord of curvature perpendicular to the radius vector = PB

$$= PD \cos \phi$$

$$= 2\rho \cos \phi$$

***Example* 26:** Find the Chord of curvature through the pole at the cardioid $r = a(1 + \cos \theta)$.

***Solution*:** The given curve be $r = a(1 + \cos \theta)$ .....(i)

Differentiating w.r.t θ, we get

$$\frac{dr}{d\theta} = -a \sin \theta$$

$\therefore \qquad \tan \phi = r \frac{dr}{d\theta} = \frac{a(1 + \cos\theta)}{-a \sin\theta} = -\cot \frac{\theta}{2} = \tan\left(\frac{\pi}{2} + \frac{\theta}{2}\right)$

or $\quad \phi = \frac{\pi}{2} + \frac{\theta}{2}$ .....(ii)

Now, the pedal equation of the given curve is

$$P = r \sin \phi$$

$$= r \sin \left(\frac{\pi}{2} + \frac{\theta}{2}\right) = r \cos \frac{\theta}{2}$$ .....(iii)

or $\quad 2P^2 = 2 r^2 \cos^2 \frac{\theta}{2}$

$$= r^2 (1 + \cos \theta)$$

$$= r^2 \frac{r}{a}, \qquad \text{from (i)}$$

or $2P^2 a = r^3$

Differentiating w.r.t r, we get

$$4ap \, \frac{dp}{dr} = 3r^2$$

or $$\frac{dp}{dr} = \frac{3r^2}{4ap}$$

$\therefore$ Radius of curvature

$$\rho = r \frac{dr}{dp}$$

$$= r.\frac{4ap}{3r^2} = \frac{4ap}{3r}$$

$\therefore$ Required chord of curvature through the pole $= 2\,\rho \sin \phi$

$$= 2.\left(\frac{4ap}{3r}\right) \sin\left(\frac{\pi}{2}+\frac{\theta}{2}\right), \qquad\qquad \text{from (ii)}$$

$$= \frac{\delta ap}{3r}.\cos\frac{\theta}{2}$$

$$= \frac{\delta a}{3r}.r\cos\frac{\theta}{2}\cos\frac{\theta}{2}, \qquad\qquad \text{from (iii)}$$

$$= \frac{\delta a}{3}\cos^2\frac{\theta}{2}$$

$$= \frac{4a}{3}\left(1+\cos\theta\right) = \frac{4r}{3}, \qquad\qquad \text{from (i)}$$

***Example* 27:** In the curve $y = a \log \sec\left(\dfrac{x}{a}\right)$, prove that the chord of curvature parallel to the axis of y is of constant length.

***Solution*:** The given curve be $y = a \log \sec\left(\dfrac{x}{a}\right)$

Differentiating w.r.t x, we get

$$\frac{dy}{dx} = a.\frac{1}{\sec\left(\frac{x}{a}\right)}\sec\left(\frac{x}{a}\right)\tan\left(\frac{x}{a}\right)\frac{1}{a} = \tan\left(\frac{x}{a}\right)$$

and $$\frac{d^2y}{dx^2} = \frac{1}{a}\sec^2\left(\frac{x}{a}\right)$$

$$\therefore \quad \tan \psi = \frac{dy}{dx} = \tan\left(\frac{x}{a}\right) \quad \Rightarrow \quad \psi = \frac{x}{a} \qquad \dots\dots(i)$$

$\therefore$ Radius of curvature

$$\rho = \frac{\left[1+\left(\dfrac{dy}{dx}\right)^2\right]^{3/2}}{\dfrac{d^2y}{dx^2}}$$

$$= \frac{\left[1+\tan^2 x/a\right]^{3/2}}{\dfrac{1}{a}\sec^2 x/a}$$

$$= \frac{a\sec^3 x/a}{\sec^2 x/a} = a\sec\left(x/a\right) \qquad \dots\dots(ii)$$

$\therefore$ Required chord of curvature parallel to y-axis

$$= 2\,\rho\cos\psi$$

$$= 2\,.\,a\sec\left(x/a\right)\cos\left(x/a\right),\text{ from (i) and (ii)}$$

$$= 2a,\text{ which is constant}$$

***Example* 28:** Find the chord of curvature through the pole of the curve $r^n = a^n \cos n\theta$.

***Solution*:** The given curve be $r^n = a^n \cos n\,\theta$ $\qquad \dots\dots(i)$

Taking log on both sides, we get

$$n \log r = n \log a + \log \cos n\,\theta$$

Differentiating w.r. t θ, we get

$$\frac{n}{r}\frac{dr}{d\theta} = -n \tan n\theta$$

or $\quad \tan\phi = r\dfrac{d\theta}{dr} = -\cot n\theta = \tan\left(\pi/2 + n\theta\right)$

$\therefore \qquad \phi = \pi/2 + n\theta$ $\qquad \dots\dots(ii)$

$\therefore$ The pedal equation of eqn (i) is

$$P = r \sin\phi$$

$$= r\left(r/a\right)^n, \qquad \sin\phi = \left(r/a\right)^n \qquad \dots\dots(iii)$$

'or' $Pa^n = r^{n+1}$

Differentiating w.r.t r, we get

$$a^n \frac{dP}{dr} = (n+1)r^n$$

∴ Radius of curvature

$$\rho = r\frac{dr}{dP} = \frac{a^n}{(n+1)r^{n-1}}$$

∴ Required chord of curvature through the pole

$$= 2\rho \sin \phi$$

$$= 2 \cdot \frac{a^n}{(n+1)r^{n-1}} \left(\frac{r}{a}\right)^n \qquad\qquad \text{from (iii)}$$

$$= \frac{2r}{n+1}$$

***Example* 29:** Show that the chord of curvature through the pole of the curve $r = a\,e^{n\theta}$ is 2r.

***Solution*:** The given curve be $r = a\,e^{n\theta}$

Differentiating w.r.t θ, we get

$$\frac{dr}{d\theta} = ane^{n\theta} = nr$$

∴ $\tan \phi = r\dfrac{d\theta}{dr} = \dfrac{1}{n}$

and $\sin\phi = \dfrac{1}{\sqrt{1+\cot^2 \phi}} = \dfrac{1}{\sqrt{1+n^2}}$

∴ Pedal equation of the curve be $P = r \sin \phi$

$$= \frac{r}{\sqrt{1+n^2}}$$

Hence $\dfrac{dP}{dr} = \dfrac{1}{\sqrt{1+n^2}}$

∴ Radius of curvature $\rho = r\dfrac{dr}{dP} = r\sqrt{1+n^2}$

Hence the required chord of curvature $= 2\,\rho \sin \phi$

$$= 2r\sqrt{1+n^2} \cdot \frac{1}{\sqrt{1+n^2}}$$

$$= 2r$$

Hence proved.

7.9 Centre of Curvature

Let P(x, y) and Q (x + δx, y + δy) be two points on the given curve.

Let C be the centre of curvature (Say (α, β))

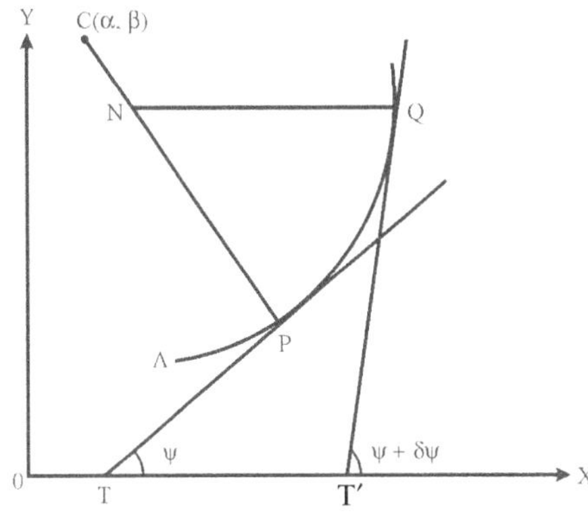

Now the equation of normal at point P is

$$(Y - y)\, \phi(x) + (X - x) = 0 \qquad \dots(1)$$

where $\phi(x) = \dfrac{dy}{dx}$

and the equation of normal at point Q is

$$[Y - (y + \delta y)]\, \phi(x + \delta x) + [X - (x + \delta x)] = 0 \qquad \dots(2)$$

Subtracting (i) from (ii), we get

$$(Y - y)\, \{\phi(x + \delta x) - \phi(x)\}\, \phi(x + \delta x)\, \delta y - \delta x = 0$$

'or' $$(Y - y) \left(\frac{\phi(x + \delta x) - \phi(x)}{\delta x} \right) - \phi(x + \delta x) \frac{\delta y}{\delta x} - 1 = 0 \qquad \dots(3)$$

Now, taking limit as $Q \to P$, $\delta x \to 0$ and $N \to C$, $y \to \beta$, we get

$$\frac{\phi(x + \delta x) - \phi(x)}{\delta x} \to \frac{d}{dx} \phi(x)$$

i.e., $$\frac{d^2 y}{dx^2} \text{ as } \phi(x) = \frac{dy}{dx}$$

and $$\frac{\delta y}{\delta x} \to \frac{dy}{dx}$$

$\therefore$ As $Q \to P$, we have from equation (3)

$$(\beta - y) \frac{d^2 y}{dx^2} - \phi(x) \frac{dy}{dx} - 1 = 0$$

or $$(\beta - y) \frac{d^2 y}{dx^2} - \left(\frac{dy}{dx} \right)^2 - 1 = 0$$

or $$\beta = y + \frac{1 + \left(\dfrac{dy}{dx} \right)^2}{\dfrac{d^2 y}{dx^2}}$$

Also as C(α, β) lies on the normal at P, therefore from (1), we get

$$(\beta - y)\frac{dy}{dx} + (\alpha - x) = 0$$

or $\quad \alpha = x - (\beta - y)\dfrac{dy}{dx}$

$$= x - \frac{\left[1 + \left(\frac{dy}{dx}\right)^2\right]\frac{dy}{dx}}{\frac{d^2y}{dx^2}}$$

Note: Equation of circle of curvature is $(x - \alpha)^2 + (y - \beta)^2 = \rho^2$.

7.10   The Evolute of a Curve

To find

(i)   The centre of curvature (α, β) is given by

$$\alpha = x - \frac{y_1\left(1 + y_1^2\right)}{y_2}, \quad \beta = y + \frac{1 + y_1^2}{y_2} \quad \text{where } y_1 = \frac{dy}{dx}, y_2 = \frac{d^2y}{dx^2}$$

(ii)   Eliminating the parameters x and y to find a relation between α and β.

(iii)   Generalizing α and β, we get the equation of the evalute.

Example 30: Find the centre of curvature of the rectangular hyperbola $xy = a^2$.

Solution: The given curve be $y = \dfrac{a^2}{x}$ $\hfill$ .....(i)

$$\therefore \quad \frac{dy}{dx} = \frac{-a^2}{x^2} \text{ and } \frac{d^2y}{dx^2} = \frac{2a^2}{x^3}$$

$$\therefore \quad \alpha = x - \frac{y_1\left(1 + y_1^2\right)}{y_2}$$

$$= x - \frac{\left(-a^2/x^2\right)\left(1 + a^4/x^4\right)}{2a^2/x^3}$$

$$= x + \frac{x^4 + a^4}{2x^3} = \frac{3x^4 + x^2y^2}{2x^3} \qquad \because a^4 = x^2 y^2 \qquad \text{(given curve)}$$

$$\alpha = \frac{3x}{2} + \frac{y^2}{2x}$$

and $\beta = y + \dfrac{1 + y_1^2}{y_2}$

$$= y + \dfrac{1 + \dfrac{a^4}{x^4}}{\dfrac{2a^2}{x^3}}$$

$$= y + \dfrac{x^4 + a^4}{2a^2 x}$$

$$= \dfrac{3y^2 + x^2}{2y} = \dfrac{3y}{2} + \dfrac{x^2}{2y}$$

Hence the required centre of curvature be $\left(\dfrac{3x}{2} + \dfrac{y^2}{2x}, \ \dfrac{3y}{2} + \dfrac{x^2}{2y} \right)$

Example **31:** Find the centre of curvature at the point 't' on the ellipse $x = a \cos t, \ y = b \sin t$.

Solution: The given curve be $x = a \cos t, y = b \sin t$

$\therefore \quad \dfrac{dx}{dt} = -a \sin t, \ \dfrac{dy}{dt} = b \cos t$

Hence $\quad \dfrac{dy}{dx} = \dfrac{dy/dt}{dx/dt} = \dfrac{-b}{a} \cot t$

and $\quad \dfrac{d^2 y}{dx^2} = \dfrac{b}{a} \operatorname{cosec}^2 t \ \dfrac{dt}{dx} = \dfrac{-b}{a^2} \operatorname{cosec}^3 t$

$\therefore \quad \alpha = x - \dfrac{y_1 \left(1 + y_1^2 \right)}{y_2}$

$$= x - \dfrac{\left(\dfrac{-b}{a} \cot t \right) \left(1 + \dfrac{b^2}{a^2} \cot^2 t \right)}{\dfrac{-b}{a^2} \operatorname{cosec}^3 t}$$

$$= a \cos t - \dfrac{1}{a} \left(a^2 \sin^2 t + b^2 \cos^2 t \right) \cos t$$

$$= \dfrac{a^2 \left(\cos t - \sin^2 t \cos t \right) - b^2 \cos^3 t}{a} = \dfrac{a^2 - b^2}{a} \cos^3 t$$

and $\beta = y + \dfrac{1 + y_1^2}{y_2}$

$$= y + \dfrac{1 + \dfrac{b^2}{a^2}\cot^2 t}{-\dfrac{b}{a^2}\cos ec^3 t}$$

$$= b \sin t - \dfrac{\left(a^2 \sin^2 t + b^2 \cos^2 t\right)\sin t}{b}$$

$$= \dfrac{1}{b}\sin t\,(b^2 - a^2 \sin^2 t - b^2 \cos^2 t)$$

$$= -\left(\dfrac{a^2 - b^2}{b}\right)\sin^3 t$$

Hence the required centre of curvature be

$$\left(\dfrac{a^2 - b^2}{b}\cos^3 t,\ \dfrac{b^2 - a^2}{b}\sin^3 t\right)$$

***Example* 32:** Find the co-ordinates of the centre of curvature for the point (x, y) on the parabola $y^2 = 4ax$ and show that their locus is $27ay^2 = 4(x - 2a)^3$.   (RGPV June 2003)

***Solution*:** The given curve is $y^2 = 4ax$

Differentiating, we get

$$\dfrac{dy}{dx} = \dfrac{2a}{y} = \dfrac{2a}{\sqrt{4ax}} = a^{1/2}x^{-1/2}$$

and $\dfrac{d^2y}{dx^2} = -\dfrac{1}{2}a^{1/2}x^{-3/2}$

$\therefore$ $\alpha = x - \dfrac{y_1\left(1 + y_1^2\right)}{y_2}$

$$= x - \dfrac{\left(1 + \dfrac{a}{x}\right)\sqrt{\dfrac{a}{x}}}{-\dfrac{1}{2}a^{1/2}x^{-3/2}}$$

$\alpha = x + 2(x + a) = 3x + 2a$ .....(i)

and $\beta = y + \dfrac{1 + y_1^2}{y_2}$

$$= y + \frac{1 + a/x}{-\dfrac{1}{2} a^{1/2} x^{3/2}}$$

$$= y - \frac{1}{2} a^{-1/2} x^{-3/2} (x + a)$$

$$= 2a^{1/2} x^{1/2} - 2a^{-1/2} x^{1/2} (x + a)$$

$$\beta = 2a^{-1/2} x^{1/2} \left[a - (x + a) \right] = -2a^{-1/2} x^{3/2} \qquad \qquad .....(ii)$$

Hence the required centre of curvature is $(3x + 2a, \ -2a^{-1/2} x^{3/2})$

To find evolute from equation (i)

$$x = \frac{\alpha - 2a}{3} \qquad \qquad .....(iii)$$

from equation (ii) $x^3 = \dfrac{\alpha \beta^2}{4} \qquad \qquad .....(iv)$

Putting the value of x from (iii) in (iv), we get

$$\left(\frac{\alpha - 2a}{3} \right)^3 = \frac{\alpha \beta^2}{4}$$

or $\quad 27\, \alpha\, \beta^2 = 4(\alpha - 2a)^3$

$\therefore$ the locus of (α, β) i.e., the equation of required evolute is

$$27xy^2 = 4(x - 2a)^3$$

***Example* 33:** Find the equation of circle of curvature at $\left(\dfrac{a}{4}, \dfrac{a}{4} \right)$ of the curve $\sqrt{x} + \sqrt{y} = \sqrt{a}$.

***Solution*:** Let the curve be $\sqrt{x} + \sqrt{y} = \sqrt{a}$

Differentiating, we get

$$\frac{dy}{dx} = -y^{1/2} x^{-1/2}$$

and $\quad \dfrac{d^2y}{dx^2} = \dfrac{\sqrt{a}}{2x\sqrt{x}}$

at the point $\left(\dfrac{a}{4}, \dfrac{a}{4} \right)$, we have $\dfrac{dy}{dx} = -1$ and $\dfrac{d^2y}{dx^2} = \dfrac{4}{a}$ and $\rho = \dfrac{a}{\sqrt{2}}$

$$\therefore \quad \alpha = x - \frac{y_1\left(1+y_1^2\right)}{y_2}$$

$$= \frac{a}{4} - \frac{(-1)(1+1)}{4\!\!\big/_a} = \frac{3a}{4}$$

and $\quad \beta = y + \dfrac{1+y_1^2}{y_2}$

$$= \frac{a}{4} + \frac{1+1}{4\!\!\big/_a} = \frac{3a}{4}$$

Hence the equation of circle of curvature at $\left(\dfrac{a}{4}, \dfrac{a}{4}\right)$ is

$$(x-\alpha)^2 + (y-\beta)^2 = \rho^2$$

or $\quad \left(x - \dfrac{3a}{4}\right)^2 + \left(y - \dfrac{3a}{4}\right)^2 = \dfrac{a^2}{2}$

or $\quad \delta(x^2+y^2) - 12\,a\,(x+y) + 5a^2 = 0$

***Example* 34:** Find the evolute of the curve $x = a(\theta + \sin\theta)$, $y = a(1 - \cos\theta)$.

***Solution*:** The given curve be $x = a(\theta + \sin\theta)$, $y = a(1 - \cos\theta)$

Differentiating, we get

$$\frac{dx}{d\theta} = a\left(1+\cos\theta\right), \; \frac{dy}{d\theta} = a\sin\theta$$

$$\therefore \quad \frac{dy}{dx} = \frac{a\sin\theta}{a\left(1+\cos\theta\right)} = \tan\frac{\theta}{2}$$

and $\quad \dfrac{d^2y}{dx^2} = \dfrac{1}{4a\cos^4\left(\frac{\theta}{2}\right)}$

$$\therefore \quad \alpha = x - \frac{y_1\left(1+y_1^2\right)}{y_2}$$

$$= a\left(\theta+\sin\theta\right) - \frac{\tan\frac{\theta}{2}\left(1+\tan^2\frac{\theta}{2}\right)}{\dfrac{1}{4a\cos^4\frac{\theta}{2}}}$$

$$= a\left(\theta+\sin\theta\right) - \frac{\sin\frac{\theta}{2}}{\cos\frac{\theta}{2}}\cdot\frac{1}{\cos^2\frac{\theta}{2}}\cdot 4a\cos^4\frac{\theta}{2}$$

$$= a(\theta + \sin\theta) - 4a\sin\frac{\theta}{2}\cos\frac{\theta}{2}$$

$$= a(\theta + \sin\theta) - 2a\sin\theta = a(\theta - \sin\theta)$$

and $\quad \beta = y + \dfrac{1 + y_1^2}{y_2}$

$$= a(1 - \cos\theta) + \frac{1 + \tan^2\frac{\theta}{2}}{\dfrac{1}{4a\cos^4\frac{\theta}{2}}}$$

$$= a(1 - \cos\theta) + 4a\cos^2\frac{\theta}{2}$$

$$= a(1 - \cos\theta) + 2a(1 + \cos\theta) = a(3 + \cos\theta)$$

Hence, the equation of evolute is $x = a(\theta - \sin\theta)$, $y = a(3 + \cos\theta)$

Note: The equation of evolute of cycloid is an equal cycloid since the radius of the generating circle for both cycloid is a.

Practice Problems

1. Find the radius of curvature at (s, ψ) on the catenary $s = c\tan\psi$.

   [**Ans.** $c\,\mathrm{Sec}^2\psi$]

2. Find the radius of curvature at (x, y) on the curve $a^2 y - x^3 - a^3$.

   [**Ans.** $\dfrac{\left(a^4 + 9x^4\right)^{3/2}}{6xa^4}$]

3. Find the radius of curvature at (x, y) on $y = c\log\sec\frac{x}{c}$.

   [**Ans.** $c\sec\frac{x}{c}$]

4. In the curve $y = ae^{x/a}$, Prove that $\rho = a\sec^2\theta\,\csc\theta$, where $\theta = \tan^{-1}\frac{y}{a}$.

5. Find the radius of curvature at the point 't' on the curve $x = a\cos t$, $y = b\sin t$.

   [**Ans.** $\dfrac{\left[a^2\sin^2 t + b^2\cos^2 t\right]^{3/2}}{ab}$]

6. If ρ_1 and ρ_2 be the radii of curvature at the extremities of a focal chord of a parabola $y^2 = 4ax$, prove that $(\rho_1)^{-2/3} + (\rho_2)^{-2/3} = (2a)^{-2/3}$.

7. Find the radius of curvature at the point (P, r) on the ellipse $\dfrac{1}{P^2} = \dfrac{1}{a^2} + \dfrac{1}{b^2} - \dfrac{r^2}{a^2 b^2}$.

$$[\text{Ans. } a^2 b^2 / P^3]$$

8. Find the radius of curvature at the point (P, r) on the cardioid $\qquad r^3 = 2ap^2$

$$[\text{Ans. } 2\sqrt{2ar}/3]$$

9. Find the radius of curvature at any point (r, θ) on the cardioid $r = a(1 + \cos \theta)$.

$$[\text{Ans. } \dfrac{4a}{3} \cos \theta/2]$$

10. Find the radius of curvature of the rectangular hyperbola $r^2 \cos 2\theta = a^2$.

$$[\text{Ans. } r^3 / a^2]$$

11. Find the radius of curvature at (r, θ) on the curve $\dfrac{l}{r} = 1 + \rho \cos \theta$.

$$[\text{Ans. } \dfrac{l \left[1 + e^2 + 2e\cos\theta \right]^{3/2}}{(1 + e\cos\theta)^3}]$$

12. Find the radius of curvature of the curve $\dfrac{2a}{r} = 1 + \cos\theta$, hence show that the square of the radius of curvature varies as the cube of the focal distance.

$$[\text{Ans. } \rho = 2r\sqrt{\dfrac{r}{a}}]$$

13. For any curve prove that $\dfrac{r}{\rho} = \sin\phi \left(1 + \dfrac{d\phi}{d\theta} \right)$, where ρ is the radius of curvature and $\tan\phi = r\dfrac{d\theta}{dr}$.

14. Find the radius of curvature for the curve $x = c \log \left\{ s + \sqrt{s^2 + c^2} \right\}, y = \sqrt{s^2 + c^2}$.

$$[\text{Ans. } y^2 / c]$$

15. Find the radius of curvature at the origin of the curve $y = x^3 + 5x^2 + 6x$. $[\text{Ans. } \dfrac{37\sqrt{37}}{10}]$

16. Find the curvature at the origin for the curve $5x^3 + 7y^3 + 4x^2 y + xy^2 + 2x^2 + 3xy + y^2 + 4x = 0$.

$$[\text{Ans. } 2]$$

17. Find the radii of curvature at the origin of the two branches of the curve given by the equations $x = 1 - t^2$, y $t - t^3$.

[**Ans.** $2\sqrt{2}$]

18. Show that the chord of curvature through the pole for the curve $P = f(r)$ is $2f(r)/\ f'(r)$.

19. Find the chord of curvature through the pole of the curve $r^2 \cos 2\theta = a^2$.

[**Ans.** 2r]

20. Show that the chord of curvature through the pole of the equi-angular spiral $r = ae^{\theta\cot\alpha}$ is 2r.

21. Show that in a parabola the chord of curvature through the focus of a parabola is four times the focal distance of the point, and chord of curvature parallel to the axis has the same length.

[Hint: eqn of parabola is $\dfrac{2a}{r} = 1 + \cos\theta$]

22. Find the centre of curvature of the catenary $y = c\cosh\left(x/c\right)$.

[**Ans.** 2y]

23. Find the co-ordinate of centre of curvature of the curve $a^2 y = x^3$.

$$\left[\textbf{Ans. } x/2\left(1 - \frac{9x^4}{a^4}\right), \frac{5x^3}{2a^2} + \frac{a^2}{6x}\right]$$

24. Find the centre of curvature at the point $(1, -1)$ of the curve

$y = x^3 + 6x^2 + 3x + 1$. $\left[\textbf{Ans. } \left(-36, \dfrac{-43}{6}\right)\right]$

25. Find the circle of curvature at the origin for the curve $x + y = x^2 + y^2 + x^3$.

[**Ans.** $x^2 + y^2 = x + y$]

26. Find the evolute of the curve $x = a(\cos t + \log \tan t/2)$, $y = a \sin t$.

$$\left[\textbf{Ans. } (a \log \tan t/2, a/\sin t), y = a \cos h\left(x/a\right)\right]$$

27. Find the evolute of the curve $x = a(\cos\theta + \theta\sin\theta)$, $y = a(\sin\theta - \theta\cos)$

[**Ans.** $(a\cos\theta, a\sin\theta)$, $x^2 + y^2 = a^2$]

28. Prove that the circle of curvature of the curve $x^3 + y^3 = 3xy$ at $\left(3/2, 3/2\right)$ is

$$\left(x - \frac{21}{16}\right)^2 + \left(y - \frac{21}{16}\right)^2 = \left(\frac{3}{8\sqrt{2}}\right)^2.$$

CHAPTER – 8

Curve Tracing

The process of determining the shape, location, singular points and other particulars of a curve from its equations is called curve tracing.

This will help us to know the area, the length, the volume, centre of gravity etc., of plane curves.

8.1 Rules for Tracing of Cartesian Curves

I   **Detect symmetry with the help of following rules:**

    (i)   A curve is symmetrical about x-axis if its equation has only even powers of y (i.e., if by putting $-y$ for y the equation remains unchanged) e.g., $y^2 = 4ax$ is symmetrical about x-axis.

    (ii)   A curve is symmetrical about y-axis if its equation contains only even powers of x, e.g., $x^2 = 4ay$ is symmetrical about y-axis.

    (iii)   If the equation of the curve contains only even powers of x and y, it is symmetrical about both axes, e.g., $x^2 + y^2 = a^2$ is symmetrical about both axes.

    (iv)   If on interchanging x and y the equation of the curve remains unchanged, it is symmetrical about the line $y = x$, e.g., $x^3 + y^3 = 3axy$ is symmetrical about the line $y = x$.

II   **Check whether the curve passes through the origin**

For this put $x = 0$, $y = 0$ in the equation of the curve; if the equation is satisfied, the curve passes through the origin, e.g., $x^2 = 4ay$ passes through the origin.

III   **Find the equation of the tangent or tangents at origin**

If the curve passes through the origin, find the equation of the tangent or tangents at origin. For this, equate the lowest-degree term in the equation to zero, e.g., in the curve $x^2 = 4ay$, the lowest-degree term is 4ay. Equating it to zero, we get $y = 0$, i.e., x-axis is the tangent at origin.

IV   Find the point of intersection of the curve and the axes

To know the point of intersection of the curve and the x-axis, put $y = 0$ in the equation of the curve. To know the point of intersection of the curve with y-axis, put $x = 0$ in the equation of the curve.

V   Find the regions of the curve

For this, solve the equation for y. Find such values of x for which y becomes imaginary. For example, if the curve is

$$y^2 (2a - x) = x^3 \text{ then } y^2 = \frac{x^3}{2a - x}$$

If $x < 0$, y is imaginary and if $x > 2a$, y is again imaginary. So, no part of the curve is either in the left of the origin or in the right of $x = 2a$.

See the behavior of y (or x) for different values of x (or y) giving particular importance to those values for which y (or x) tends to infinity or zero.

VI   Find $\dfrac{dy}{dx}$ and the points where tangents are parallel to axes.

VII Find the asymptote or asymtotes, if any

Equating to zero the coefficient of highest power of y we get asymptote parallel to y-axis and equating the coefficient of highest power of x to zero we get asymptote parallel to x-axis.

For example, if the curve be of nth degree and the term containing y^n be absent then the coefficient of y^{n-1} equated to zero will give asymptote parallel to y-axis.

If the equation of the curve be of nth degree and the coefficient of x^n is not zero then there will be no asymptote parallel to x-axis. Similarly, if the coefficient of y^n is not zero then there will be no asymptote parallel to y-axis. For example, $x^3 + y^3 = 3axy$ will have no asymptote parallel either to x-axis or y-axis.

Problems

***Example* 1: Trace the curve $y^2 (2a - x) = x^3$.**

    (i)   Since only even power of y occurs, so the curve is symmetrical about x-axis.

    (ii)   Since $x = 0$, $y = 0$ satisfies the equation, so the curve passes through origin.

    (iii) Equating the lowest-degree term to zero, the equation of tangent at origin is $2ay^2 = 0$, i.e., $y^2 = 0$, i.e., two coincident tangents at origin. So, origin is a cusp.

    (iv) Putting $x = 0$, we get $y = 0$ and putting $y = 0$ we get $x = 0$. So, the curve does not cut the axes at any point other than the origin.

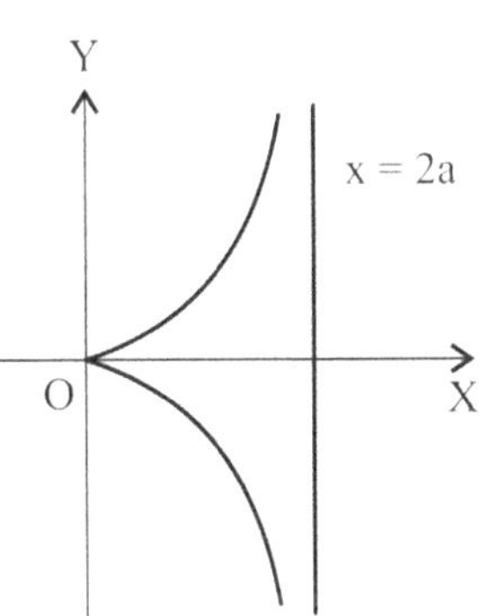

(v)   Solving the equation of the curve,

$$y^2 = \frac{x^3}{2a - x}$$

Clearly, y^2 is –ve, i.e., y is imaginary if $x > 2a$

So, no portion of curve is beyond the line $x = 2a$

Also y is imaginary if x is –ve. So, no portion of the curve is in the left of $x = 0$, (i.e., y-axis).

At $x = 2a$, $y = \pm\,\infty$, so the curve goes to infinity

(vi)   Equating to zero the coefficient of the highest power of y we get $2a - x = 0$, i.e., $x = 2a$. So the asymptote parallel to y-axis is given by $x = 2a$.

Hence the curve is as shown in the figure. The curve is called a cissoid.

***Example* 2: Trace the curve $y^2\,(a - x) = x^2\,(a + x)$.**

(i)   The curve is symmetrical about x-axis since only even power of y occurs.

(ii)   Putting $x = 0$, $y = 0$, the equation is satisfied. So, it passes through the origin.

(iii)   Equating lowest-degree term to zero, we get $y^2 - x^2 = 0$, i.e., $y = \pm\,x$. So, $y = \pm\,x$ are two real and distinct tangents at origin. So, origin is a node.

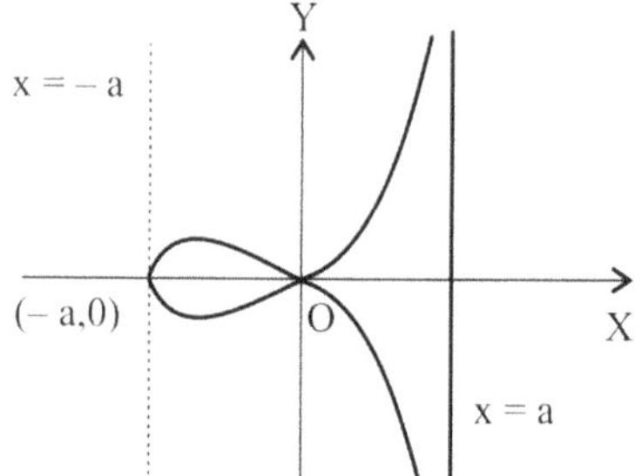

(iv)   Putting $y = 0$ in the equation of the curve we get $x^2(a + x) = 0$; $\therefore$ $x = 0$ and $x = -a$. So, the curve cuts x-axis at $(-a, 0)$ and $(0, 0)$. Again putting $x = 0$, we get $y = 0$. So, the curve cuts y-axis at $(0, 0)$

(v)   Solving for y, we get

$$y^2 = \frac{x^2\,(a + x)}{a - x};$$

$$\therefore \qquad y = x\sqrt{\frac{a + x}{a - x}}$$

Clearly y is imaginary when $x < -\,a$ and when $x > a$. So, no portion of the curve is in the left of the line $x = -\,a$, and in the right of $x = a$. Also $y = 0$ when $x = 0$ and $x = -\,a$, and $y^2 > 0$ if $-\,a < x < 0$, so there is a loop between $x = 0$ and $x = -\,a$.

(vi)   Equating the coefficient of the highest power of y to zero, $a - x = 0$, $x = a$ is the equation of the asymptote parallel to y-axis.

(vii)   At $(-\,a, 0)$, $\dfrac{dy}{dx} = \infty$, i.e., tangent at $(-\,a, 0)$ is at right angles to the axis of x.

Hence the curve is as shown.

***Example* 3: Trace the curve $y^2\,(a + x) = x^2\,(3a - x)$.**

(i)   Since only even powers of y occurs, the curve is symmetrical about x-axis.

(ii)   The curve passes through origin and tangent at origin is given by $y^2 = 3x^2$, which gives two real and distinct tangents, their equations being

$$y = \pm\sqrt{3}x .$$

(iii)   Putting $y = 0$, $x^2(3a - x) = 0$; $\therefore$ $x = 0$, 3a. So, the curve cuts x-axis at $(0, 0)$ and $(3a, 0)$. Putting $x = 0$, we get $y = 0$, so the curve cuts y-axis only at $(0, 0)$.

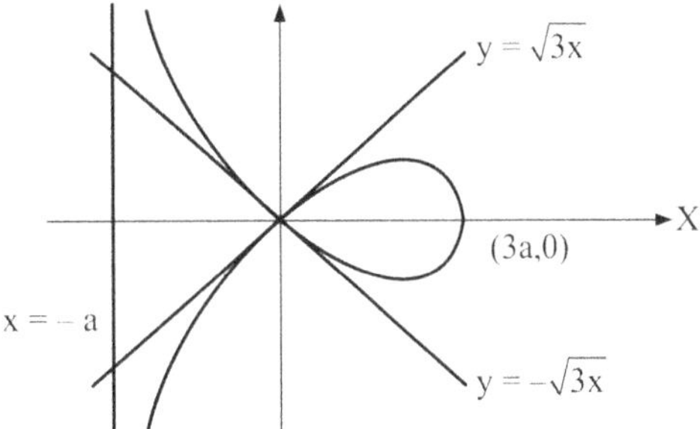

(iv)   Equating the coefficient of the highest power of y to zero, the asymptote parallel to y-axis is $x = -a$.

(v)   Solving for y, $y = \pm x \sqrt{\dfrac{3a - x}{a + x}}$. If $x > 3a$ or $x < -a$, y is imaginary. So, no portion of the curve is in the right of $x = 3a$ or in the left of $x = -a$.

Hence the curve is as shown.

***Example* 4: Trace the curve $x^3 + y^3 = 3axy$.**

(Folium of Descartes)

(i)   Interchanging x and y, the equation remains unaltered. So, the curve is symmetrical about the line $y = x$.

(ii)   There is no constant term in the equation of the curve, so it passes through the origin.

(iii)   Equating the lowest-degree term to zero we get $3axy = 0$; i.e., $x = 0$, $y = 0$. So, at origin both axes are tangent. So origin is node.

(iv)   Putting $y = 0$ in the equation of the curve, we get $x = 0$, and putting $x = 0$ we get $y = 0$. So, the curve cuts the axes at origin only.

(v)   x and y cannot both be negative since then LHS = $-$ve and RHS = $+$ve. Hence, no portion of the curve is in the 3rd quadrant (where x and y have negative values).

(vi)   Highest degree is 3 (i.e., $n = 3$)

Putting $x = 1$, $y = m$ in third-degree terms, we have $\varphi_3(m) = 1 + m^3$, $\varphi_2(m) = -3am$ (putting $x = 1$, $y = m$ in 2nd degree terms).

$\varphi_3(m) = 0$ gives $1 + m^2 = 0$, i.e., $(1 + m)(1 - m + m^2) = 0$

$\therefore$ $m = -1$, is the only real value of m

Now $c = \dfrac{\varphi_{n-1}(m)}{\varphi_n'(m)} = -\dfrac{\varphi_2(m)}{\varphi_3'(m)} = -\dfrac{-3am}{3m^2} = \dfrac{a}{m} = -a$, for $m = -1$

$\therefore$ the only asymptote is $y = -x - a$, i.e., $x + y + a = 0$

(vii)   Putting x = y in the equation of the curve we find that the line y = x cuts the curve

at the point $\left(\dfrac{3a}{2}, \dfrac{3a}{2}\right)$

(viii)   Differentiating the equation of the curve,

$$3x^2 + 3y^2 \frac{dy}{dx} = 3a\left(y + x\frac{dy}{dx}\right)$$

or $\quad 3\left(y^2 - ax\right)\dfrac{dy}{dx} = 3ay - 3x^2$

$\therefore \qquad \dfrac{dy}{dx} = \dfrac{ay - x^2}{y^2 - ax}$

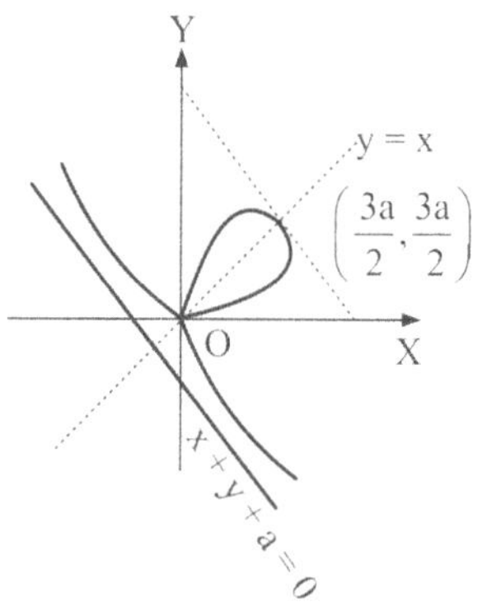

At $\quad \left(\dfrac{3a}{2}, \dfrac{3a}{2}\right), \dfrac{dy}{dx} = \dfrac{a \cdot \dfrac{3a}{2} - \dfrac{9a^2}{4}}{\dfrac{9a^2}{4} - a\dfrac{3a}{2}}$

$$= \dfrac{\dfrac{3a^2}{2} - \dfrac{9a^2}{4}}{\dfrac{9a^2}{4} - \dfrac{3a^2}{2}} = -1$$

So, the tangent at this point makes an angle 135° with the positive direction of x-axis.

Hence the curve is as shown.

Example 5: Trace the curve $x^4 + y^4 = a^2 (x^2 - y^2)$.

(i)   Since only even powers of x and y occur, the curve is symmetrical about both axes.

(ii)   x = 0, y = 0 satisfy the equation of the curve, so the curve passes through the origin.

(iii)   Equating the term of lowest degree to zero we get $x^2 - y^2 = 0$, i.e., $y = \pm x$.

So the equations of tangent at origin are y = x, y = –x.

(iv)   Putting y = 0, $x^4 - a^2 x^2 = 0$, i.e., $x^2(x^2 - a^2) = 0$, i.e., x = 0, + a.

So, the curve cuts x-axis at three points (0, 0), (a, 0), (0, –a).

Again by putting x = 0, $y^4 = -a^2 y^2$ or $y^4 + a^2 y^2 = 0$

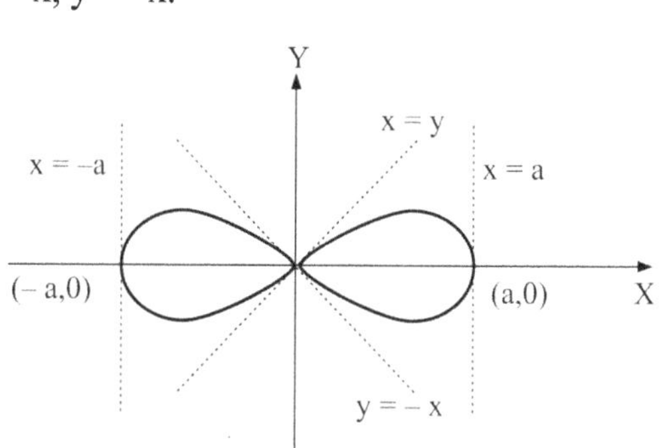

$\therefore$ y = 0, i.e., the curve cuts y-axis at the origin only

(v)   The curve has no asymptote.

(vi)   Solving the equation for y,

$$y^4 + a^2 y^2 = a^2 x^2 - x^4 = x^2(a^2 - x^2)$$

If $x > a$ or $x < -a$, y is imaginary

So, no portion of the curve is in the right of $x = a$ or in the left of $x = -a$.

(vii)   Differentiating the equation of the curve,

$$4x^3 + 4y^3 \frac{dy}{dx} = a^2 \left(2x - 2y\frac{dy}{dx} \right)$$

$$\therefore \quad \frac{dy}{dx} = \frac{2a^2 x - 4x^3}{4y^3 + 2a^2 y}$$

$$\therefore \quad \text{at } (a, 0),\ \frac{dy}{dx} = \infty \ \text{ and}$$

also at $(-a, 0)$, $\dfrac{dy}{dx} = \infty$.

So, at $(a, 0)$ and at $(-a, 0)$ tangent is perpendicular to x-axis. The curve is as shown.

Example 6: **Trace the curve $y^2 (a^2 + x^2) = x^2 (a^2 - x^2)$.**

(i)   The curve is symmetrical about both the axes, since there occur even powers of x and y both. On changing signs of x and y both the equation remains unaltered, so there is symmetry in opposite quadrants.

(ii)   The curve passes through the origin as by putting $x = 0$ and $y = 0$, the equation is satisfied.

(iii)   Equating the lowest-degree term to zero, the tangents at origin are given by $y^2 - x^2 = 0$, i.e., $y = \pm x$.

Hence the origin is a node.

(iv)   Putting $y = 0$, we have $x^2 (a^2 - x^2) = 0$, i.e., $x = 0$, $x = \pm a$. So, the curve cuts the x-axis at three points $(0, 0)$, $(a, 0)$ and $(-a, 0)$.

The curve does not cut y-axis at any point other than $(0, 0)$ since by putting $x = 0$, we get $y = 0$.

(v)   Equating the coefficient of highest power of y to zero. $a^2 + x^2 = 0$, giving asymptotes parallel to y-axis which are clearly imaginary. There is also no other real asymptote.

(vi)   Solving for y, we get

$$y^2 = \frac{x^2 \left(a^2 - x^2 \right)}{a^2 + x^2}$$

If $x > a$, y is imaginary. Also if $x < -a$, y is imaginary. So, no portion of the curve lies beyond $x = a$ and $x = -a$.

(vii)   Differentiating the equation of the curve w.r.t. x,

$$2y\frac{dy}{dx} = \frac{\left(2a^2 x - 4x^3\right)\left(a^2 + x^2\right) - 2x.x^2 \left(a^2 - x^2\right)}{\left(a^2 + x^2\right)^2}$$

Or $\quad 2\dfrac{dy}{dx} = \dfrac{2a^4x + 2a^2x^3 - 4a^2x^3 - 4x^5 - 2x^3a^2 + 2x^5}{\dfrac{x\left(a^2 - x^2\right)^{1/2}}{\left(a^2 + x^2\right)^{1/2}} \cdot \left(a^2 + x^2\right)^2}$

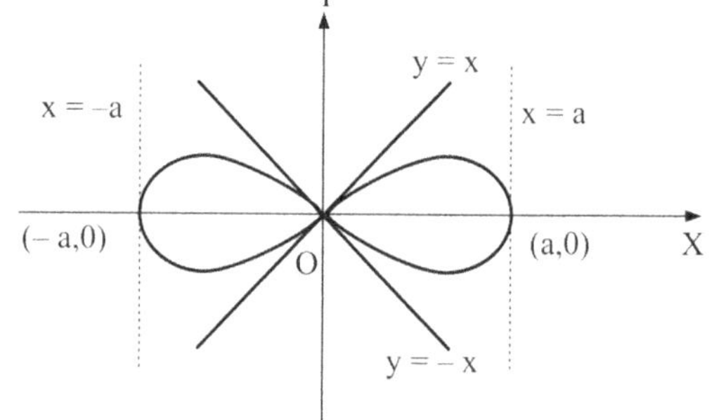

$\qquad\qquad = \dfrac{2x\left(a^4 - 2a^2x^2 - x^4\right)}{x\left(a^2 - x^2\right)^{1/2}\left(a^2 + x^2\right)^{3/2}}$

$\therefore \quad \dfrac{dy}{dx} = \dfrac{a^4 - 2a^2x^2 - x^4}{\left(a^2 - x^2\right)^{1/2}\left(a^2 + x^2\right)^{3/2}}$

$\qquad\qquad = \infty \quad \text{when } x = \pm a$

Thus at the points (a, 0) and (–a, 0), tangents are perpendicular to x-axis. Hence, the curve is as shown.

***Example* 7: Trace the curve** $\dfrac{a^2}{x^2} - \dfrac{b^2}{y^2} = 1$.

The curve is $a^2 y^2 - b^2 x^2 = x^2 y^2$

(i) Since even powers of x and y occur, so the curve is symmetrical about both axes.

(ii) Putting x = 0, y = 0, the equation of the curve is satisfied, so the curve passes through the origin.

(iii) Equating the lowest-degree term to zero, we get the equation of tangent at the origin as $a^2y^2 - b^2x^2 = 0$, i.e., $ay = \pm bx$

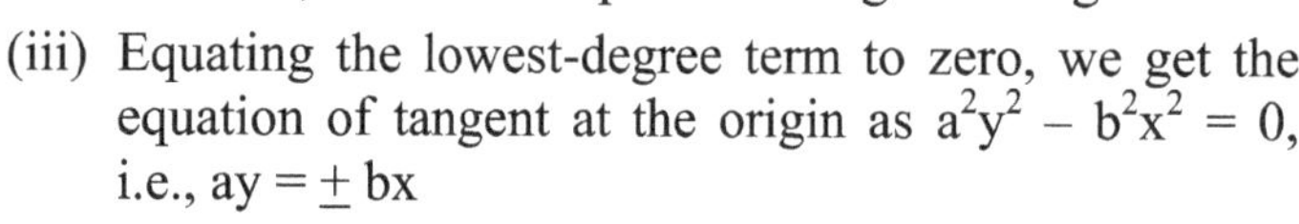
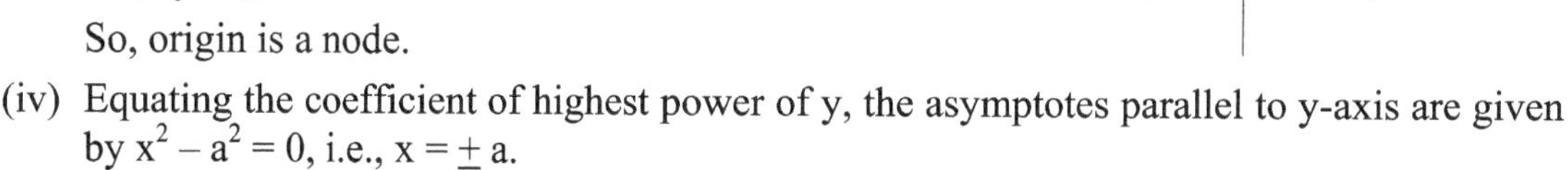

So, origin is a node.

(iv) Equating the coefficient of highest power of y, the asymptotes parallel to y-axis are given by $x^2 - a^2 = 0$, i.e., $x = \pm a$.

Equating the coefficient of highest power of x, $b^2 + y^2 = 0$, i.e., $y^2 = -b^2$, so asymptotes parallel to x-axis are imaginary.

(v) The curve does not intersect the axes at any point other than (0, 0).

(vi) Solving for y, $y^2\left(a^2 - x^2\right) = b^2 x^2$ $\quad$ or $\quad y^2 = \dfrac{b^2 x^2}{a^2 - x^2}$

$\therefore \qquad\qquad\qquad y = \pm \dfrac{bx}{\sqrt{a^2 - x^2}}$

So, y is real only when x lies between a and – a. So, the whole curve lies between x = a and x = – a.

As $x \to \pm a$, $y \to \pm \infty$. The origin is the point of inflexion as $\dfrac{d^2y}{dx^2} = 0$ at origin and $\dfrac{d^3y}{dx^3}$ is not zero at origin. So, the curve is as shown.

Example 8: Trace the curve $xy^2 + (x + a)^2 (x + 2a) = 0$.

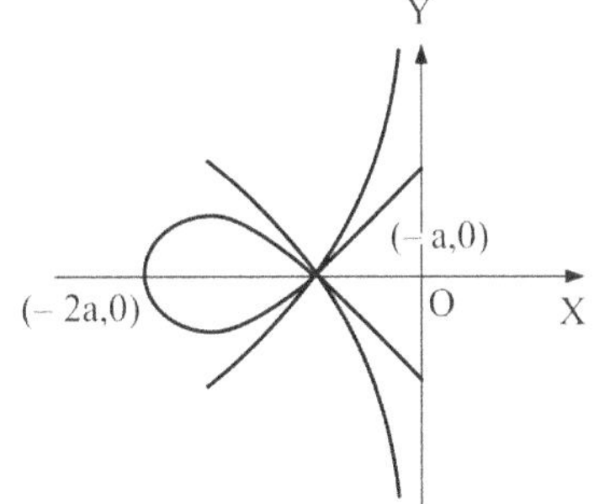

(i) The curve is symmetric about x-axis as only even powers of y occur.

(ii) The curve does not pass through the origin as $x = 0$, $y = 0$ do not satisfy the equation.

(iii) Putting $x = 0$, we get $2a^3 = 0$, which is impossible. So the curve never cuts the y-axis. Putting $y = 0$, we get $(x + a)^2 (x + 2a) = 0$.

∴ $x = -a$ and $x = -2a$. So the curve cuts x-axis at $(-a, 0)$ and $(-2a, 0)$.

(iv) Equating to zero the coefficient of highest power of y, i.e., of y^2, we get $x = 0$. So, the asymptote parallel to y-axis is $x = 0$.

(v) Solving the equation of the curve for y,

$$y^2 = \frac{(x+a)^2 (x+2a)}{x}$$

Clearly, y is imaginary if $x > 0$. Hence the curve lies in the left of $x = 0$, i.e., the y-axis. Also y is imaginary if $x < -2a$. So no portion of the curve lies in the left of $x = -2a$.

Thus using (iii), we find that there is a loop between $x = -2a$ and $x = -a$.

Tangents at $(-a, 0)$ are parallel to $y = \pm x$. The curve is as shown.

Practice Problems

1. Trace the curve (strophoid)

   $y^2 (a + x) = x^2 (a - x)$ or $(x^2 + y^2) x - a (x^2 - y^2) = 0$.

   [*Hint*: Use a in place of 3a in example 3]

2. Trace the curve $y^2 (x^2 + y^2) + a^2 (x^2 - a^2) = 0$.

   [*Hint*: Proceed just as example 6]

3. Trace the curve $a^2 y^2 = x(x - a)^2$

4. Trace the curve $ay^2 = x^2 (a - x)$.

5. Trace the cuve $a^2 y^2 = x^2 (a^2 - x^2)$.

6. Trace the semicubical parabola $y^2 = x^3$.

7. Trace the curve $(a - x) y^2 = a^2 x$.

   Note: Tracing itself contains hint. You have to only justify it.

Answers

1.

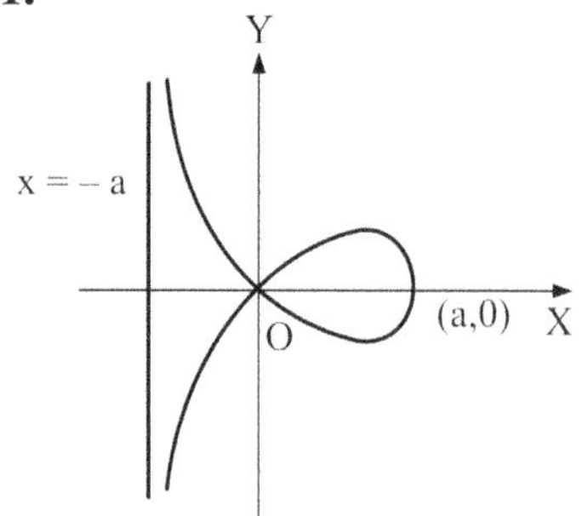

2.

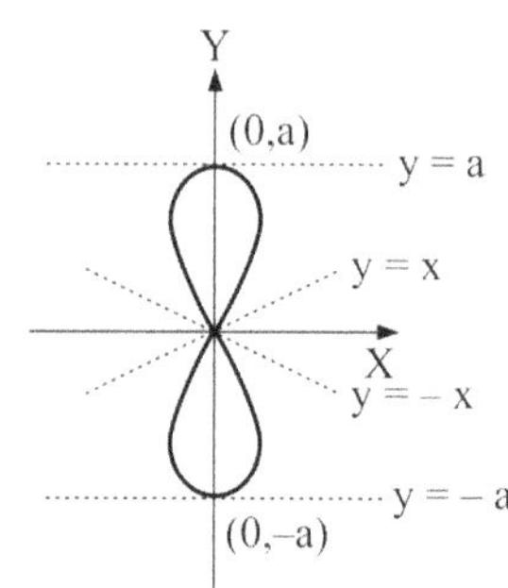

3.

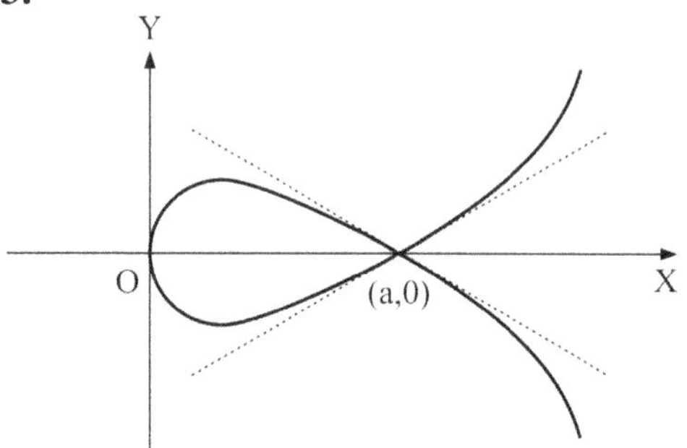

4.

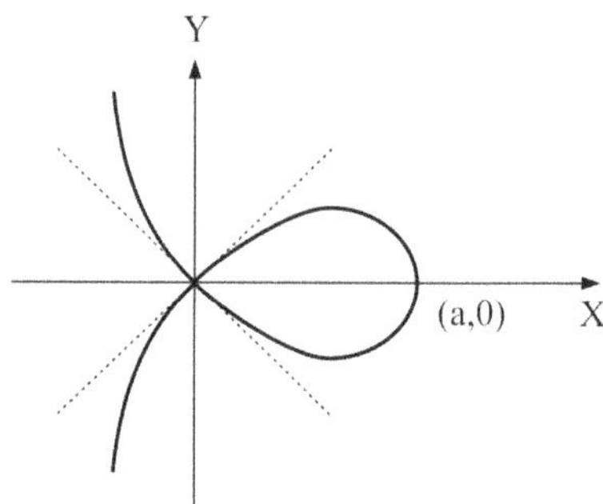

5.

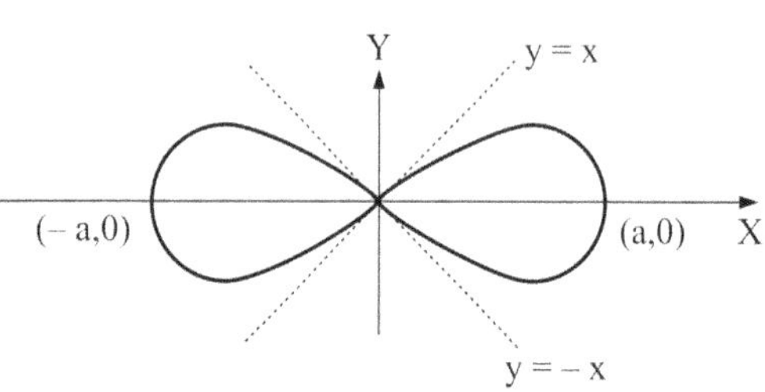

6.

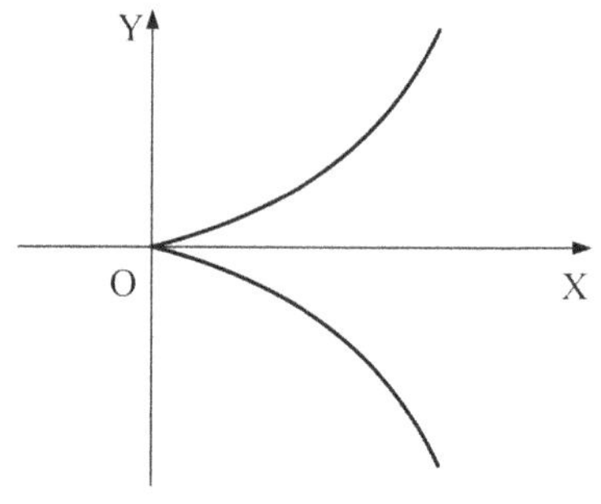

7.

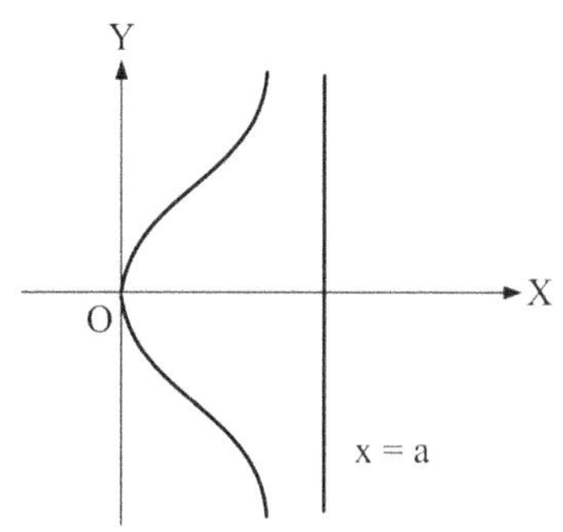

8.2 Tracing of Polar Curves

I.   Symmetry

    (i)   Replace θ by $-\theta$; if the equation of the curve remains unchanged, the curve is symmetrical about the initial line.

    (ii)   Replace r by $-$ r; if the equation of the curve is unchanged, i.e., if only even powers of r occur, the curve is symmetrical about the pole.

    (iii)   Replace θ by $\pi - \theta$; if the equation of the curve is unchanged, the curve is symmetrical about the line $\theta = \dfrac{\pi}{2}$ (i.e., symmetrical about OX).

    (iv)   Replace θ by $\dfrac{\pi}{2} - \theta$; if the equation remains unchanged, the curve is symmetrical about the line $\theta = \dfrac{\pi}{4}$.

II.   Whether the curve passes through the pole

If r = 0 for some real value of θ, the curve passes through the pole.

If the curve passes through the pole, the value of θ for which r = 0 gives the equation of the tangent at pole.

If r = 0 for more than one value of θ, the curve is said to have a loop.

III.   Intersection of the curve with the initial line OX and also with OY

For this, we know r by putting $\theta = 0$ and $\theta = \dfrac{\pi}{2}$ in the equation of the curve.

IV.   Region of the curve

To find the region in which no portion of the curve lies, we determine those values of θ for which r is imaginary.

V.   Asymptotes, if any

If the curve possesses never-ending branch, find its asymptote. This is sometimes best accomplished by transforming the equation to the Cartesian form.

VI.   Some more points

Some specific points on the curve should be plotted. For this we determine corresponding values of r for some special values of θ, e.g., $\theta = 0, \dfrac{\pi}{6}, \dfrac{\pi}{4}, \dfrac{\pi}{3}, \dfrac{\pi}{2}, \ldots$ and plot them.

VII.   Sometimes tracing of the curve becomes easy by transforming its equation into cartesian form.

Problems

***Example* 9:** Trace the cardiod $r = a\,(1 + \cos\theta)$.

  (i) By changing θ to $-\theta$, we get

$$r = a\,[1 + \cos(-\theta)] = a\,(1 + \cos\theta),$$

i.e., the equation does not change. So, the curve is symmetrical about the initial line.

  (ii) When $\theta = \pi$, $r = 0$, hence the curve passes through the pole and the equation of tangent at the pole is $\theta = \pi$, i.e., the initial line.

  (iii) Now we put some points on the curve. When $\theta = 0$, $r = 2a$; when $\theta = \dfrac{\pi}{3}$, $r = \dfrac{3}{2}a$; when $\theta = \dfrac{\pi}{2}$, $r = a$; when $\theta = \dfrac{2\pi}{3}$, $r = \dfrac{a}{2}$, when $\theta = \pi$, $r = 0$.

  (iv) When θ increases from 0 to π, r decreases from 2a to 0 as shown in (iii)

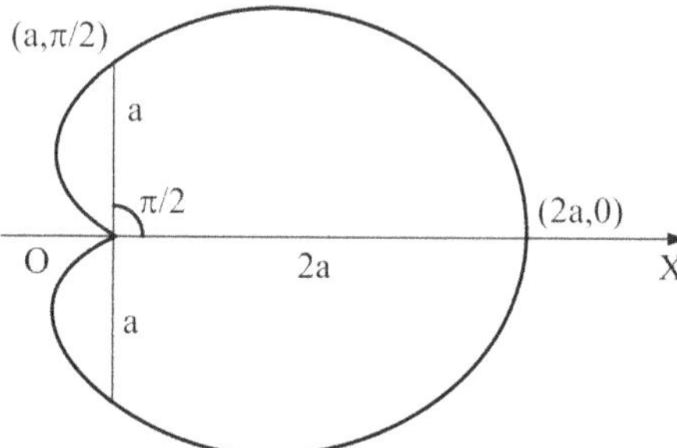

Also when θ increases from π to 2π, increases from 0 to 2a as the curve is symmetrical about the initial line. Hence, the curve is as shown.

***Example* 10:** Trace the curve $r = a + b\cos\theta$ (limacon). The following cases may arise:

Case I If $a > b$

  (i) The curve is symmetrical about the initial line since by changing θ to $-\theta$ the equation of the curve remains unchanged.

  (ii) r is always positive and is never greater than $a + b$.

  (iii) Since $a > b$, r can never be zero. So, the curve does not pass through the pole.

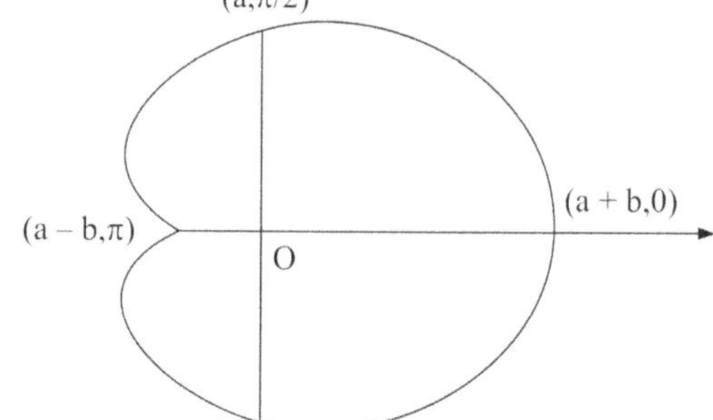

  (iv) Some points on the curve are evaluated. When $\theta = 0$, $r = (a + b)$; when $\theta = \dfrac{\pi}{2}$, $r = a$; when $\theta = p$, $r = (a - b)$; when $\theta = \dfrac{3\pi}{2}$, $r = a$; when $\theta = 2\pi$, $r = a + b$.

  (v) It is clear from (iv) that when θ increases from 0 to π, r decreases from $(a + b)$ to $(a - b)$. Also when θ increases from π to 2π, r increases from $(a - b)$ to $(a + b)$. Hence the curve is as shown.

Case II If $a = b$

In this case the curve becomes a cardoid which has been traced in the above sum.

Case III If $a < b$

(i)   The curve is symmetrical about the initial line.

(ii) r is never greater than $a + b$.

(iii) $r = 0$ when $\theta = \cos^{-1}\left(-\dfrac{a}{b}\right)$. So, the curve passes through the pole. But $\cos^{-1}\left(-\dfrac{a}{b}\right)$ has

two values in the region $\theta = 0$ to $\theta = 2\pi$, one value being in the second and the other in the third quadrant. If these values be θ_1, and θ_2 then the equation of tangents at the pole are $\theta = \theta_1$ and $\theta = \theta_2$. Here is a loop between $\theta = \theta_1$ and $\theta = \theta_2$.

(iv) Some points on the curve are taken below.

When $\theta = 0$, $r = (a + b)$;

when $\theta = \pi$, $r = -(b - a)$;

when $\theta = \dfrac{3\pi}{2}$, $r = a$;

when $\theta = 2\pi$, $r = (a + b)$

The curve is as shown

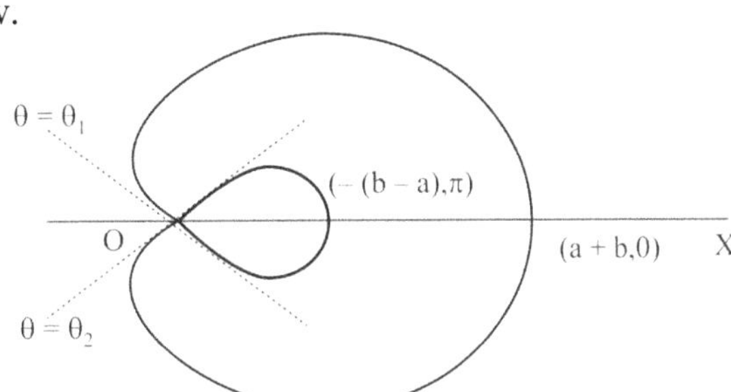

***Example* 11:** Trace the curve $r^2 = a^2 \cos 2\theta$. (lemnislate of Bernoulli)

(i)   (a)   Replacing θ by $-\theta$, the equation of the curve remains unchanged, so the curve is symmetrical about the initial line.

     (b)   Replacing r by $-r$, the equation of the curve remains unchanged, so the curve is symmetrical about the pole.

     (c)   Putting $\pi - \theta$ for θ, we get

$$r^2 = a^2 \cos^2(\pi - \theta) = a^2 \cos 2\theta,$$

     i.e., the equation of the curve remains unchanged, so the curve is symmetrical about the line $\theta = \dfrac{\pi}{2}$

(ii) $r = 0$ when $\cos 2\theta = 0$, i.e., $2\theta = \dfrac{\pi}{2}, \dfrac{3\pi}{2}$

$$\therefore \qquad \theta = \dfrac{\pi}{4}, \dfrac{3\pi}{4}$$

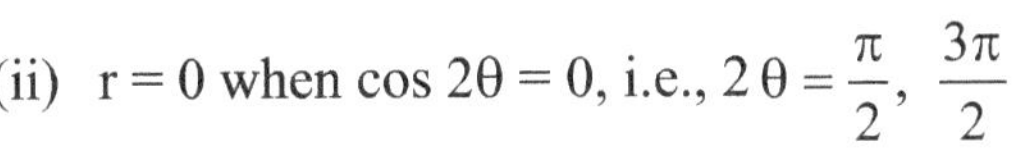

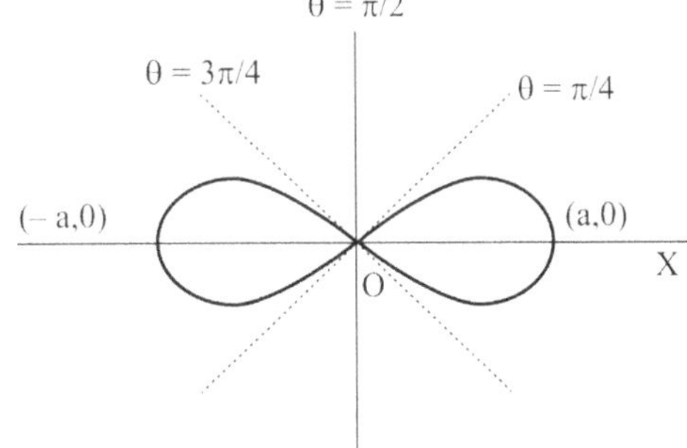

So, the curve passes through the pole and the equation of tangents at the pole are

$$\theta = \dfrac{\pi}{4}, \quad \theta = \dfrac{3\pi}{4}$$

(iii) When $\theta = 0$, $r^2 = a^2$, $\therefore$ $r = \pm\, a$. So, the points $(a, 0)$ and $(-a, 0)$ lie on the curve.

(iv)   When θ increases from 0 to $\dfrac{\pi}{4}$, r^2 decreases from a^2 to 0. Again when θ increases

from $\dfrac{3\pi}{4}$ to π, r^2 increases from 0 to a^2. But when θ lies between $\dfrac{\pi}{4}$ and $\dfrac{3\pi}{4}$ then

$r^2 = -$ve. So, r is imaginary. Hence no portion of the curve lies between $\theta = \pi/4$

and $\theta = \dfrac{3\pi}{4}$. The curve is as shown.

Note: Changing the equation in Cartesian form,

$r^2 = a^2 (\cos^2 \theta - \sin^2 \theta)$ or $r^4 = a^2 (r^2 \cos^2 \theta - r^2 \sin^2 \theta)$ i.e., $(x^2 + y^2)^2 = a^2 (x^2 - y^2)$.
Now the curve can be traced as Cartesian curve.

Example 12: Trace the curve $r = ae^{m\theta}$. (equiangular spiral)

(i) r is always positive and increases with θ
and when $\theta \to \infty$, $r \to \infty$.

(ii) r is never zero for any finite value of θ,
hence pole does not lie on the curve.

(iii) Some points on the curve are as follows:
When $\theta = 0$, $r = a$;

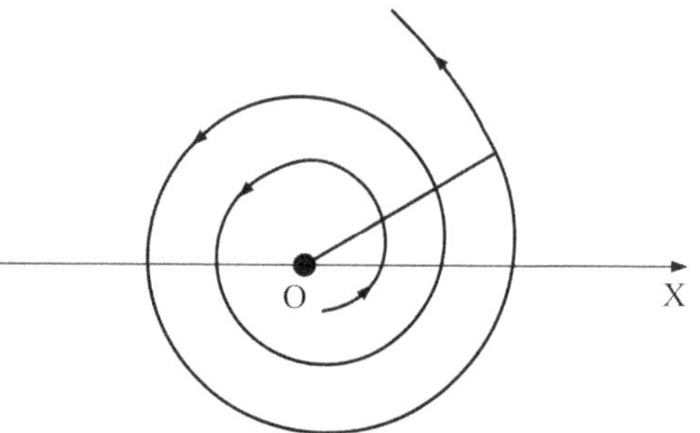

when $\theta = \dfrac{\pi}{2}$, $r = ae^{\frac{m\pi}{2}}$; when $\theta = \pi$,

$r = ae^{m\pi}$

Thus as θ increases from 0 to ∞, r increases from a to ∞.

(iv) Differentiating the equation of the curve w.r.t. θ, we get

$$\frac{dr}{d\theta} = ae^{m\theta} m = mr$$

$\therefore$ $$\tan \varphi = r\frac{d\theta}{dr} = \frac{r}{mr} = \frac{1}{m}$$

$\therefore$ $$\varphi = \tan^{-1}\left(\frac{1}{m}\right) = \text{constant}$$

So, at every point the angle between the radius vector and the tangent is constant. This is
way the curve is called equiangular spiral. The shape of the curve is as shown.

Example 13: Trace the curve $r = a \sin 3\theta$.

(i) The curve is symmetrical about the line $\theta = \pi/2$ since the equation remains unchanged
when θ is replaced by $\pi - \theta$.

(ii) $r = 0$ when $\sin 3\theta = 0$, i.e., when $3\theta = n\pi$

$\therefore \ \theta = \dfrac{n\pi}{3}$, i.e., r = 0 when $\theta = 0, \ \dfrac{\pi}{3}, \ \dfrac{2\pi}{3}, \ \pi,....$

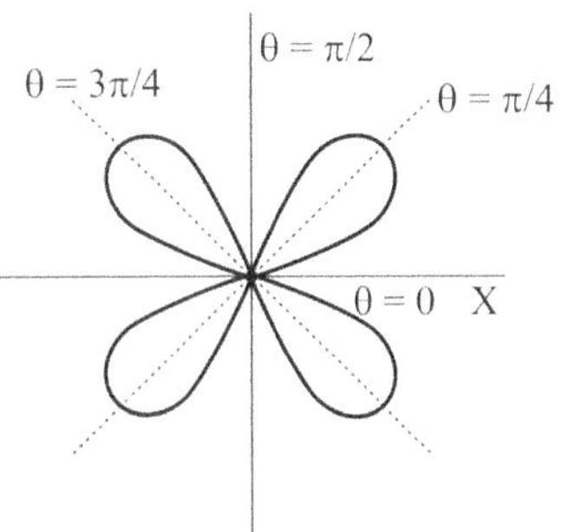

So, the curve passes through the pole and the equations of tangents at pole are

$$\theta = 0, \ \theta = \pi/3, \ \theta = \dfrac{2\pi}{3}$$

Also when $\theta = \pi/6$, r = a; when $\theta = \pi/2$, r = –a; when

$\theta = \dfrac{5\pi}{6}$, r = a.

(a) We find that as θ increases from 0 to $\pi/6$, r increases from 0 to a. As θ increases from $\pi/6$ to $\pi/3$, r decreases from a to 0. Thus there is a loop between $\theta = 0$ and $\theta = \pi/3$.

(b) when θ increases from $\pi/3$ to $\pi/2$, r becomes – a from 0. When θ increases from $\pi/2$ to $2\pi/3$, r becomes 0 from – a. Thus, there is second loop between $\theta = \pi/3$ and $\theta = 2\pi/3$.

(c) Similarly, there is third loop between $\theta = 2\pi/3$ and $\theta = \pi$. Thus, the curve is as shown.

***Example* 14:** Trace the curve r = a sin 2θ

(i) No part of the curve lies beyond the region r = a as the maximum value of r is a.

(ii) The curve is symmetrical about the line $\theta = \dfrac{\pi}{2}$.

(iii) The curve is symmetrical about $\theta = \dfrac{\pi}{4}$ and $\theta = \dfrac{3\pi}{4}$.

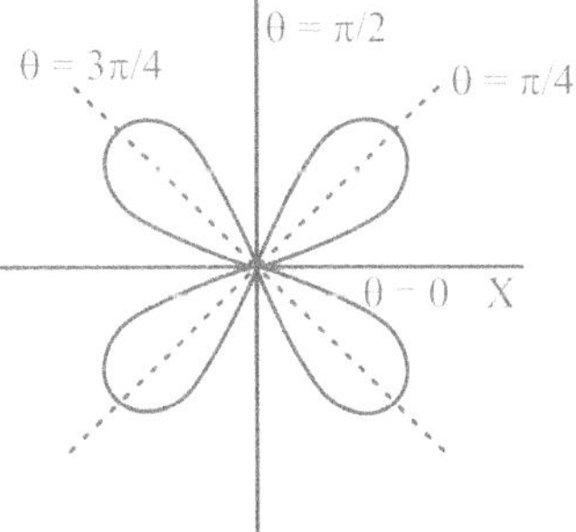

(iv) r = 0 when $\theta = 0, \ \dfrac{\pi}{2}, \ \pi,...$

The curve passes through the pole and tangents at pole are $\theta = 0$ and the $\theta = \dfrac{\pi}{2}$, the other values of θ giving the same tangents.

(v) When $\theta = \dfrac{\pi}{4}$, r = a; when $\theta = \dfrac{3\pi}{4}$, r = – a; when $\theta = \dfrac{5\pi}{4}$, r = a. Thus r increases from 0 to a as θ increases from 0 to $\dfrac{\pi}{4}$ and r decreases from a to 0 as θ increases from $\dfrac{\pi}{4}$ to $\dfrac{\pi}{2}$.

Thus there is a loop in the first quadrant. Similarly, there are loops each in 2nd, 3rd and 4th quadrant. Hence the curve is as shown.

Note: It may be noted that for r = a sin nθ or r = a cos nθ, there are n or 2n loops according as n is odd or even.

8.3   Some well-known Curves

The following curves should always be kept in mind.

Folium of Descartes

Its equation is $x^3 + y^3 = 3axy$. It is symmetrical about the line $y = x$. The axes are tangents at the origin and there is a loop in the first quadrant. It has an asymptote $x + y + a = 0$.

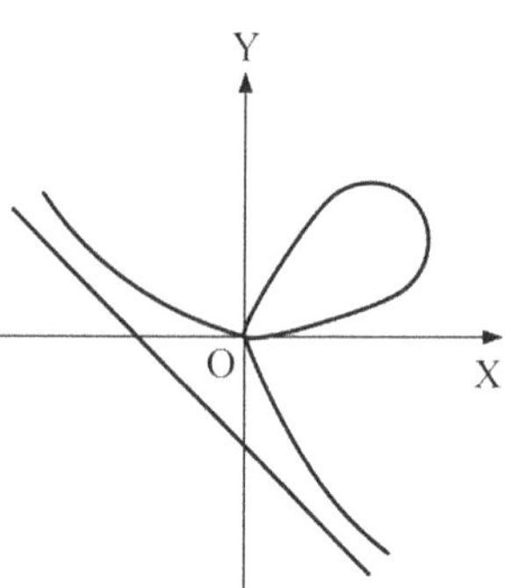

Cissoid of Diocles

Its equation is $y^2(2a - x) = x^3$ or $y^3 = \dfrac{x^3}{2a - x}$; $x = 2a$ is an asymptote.

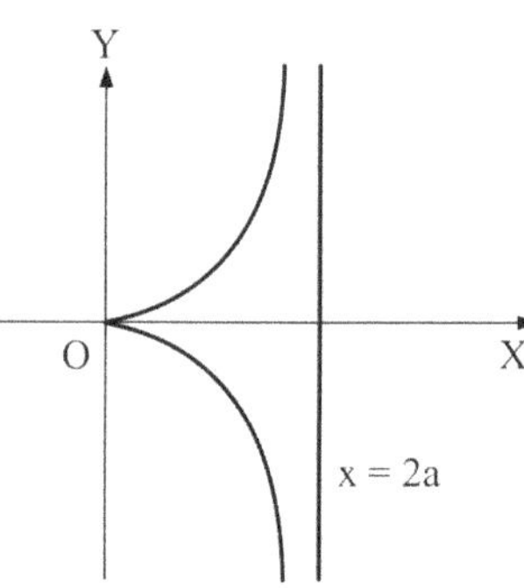

Strophoid

Its equation is $y^2 = x^2 \dfrac{a + x}{a - x}$ or $y(a - x) = x^2(a + x)$. OABCO is a loop $OD = OB = a$. $x = a$ is an asymptote.

The curve $y^2 = x^2 \dfrac{a - x}{a + x}$ is similar and just reverse of strophoid.

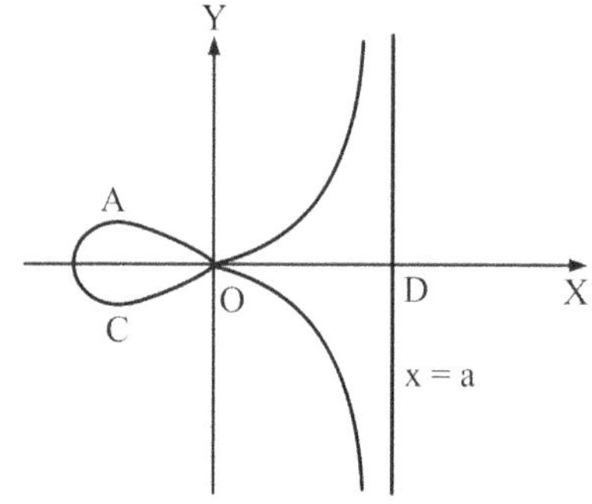

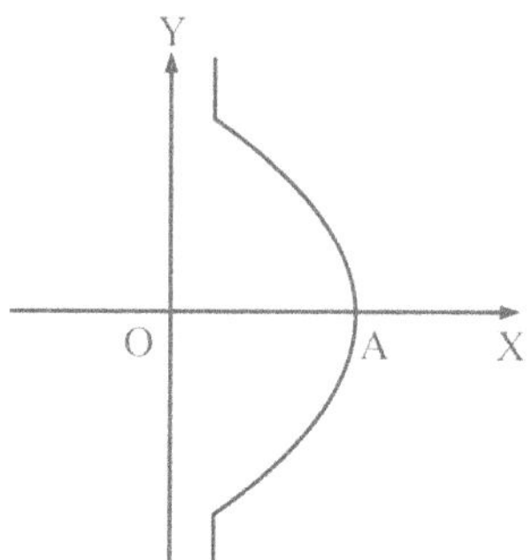

Witch of Agnesi

Its equation is $xy^2 = 4a^2(2a - x)$. Here $OA = 2a$.

Astroid

Its equation is $x^{2/3} + y^{2/3} = a^{2/3}$ or $x = a\cos^3 t$, $y = a\sin^3 t$. Here $OA = OB = OA' = OB' = a$. The whole figure lies completely within a circle of radius a and centre O. The points A, A', B' are called cusps. It is a special case of a four-cusped hypocycloid described below.

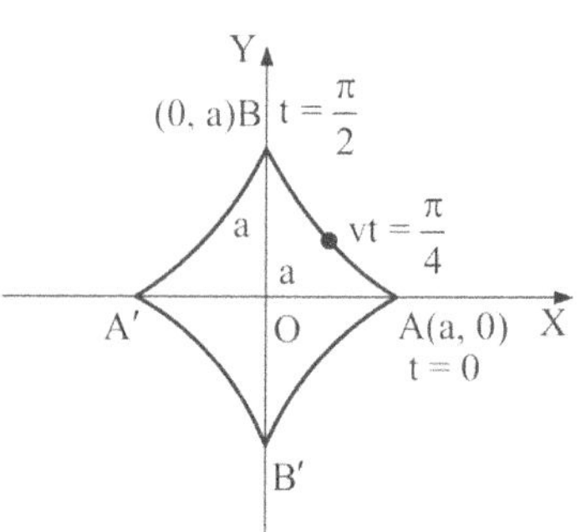

For cusp A, t = 0; for B, t = $\pi/2$;

For vertex V, t = $\pi/4$.

Four Cusped hypocycloid

Its equation is $\left(\dfrac{x}{a}\right)^{2/3} + \left(\dfrac{y}{b}\right)^{2/3} = 1$

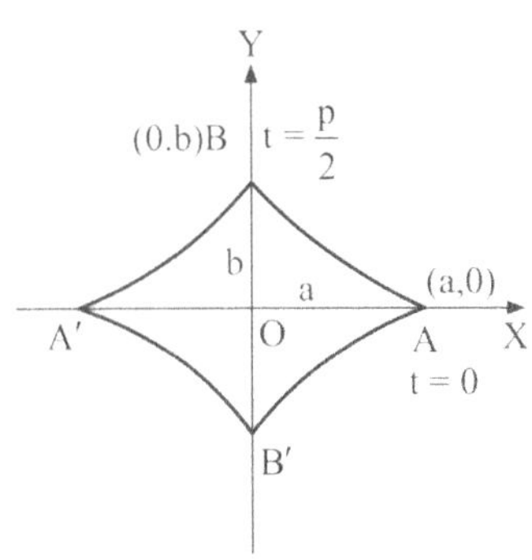

or $x = a\cos^2 t,\ y = b\sin^3 t$.

Here OA = OA′ = a; OB = OB′ = b . When a = b, it is astroid.

Cycloid

The cycloid is the curve traced out by a point on the circumference of a circle which rolls (without sliding) on a straight line. The parametric equations of the cycloid with the starting point as the origin and the line on which the circle rolls, called the base, as x-axis are

$$x = a(\theta - \sin\theta),\quad y = a(1 - \cos\theta)$$

For O and O′, y = 0, $\therefore$ cos θ = 1, $\therefore \theta = 0$ and 2π and x = 0 and $2a\pi$.

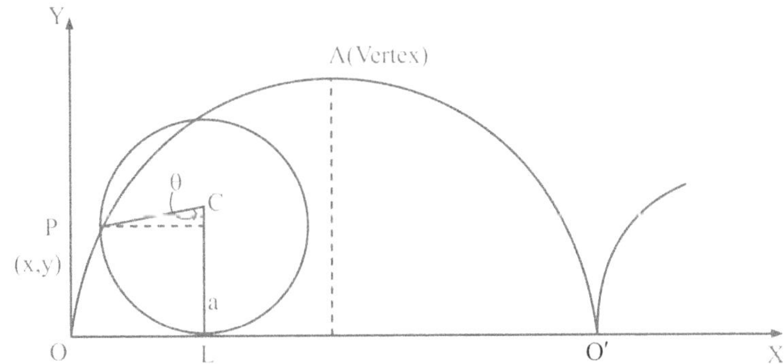

When the circle rolls on, arches like OAO′ are generated over and over again and any single arch is called a cycloid.

The equation of the cycloid with the vertex as origin and tangent at the vertex as x-axis is

$$x = a(\theta + \sin\theta),\ y = a(1 - \cos\theta)$$

For vertex A, θ = 0; for O. θ = π and for O′, θ = $-\pi$

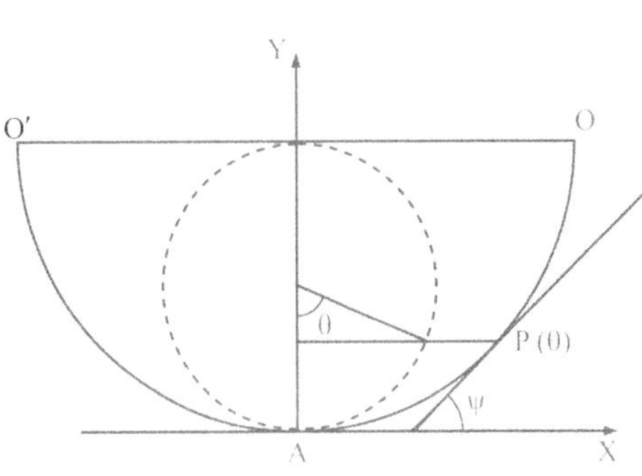

Catenary

The catenary is the curve in which a uniform heavy string hangs under the action of gravity when suspended freely from two points.

Its equation is $y = c \cos h \dfrac{x}{c} = \dfrac{c}{2}\left(e^{x/c} + e^{-x/c}\right)$.

C is called vertex, OC = c, OX is called directrix.

Logarithmic and Exponential Curves

In $y = \log x$, x is always positive; $y = 0$ when $x = 1$ and as x becomes smaller and smaller, y, being negative, becomes numerically larger and larger. For $x > 0$ the curve is continuous.

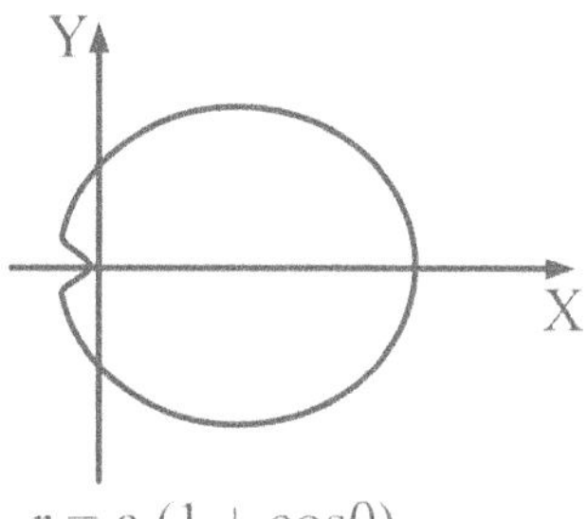

$$r = a\,(1 + \cos\theta)$$

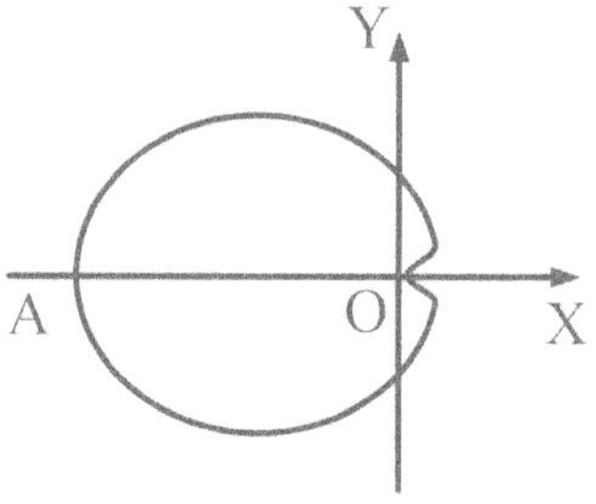

$$r = a\,(1 - \cos\theta)$$

$y = e^x$, x may be positive or negative but y is always positive. y becomes smaller and smaller as x, being negative, becomes numerically larger and larger. The curve is continuous for all values of x.

Limacon

Its equation is $r = a + b\cos\theta$. When $a > b$, we have the outer curve when $a < b$ we have the inner curve with a loop. When $a = b$ the curve becomes a cardioide.

Lemniscate

Its equation is $r^2 = a^2 \cos 2\theta$, or $(x^2 + y^2) = a^2(x^2 - y^2)$. It consists of two equal loops. Each loop is symmetrical about the initial line, which divides each loop into two equal halves. $OA = OA' = a$. The tangent at origin are $y = \pm x$. For the upper half of the right-hand loop, θ varies from 0 to $\pi/4$.

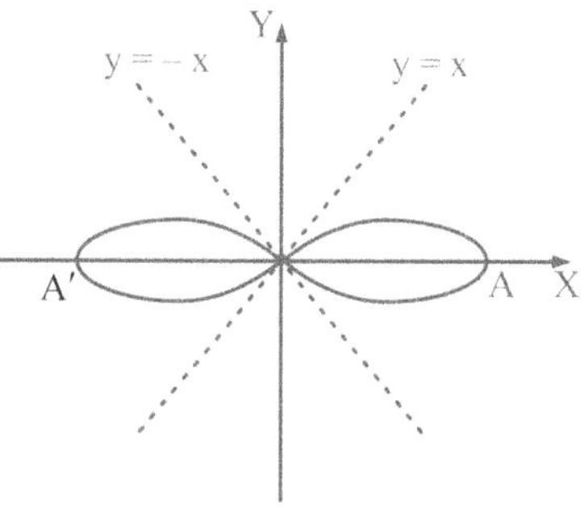

Rose or Rose Petals

Its equation in general form is

$$r = a \sin n\theta,$$

or
$$r = a \cos n\theta.$$

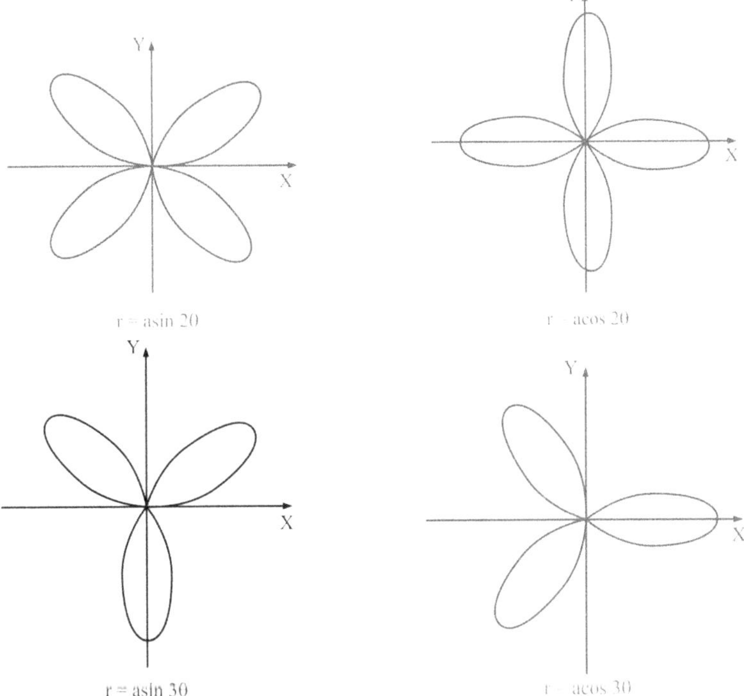

If n is odd, the number of loops is n and if n is even, the number of loops is 2n.

Practice Problems

1. Trace the curve $r = a(1 - \cos \theta)$. (*Hint*: Proceed as example 1]

2. Trace the curve $r^2 = 4 \cos 2\theta$.(*Hint*: Put a = 2 in example 3, curve shape is as in example 3)

3. Trace the curve $r = a \cos 2\theta$. (*Hint*: Proceed as example 6)

4. Trace the curve $r = a \cos 3\theta$. (*Hint*: Proceed as example 5)

Answers

1.

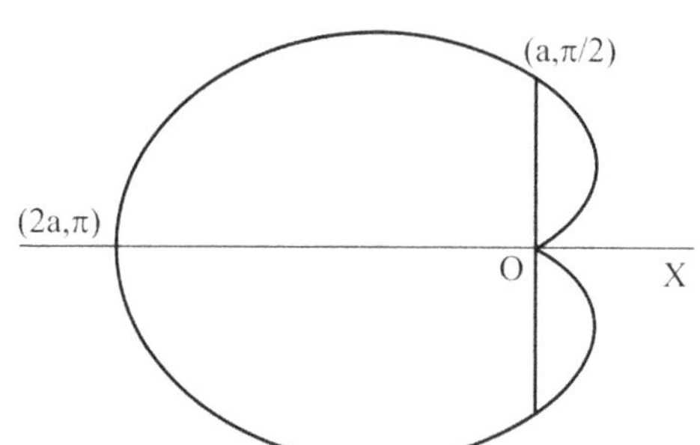

3.

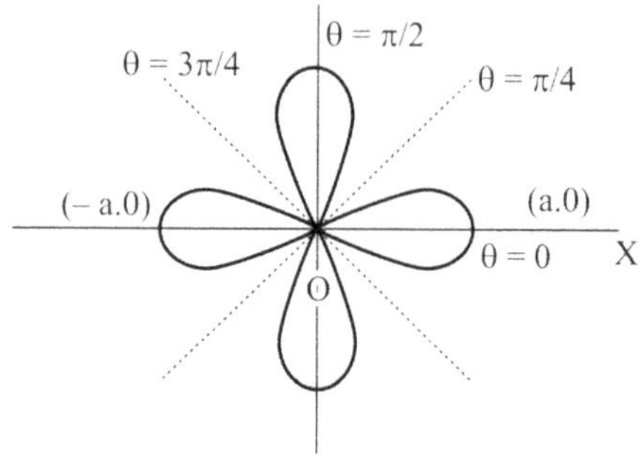

4.

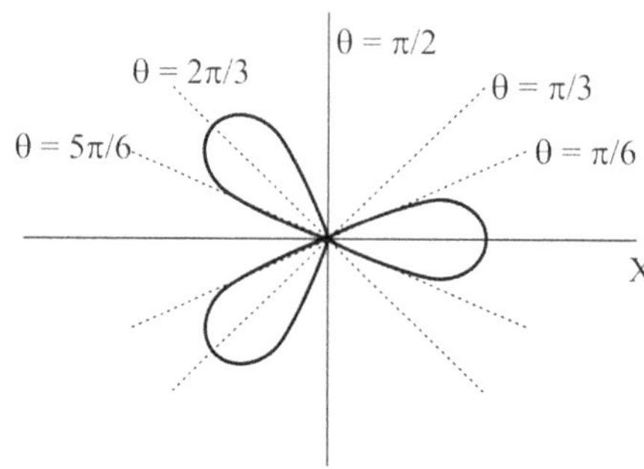

Unit – III

Partial Derivatives and Applications

- Partial Differentiation
- Jacobian
- Approximations and Errors

CHAPTER – 9

Partial Differentiation

9.1 Introduction

Let u be a function of two independent variables x and y i.e., u = f(x, y). The partial derivative of u with respect to x is defined as $\dfrac{\partial f}{\partial x} = \displaystyle\lim_{\delta x \to 0} \dfrac{f(x+\delta x, y) - f(x,y)}{\delta x}$ provided this limit exists and written as f_x or $D_x f$. Therefore the partial derivative of f(x, y) with respect to x is the ordinary differential coefficient of f(x, y) when y is regarded as constant.

Similarly the partial derivative of f(x, y) with respect to y is defined as $\dfrac{\partial f}{\partial y} = \displaystyle\lim_{\delta y \to 0} \dfrac{f(x, y+\delta y) - f(x,y)}{\delta y}$ provided this limit exists and written as f_y or $D_y f$. Therefore the partial derivative of f(x, y) w.r.f y is the ordinary differential coefficient of f(x, y) when x is regarded as constant.

The second order derivatives are

$$\frac{\partial^2 f}{\partial x^2}, \frac{\partial^2 f}{\partial y \partial x}, \frac{\partial^2 f}{\partial x \partial y} \text{ and } \frac{\partial^2 f}{\partial y^2}$$

and the notations are f_{xx}, f_{xy}, f_{yx} and f_{yy} respectively.

Illustrative Examples

Example 1: If u = log(x sin y + y sin x) then prove that $\dfrac{\partial^2 u}{\partial y \partial x} = \dfrac{\partial^2 u}{\partial x \partial y}$.

Solution: Let u = log (x sin y + y sin x)   .....(i)

Differentiating (i) partially w.r.t x, we get

$$\frac{\partial u}{\partial x} = \frac{\sin y + y \cos x}{x \sin y + y \sin x} \qquad \text{.....(ii)}$$

Differentiating (i) partially w.r.t y, we get

$$\frac{\partial u}{\partial y} = \frac{x \cos y + \sin x}{x \sin y + y \sin x} \qquad .....(iii)$$

Differentiating (ii) partially w.r.t y, we get

$$\frac{\partial^2 u}{\partial y \partial x} = \frac{\partial}{\partial y}\left[\frac{\sin y + y \cos x}{x \sin y + y \sin x}\right]$$

$$= \frac{(x \sin y + y \sin x)(\cos y + \cos x) - (\sin y + y \cos x)(x \cos y + \sin x)}{(x \sin y + y \sin x)^2} \qquad .....(iv)$$

Differentiating (iii) partially w.r.t x, we get

$$\frac{\partial^2 u}{\partial x \partial y} = \frac{\partial}{\partial x}\left[\frac{x \cos y + \sin x}{x \sin y + y \sin x}\right]$$

$$= \frac{(x \sin y + y \sin x)(\cos y + \cos x) - (x \cos y + \sin x)(\sin y + y \cos x)}{(x \sin y + y \sin x)^2} \qquad .....(v)$$

From (iv) and (v), we get

$$\frac{\partial^2 u}{\partial y \partial x} = \frac{\partial^2 u}{\partial x \partial y}$$

***Example* 2:** If $u = f\left(\frac{y}{x}\right)$, show that $x\dfrac{\partial u}{\partial x} + y\dfrac{\partial u}{\partial y} = 0$.

***Solution*:** Let $u = f\left(\frac{y}{x}\right)$

Differentiating partially w.r.t x, we get

$$\therefore \qquad \frac{\partial u}{\partial x} = f'\left(\frac{y}{x}\right)\left(-\frac{y}{x^2}\right)$$

$$\text{or} \qquad x\frac{\partial u}{\partial x} = -f'\left(\frac{y}{x}\right)\frac{y}{x} \qquad .....(i)$$

Differentiating partially w.r.t y, we get

$$\frac{\partial u}{\partial y} = f'\left(\frac{y}{x}\right)\left(\frac{1}{x}\right)$$

$$\text{or} \qquad y\frac{\partial u}{\partial y} = f'\left(\frac{y}{x}\right)\left(\frac{y}{x}\right) \qquad .....(ii)$$

Adding (i) and (ii), we get

$$x\frac{\partial u}{\partial x} + y\frac{\partial u}{\partial y} = 0$$

Hence proved

Example 3: If $u = \sin^{-1}\left(x\!\!\!\Big/\!\!\!y\right) + \tan^{-1}\left(y\!\!\!\Big/\!\!\!x\right)$, then prove that $x\dfrac{\partial u}{\partial x} + y\dfrac{\partial u}{\partial y} = 0$.

Solution: Let $u = \sin^{-1}\left(x\!\!\!\Big/\!\!\!y\right) + \tan^{-1}\left(y\!\!\!\Big/\!\!\!x\right)$ .....(i)

Differentiating (i) partially w.r.t x, we get

$$\frac{\partial u}{\partial x} = \frac{1}{\sqrt{1 - \dfrac{x^2}{y^2}}} \cdot \frac{1}{y} + \frac{1}{1 + \dfrac{y^2}{x^2}}\left(\frac{-y}{x^2}\right)$$

or $x\dfrac{\partial u}{\partial x} = \dfrac{x}{\sqrt{y^2 - x^2}} - \dfrac{xy}{x^2 + y^2}$ .....(ii)

Differentiating (i) partially w.r.t y, we get

$$\frac{\partial u}{\partial y} = \frac{1}{\sqrt{1 - \dfrac{x^2}{y^2}}}\left(-x\!\!\!\Big/\!\!\!y^2\right) + \frac{1}{1 + \dfrac{y^2}{x^2}} \cdot \frac{1}{x}$$

'or' $y\dfrac{\partial u}{\partial y} = -\dfrac{x}{\sqrt{y^2 - x^2}} + \dfrac{xy}{x^2 + y^2}$ .....(iii)

Adding (ii) and (iii), we get

$$x\frac{\partial u}{\partial x} + y\frac{\partial u}{\partial y} = 0.$$

Example 4: If $z = f(x + ay) + \phi(x - ay)$, then prove that $\dfrac{\partial^2 z}{\partial y^2} = a^2 \dfrac{\partial^2 z}{\partial x^2}$.

Solution: Let $z = f(x + ay) + \phi(x - ay)$ .....(i)

Differentiating (i) partially w.r.t x, we get

$$\frac{\partial z}{\partial x} = f'(x + ay) + \phi'(x - ay)$$

and $\dfrac{\partial^2 z}{\partial x^2} = f''(x + ay) + \phi''(x - ay)$ .....(ii)

Differentiating (i) partially w.r.t y, we get

$$\frac{\partial z}{\partial y} = f'(x + ay).a + \phi'(x - ay)(-a)$$

and $\dfrac{\partial^2 z}{\partial y^2} = a^2 f''(x + ay) + a^2\phi''(x - ay)$ .....(iii) $\therefore$

$\therefore$ $\dfrac{\partial^2 z}{\partial y^2} = a^2 \dfrac{\partial^2 z}{\partial x^2}$,   from (ii).

Example 5: If $F = \left(x^2 + y^2 + z^2\right)^{-\frac{1}{2}}$, then prove that $\dfrac{\partial^2 F}{\partial x^2} + \dfrac{\partial^2 F}{\partial y^2} + \dfrac{\partial^2 F}{\partial z^2} = 0$.

(RGPV June 2002, 2005)

Solution: Let $\qquad F = \left(x^2 + y^2 + z^2\right)^{-\frac{1}{2}}$ $\qquad\qquad$ .....(i)

Differentiating (i) partially w.r.t x, we get

$$\frac{\partial F}{\partial x} = -\frac{1}{2}\left(x^2 + y^2 + z^2\right)^{\frac{-3}{2}}.2x = -x\left(x^2 + y^2 + z^2\right)^{\frac{-3}{2}}$$

$\therefore\qquad$
$$\frac{\partial^2 F}{\partial x^2} = -\left[x\left\{-\frac{3}{2}\left(x^2 + y^2 + z^2\right)^{\frac{-5}{2}}.2x\right\} + \left(x^2 + y^2 + z^2\right)^{\frac{-3}{2}}\right]$$

$$= 3x^2\left(x^2 + y^2 + z^2\right)^{\frac{-5}{2}} - \left(x^2 + y^2 + z^2\right)^{\frac{-3}{2}}$$

$$= \left(x^2 + y^2 + z^2\right)^{\frac{-5}{2}}\left[3x^2 - \left(x^2 + y^2 + z^2\right)\right]$$

$$= \left(x^2 + y^2 + z^2\right)^{\frac{-5}{2}}\left(2x^2 - y^2 - z^2\right) \qquad\qquad .....(ii)$$

Similarly $\qquad \dfrac{\partial^2 F}{\partial y^2} = \left(x^2 + y^2 + z^2\right)^{\frac{-5}{2}}\left(2y^2 - x^2 - z^2\right) \qquad\qquad .....(iii)$

and $\qquad \dfrac{\partial^2 F}{\partial z^2} = \left(x^2 + y^2 + z^2\right)^{\frac{-5}{2}}\left(2z^2 - x^2 - y^2\right) \qquad\qquad .....(iv)$

Adding (ii) (iii) and (iv), we have

$$\frac{\partial^2 F}{\partial x^2} + \frac{\partial^2 F}{\partial y^2} + \frac{\partial^2 F}{\partial z^2} = 0$$

Hence proved.

Example 6: If $u = \log(x^3 + y^3 + z^3 - 3xyz)$ then prove that

(i) $\qquad \dfrac{\partial u}{\partial x} + \dfrac{\partial u}{\partial y} + \dfrac{\partial u}{\partial z} = \dfrac{3}{x + y + z}$ $\qquad\qquad$ (RGPV June 2007)

(ii) $\qquad \left(\dfrac{\partial}{\partial x} + \dfrac{\partial}{\partial y} + \dfrac{\partial}{\partial z}\right)^2 u = -\dfrac{9}{\left(x + y + z\right)^2}$ $\qquad$ (RGPV June 2005, June 2007, Jan 2008)

Solution:

(i)   Here $u = \log(x^3 + y^3 + z^3 - 3xyz)$

Differentiating partially w.r.t x, we get

$$\frac{\partial u}{\partial x} = \frac{3x^2 - 3yz}{x^3 + y^3 + z^3 - 3xyz} \qquad\qquad .....(i)$$

Similarly, $\quad \dfrac{\partial u}{\partial y} = \dfrac{3y^2 - 3xz}{x^3 + y^3 + z^3 - 3xyz} \qquad\qquad .....(ii)$

and $\qquad \dfrac{\partial u}{\partial z} = \dfrac{3z^2 - 3xy}{x^3 + y^3 + z^3 - 3xyz} \qquad\qquad ....(iii)$

Adding (i), (ii) and (iii), we get

$$\frac{\partial u}{\partial x} + \frac{\partial u}{\partial y} + \frac{\partial u}{\partial z} = \frac{3\left(x^2 + y^2 + z^2 - xy - yz - zx\right)}{x^3 + y^3 + z^3 - 3xyz}$$

$$= \frac{3\left(x^2 + y^2 + z^2 - xy - yz - zx\right)}{\left(x + y + z\right)\left(x^2 + y^2 + z^2 - xy - yz - zx\right)}$$

$$= \frac{3}{x + y + z} \qquad\qquad .....(iv)$$

(ii) Now, $\quad \left(\dfrac{\partial}{\partial x} + \dfrac{\partial}{\partial y} + \dfrac{\partial u}{\partial z}\right)^2 u = \left(\dfrac{\partial}{\partial x} + \dfrac{\partial}{\partial y} + \dfrac{\partial}{\partial z}\right)\left(\dfrac{\partial}{\partial x} + \dfrac{\partial}{\partial y} + \dfrac{\partial}{\partial z}\right) u$

$$= \left(\frac{\partial}{\partial x} + \frac{\partial}{\partial y} + \frac{\partial}{\partial z}\right)\left(\frac{\partial u}{\partial x} + \frac{\partial u}{\partial y} + \frac{\partial u}{\partial z}\right)$$

$$= \left(\frac{\partial}{\partial x} + \frac{\partial}{\partial y} + \frac{\partial}{\partial z}\right)\left(\frac{3}{x + y + z}\right), \quad \text{from (iv)}$$

$$= 3\left[\frac{\partial}{\partial x}\left(\frac{1}{x + y + z}\right) + \frac{\partial}{\partial y}\left(\frac{1}{x + y + z}\right) + \frac{\partial}{\partial z}\left(\frac{1}{x + y + z}\right)\right]$$

$$= 3\left[-\frac{1}{\left(x + y + z\right)^2} - \frac{1}{\left(x + y + z\right)^2} - \frac{1}{\left(x + y + z\right)^2}\right]$$

$$= \frac{-9}{\left(x + y + z\right)^2}$$

Example 7: If $u(x + y) = x^2 + y^2$, then prove that $\left(\dfrac{\partial u}{\partial x} - \dfrac{\partial u}{\partial y}\right)^2 = 4\left(1 - \dfrac{\partial u}{\partial x} - \dfrac{\partial u}{\partial y}\right).$

(RGPV Dec 2000)

Solution: Let $u = \dfrac{x^2 + y^2}{x + y}$

Differentiating partially w.r.t x, we get

$$\frac{\partial u}{\partial x} = \frac{(x+y)2x - (x^2 + y^2)}{(x+y)^2} = \frac{x^2 + 2xy - y^2}{(x+y)^2}$$

Similarly $\quad \dfrac{\partial u}{\partial y} = \dfrac{y^2 + 2xy - x^2}{(x+y)^2}$

$$\therefore \quad \frac{\partial u}{\partial x} - \frac{\partial u}{\partial y} = \frac{(x^2 + 2xy - y^2) - (y^2 + 2xy - x^2)}{(x+y)^2}$$

$$= \frac{2(x^2 - y^2)}{(x+y)^2}$$

$$= \frac{2(x-y)}{(x+y)}$$

$$\therefore \quad \left(\frac{\partial u}{\partial x} - \frac{\partial u}{\partial y}\right)^2 = \frac{4(x-y)^2}{(x+y)^2} \qquad \qquad \dots\dots(i)$$

Now, $\quad 1 - \dfrac{\partial u}{\partial x} - \dfrac{\partial u}{\partial y} = 1 - \dfrac{x^2 + 2xy - y^2 + y^2 + 2xy - x^2}{(x+y)^2}$

$$= 1 - \frac{4xy}{(x+y)^2}$$

$$= \frac{(x-y)^2}{(x+y)^2}$$

$$\therefore \quad 4\left(1 - \frac{\partial u}{\partial x} - \frac{\partial u}{\partial y}\right) = \frac{4(x-y)^2}{(x+y)^2} \qquad \qquad \dots\dots(ii)$$

$\therefore$ From (i) and (ii), we have

$$\left(\frac{\partial u}{\partial x} - \frac{\partial u}{\partial y}\right)^2 = 4\left(1 - \frac{\partial u}{\partial x} - \frac{\partial u}{\partial y}\right)$$

Example 8: If $u = f(r)$, where $r^2 = x^2 + y^2$, show that $\dfrac{\partial^2 u}{\partial x^2} + \dfrac{\partial^2 u}{\partial y^2} = f''(r) + \dfrac{1}{r}f'(r)$.

(RGPV Dec 2001, June 2006)

Solution: Given $r^2 = x^2 + y^2$

Differentiating partially w.r.t x, we get

$$2r\frac{\partial r}{\partial x} = 2x \quad \text{'or'} \quad \frac{\partial r}{\partial x} = \frac{x}{r}$$

and $\quad \dfrac{\partial r}{\partial y} = \dfrac{y}{r}$

Now, $\quad u = f(r)$

$\therefore \quad \dfrac{\partial u}{\partial x} = f'(r)\dfrac{\partial r}{\partial x} = f'(r)\dfrac{x}{r}$

$$\dfrac{\partial^2 u}{\partial x^2} = \dfrac{r\left[f'(r) + xf''(r)\dfrac{dr}{dx}\right] - xf'(r).\dfrac{\partial r}{\partial x}}{r^2}$$

$$= \dfrac{1}{r^2}\left[rf'(r) + x^2 f''(r) - \dfrac{x^2}{r}f'(r)\right] \qquad\qquad .....(i)$$

Similarly, $\quad \dfrac{\partial^2 u}{\partial y^2} = \dfrac{1}{r^2}\left[rf'(r) + y^2 f''(r) - \dfrac{y^2}{r}f'(r)\right] \qquad .....(ii)$

Adding (i) and (ii), we get

$$\dfrac{\partial^2 u}{\partial x^2} + \dfrac{\partial^2 u}{\partial y^2} = \dfrac{1}{r^2}\left[2rf'(r) + \left(x^2 + y^2\right)f''(r) - \dfrac{\left(x^2 + y^2\right)}{r}f'(r)\right]$$

$$= \dfrac{1}{r^2}\left[2rf'(r) + r^2 f''(r) - rf'(r)\right]$$

$$= \dfrac{1}{r^2}\left[rf'(r) + r^2 f''(r)\right]$$

$$= \dfrac{1}{r}f'(r) + f''(r)$$

Hence proved.

Example 9: If $u = e^{xyz}$, then prove that $\dfrac{\partial^3 u}{\partial x \partial y \partial z} = \left(1 + 3xyz + x^2 y^2 z^2\right)e^{xyz}$. $\qquad$ (RGPV 2001)

Solution: Let $u = e^{xyz}$

Differentiating partially w.r.t z, we get

$$\dfrac{\partial u}{\partial z} = xy\, e^{xyz}$$

Differentiating partially w.r.t y, we get

$$\dfrac{\partial^2 u}{\partial y \partial z} = \dfrac{\partial}{\partial y}\left(xy\, e^{xyz}\right)$$

$$= x[y \cdot xze^{xyz} + e^{xyz}]$$

$$= e^{xyz}\,[x^2 yz + x]$$

Differentiating partially w.r.t x, we get

$$\frac{\partial^3 u}{\partial x \partial y \partial z} = \frac{\partial}{\partial x}\left[e^{xyz}\left(x^2 yz + x\right)\right]$$

$$= e^{xyz}(2xyz + 1) + yze^{xyz}(x^2 yz + x)$$

$$= e^{xyz}(2xyz + 1 + x^2 y^2 z^2 + xyz]$$

$$= e^{xyz}(1 + 3xyz + x^2 y^2 z^2)$$

Hence Proved.

***Example* 10:** If $x^x y^y z^z = c$, then show that $\dfrac{\partial^2 z}{\partial x \partial y} = -(x \log ex)^{-1}$, here z is the function of x and y.

(RGPV Dec 2001, 2004, 2005, 2006)

***Solution*:** Given $x^x y^y z^z = c$

Taking logarithm on both sides, we get

$$x \log x + y \log y + z \log z = \log c \qquad\qquad\qquad .....(i)$$

Differentiating (i) partially w.r.t x, we get

$$\left[x.\frac{1}{x} + \log x\right] + \left[z.\frac{1}{z} + \log z\right]\frac{\partial z}{\partial x} = 0$$

'or' $\quad \dfrac{\partial z}{\partial x} = -\dfrac{(1 + \log x)}{(1 + \log z)} \qquad\qquad\qquad .....(ii)$

Similarly $\dfrac{\partial z}{\partial y} = -\dfrac{(1 + \log y)}{(1 + \log z)} \qquad\qquad\qquad .....(iii)$

Differentiating (ii) partially w.r.t x, we get

$$\frac{\partial^2 z}{\partial x \partial y} = \frac{\partial}{\partial x}\left[-\frac{1 + \log y}{1 + \log z}\right]$$

$$= -(1 + \log y)\frac{\partial}{\partial x}(1 + \log z)^{-1}$$

$$= -(1 + \log y)\left[-(1 + \log z)^{-2}.\frac{1}{z}\frac{\partial z}{\partial x}\right]$$

$$= +\frac{1 + \log y}{z(1 + \log z)^2}\left\{-\frac{1 + \log x}{1 + \log z}\right\}, \qquad \text{from (ii)}$$

Putting $x = y = z$, we have

$$\frac{\partial^2 z}{\partial x \partial y} = -\frac{(1 + \log x)^2}{z(1 + \log x)^3}$$

$$= -\frac{1}{x(1 + \log x)}$$

$$= -\frac{1}{x(\log e + \log x)} \qquad (\because \log e = 1)$$

$$= -\frac{1}{x \log ex}$$

$$= -(x \log ex)^{-1}$$

Hence proved

***Example* 11:** If $u = \left(x^2 + y^2 + z^2\right)^{-\frac{1}{2}}$, $x^2 + y^2 + z^2 \neq 0$, then prove that $x\dfrac{\partial u}{\partial x} + y\dfrac{\partial u}{\partial y} + z\dfrac{\partial u}{\partial z} = -u$

.   (RGPV Feb 2005, Jan 2007)

***Solution*:** Given $u = \left(x^2 + y^2 + z^2\right)^{-\frac{1}{2}}$

Differentiating partially w.r.t x, we get

$$\frac{\partial u}{\partial x} = -\frac{1}{2}\left(x^2 + y^2 + z^2\right)^{-\frac{3}{2}} 2x = -x\left(x^2 + y^2 + z^2\right)^{-\frac{3}{2}}$$

$$\therefore \qquad x\frac{\partial u}{\partial x} = -x^2\left(x^2 + y^2 + z^2\right)^{-\frac{3}{2}} \qquad\qquad .....(i)$$

Similarly $y\dfrac{\partial u}{\partial y} = -y^2\left(x^2 + y^2 + z^2\right)^{-\frac{3}{2}} \qquad\qquad .....(ii)$

and $\qquad z\dfrac{\partial u}{\partial z} = -z^2\left(x^2 + y^2 + z^2\right)^{-\frac{3}{2}} \qquad\qquad .....(iii)$

$\therefore$ Adding (i), (ii) and (iii), we get

$$x\frac{\partial u}{\partial x} + y\frac{\partial u}{\partial y} + z\frac{\partial u}{\partial z} = -\left(x^2 + y^2 + z^2\right)\left(x^2 + y^2 + z^2\right)^{-\frac{3}{2}}$$

$$= -\left(x^2 + y^2 + z^2\right)^{-\frac{1}{2}}$$

$$= -u$$

Hence proved.

***Example* 12:** If $u = \log\left(\dfrac{x^4 + y^4}{x + y}\right)$, then prove that $x\dfrac{\partial u}{\partial x} + y\dfrac{\partial u}{\partial y} = 3$.   (RGPV June 2006)

***Solution*:** Given $u = \log\left(\dfrac{x^4 + y^4}{x + y}\right)$

Differentiating partially w.r.t x, we get

$$\frac{\partial u}{\partial x} = \frac{1}{\left(x^4 + y^4\right)\Big/(x+y)} \cdot \frac{(x+y)4x^3 - \left(x^4 + y^4\right)}{(x+y)^2}$$

$$= \frac{3x^4 + 4x^3 y - y^4}{\left(x^4 + y^4\right)(x+y)}$$

Similarly $\dfrac{\partial u}{\partial y} = \dfrac{3y^4 + 4xy^3 - x^4}{\left(x^4 + y^4\right)(x+y)}$

$\therefore \qquad x\dfrac{\partial u}{\partial x} + y\dfrac{\partial u}{\partial y} = \dfrac{3x^5 + 4x^4 y - xy^4 + 3y^5 + 4xy^4 - x^4 y}{\left(x^4 + y^4\right)(x+y)}$

$$= \frac{3\left(x^5 + x^4 y + xy^4 + y^5\right)}{\left(x^4 + y^4\right)(x+y)}$$

$$= \frac{3\left(x^4 + y^4\right)(x+y)}{\left(x^4 + y^4\right)(x+y)} = 3.$$

***Example* 13:** If $\dfrac{x^2}{a^2 + u} + \dfrac{y^2}{b^2 + u} + \dfrac{z^2}{c^2 + u} = 1$, then prove that $u_x^2 + u_y^2 + u_z^2 = 2(xu_x + yu_y + zu_z)$.

***Solution*:** Let $\dfrac{x^2}{a^2 + u} + \dfrac{y^2}{b^2 + u} + \dfrac{z^2}{c^2 + u} = 1$ …..(i)

Differentiating (i) partially w.r.t x, we get

$$\frac{2x}{a^2 + u} - \left[\frac{x^2}{\left(a^2 + u\right)^2} + \frac{y^2}{\left(b^2 + u\right)^2} + \frac{z^2}{\left(c^2 + u\right)^2}\right]\frac{\partial u}{\partial x} = 0$$

'or' $\qquad \dfrac{\partial u}{\partial x} = \dfrac{2x\Big/\left(a^2 + u\right)}{\dfrac{x^2}{\left(a^2 + u\right)^2} + \dfrac{y^2}{\left(b^2 + u\right)^2} + \dfrac{z^2}{\left(c^2 + u\right)^2}}$

Similarly
$$\frac{\partial u}{\partial y} = \frac{2y \Big/ \left(b^2 + u\right)}{\dfrac{x^2}{\left(a^2+u\right)^2} + \dfrac{y^2}{\left(b^2+u\right)^2} + \dfrac{z^2}{\left(c^2+u\right)^2}}$$

and
$$\frac{\partial u}{\partial z} = \frac{2z \Big/ \left(c^2 + u\right)}{\dfrac{x^2}{\left(a^2+u\right)^2} + \dfrac{y^2}{\left(b^2+u\right)^2} + \dfrac{z^2}{\left(c^2+u\right)^2}}$$

$\therefore$
$$\left(\frac{\partial u}{\partial x}\right)^2 + \left(\frac{\partial u}{\partial y}\right)^2 + \left(\frac{\partial u}{\partial z}\right)^2 = \frac{4\left[\dfrac{x^2}{\left(a^2+u\right)^2} + \dfrac{y^2}{\left(b^2+u\right)^2} + \dfrac{z^2}{\left(c^2+u\right)^2}\right]}{\left[\dfrac{x^2}{\left(a^2+u\right)^2} + \dfrac{y^2}{\left(b^2+u\right)^2} + \dfrac{z^2}{\left(c^2+u\right)^2}\right]^2}$$

'or'
$$u_x^2 + u_y^2 + u_z^2 = \frac{4}{\left[\dfrac{x^2}{\left(a^2+u\right)^2} + \dfrac{y^2}{\left(b^2+u\right)^2} + \dfrac{z^2}{\left(c^2+u\right)^2}\right]^2} \qquad\qquad \dots\dots(ii)$$

Also,
$$xu_x + yu_y + zu_z = \frac{\dfrac{2x^2}{a^2+u} + \dfrac{2y^2}{b^2+u} + \dfrac{2z^2}{c^2+u}}{\dfrac{x^2}{\left(a^2+u\right)^2} + \dfrac{y^2}{\left(b^2+u\right)^2} + \dfrac{z^2}{\left(c^2+u\right)^2}} \quad \text{from (i),} \qquad \dots\dots(iii)$$

$\therefore$ from (ii) and (iii), we have $u_x^2 + u_y^2 + u_z^2 = 2(xu_x + yu_y + zu_z)$

Hence proved

9.2 Homogeneous Functions

Let $f(x, y) = a_0x^n + a_1x^{n-1} y + a_2x^{n-2} . y^2 + \dots\dots + a_ny^n$ be a function of x and y in which every term is of degree n. Such types of function is called a homogeneous function of x and y of degree n and can be written as $x^n\ f\left(y \big/ x\right)$.

9.3 Euler's Theorem

If $f(x, y)$ be a homogeneous function of x and y of degree n, then
$$x\frac{\partial f}{\partial x} + y\frac{\partial f}{\partial y} = nf$$

***Proof*:** Let f(x, y) be a homogeneous function of x and y of degree n, which can be written as

$$f(x, y) = x^n f\left(\frac{y}{x}\right) \qquad \qquad .....(i)$$

Differentiating partially w.r.t x and y, we get

$$\frac{\partial f}{\partial x} = nx^{n-1} f\left(\frac{y}{x}\right) + x^n f'\left(\frac{y}{x}\right)\left(-\frac{y}{x^2}\right)$$

$$\therefore \quad x\frac{\partial f}{\partial x} = nx^n f\left(\frac{y}{x}\right) - x^{n-1} y f'\left(\frac{y}{x}\right) \qquad \qquad .....(ii)$$

and

$$\frac{\partial f}{\partial y} = x^n f'\left(\frac{y}{x}\right) \cdot \frac{1}{x}$$

$$\therefore \quad y\frac{\partial f}{\partial y} = x^{n-1} y f'\left(\frac{y}{x}\right) \qquad \qquad .....(iii)$$

Adding (ii) and (iii), we get

$$x\frac{\partial f}{\partial x} + y\frac{\partial f}{\partial y} = nx^n\left(\frac{y}{x}\right)$$

$$= nf$$

Illustrative Problems

***Example* 14:** Verify Euler's Theorem, when $f(x, y) = ax^2 + 2h\,xy + by^2$.

***Solution*:** Given $f(x, y) = ax^2 + 2h\,xy + by^2$ $\qquad \qquad .....(i)$

which is a homogeneous function of degree 2. Hence we prove that

$$x\frac{\partial f}{\partial x} + y\frac{\partial f}{\partial y} = 2f$$

Now, differentiating partially w.r.t x and y, we get

$$\frac{\partial f}{\partial x} = 2ax + 2hy$$

$$\therefore \quad x\frac{\partial f}{\partial x} = 2ax^2 + 2hxy \qquad \qquad .....(ii)$$

and

$$\frac{\partial f}{\partial y} = 2hx + 2by$$

$$\therefore \quad y\frac{\partial f}{\partial y} = 2hxy + 2by^2 \qquad \qquad .....(iii)$$

Adding (ii) and (iii), we get

$$x\frac{\partial f}{\partial x} + y\frac{\partial f}{\partial y} = 2ax^2 + 4hxy + 2by^2$$

$$= 2(ax^2 + 2hxy + by^2)$$

$$= 2f \qquad \text{from (i)}$$

***Example* 15:** If $u = \tan^{-1}\left(\dfrac{x^3 + y^3}{x - y}\right)$, prove that

(i) $\quad x\dfrac{\partial u}{\partial x} + y\dfrac{\partial u}{\partial y} = \sin 2u$ $\qquad\qquad$ (RGPV Dec 2002, Dec 2005)

(ii) $\quad x^2\dfrac{\partial^2 u}{\partial x^2} + 2xy\dfrac{\partial^2 u}{\partial x\partial y} + y^2\dfrac{\partial^2 u}{\partial y^2} = \sin 4u - \sin 2u \; = \; 2\cos 3u \sin u$

$$\text{(RGPV Dec 2003, Jan 2006, Jan 2008, April 2009)}$$

***Solution*:**

(i) Let $\quad u = \tan^{-1}\left(\dfrac{x^3 + y^3}{x - y}\right)$

$\therefore \qquad \tan u = \dfrac{x^3 + y^3}{x - y} = f \,(\text{say})$

Then f is a homogeneous function of x and y of degree 2. Hence by Euler's Theorem, we have

$$x\frac{\partial f}{\partial x} + y\frac{\partial f}{\partial y} = 2f$$

$\therefore \qquad x\dfrac{\partial}{\partial x}\left(\tan u\right) + y\dfrac{\partial}{\partial y}\left(\tan u\right) = 2\tan u \qquad (\because \; f = \tan u)$

or $\quad x\sec^2 u \,\dfrac{\partial u}{\partial x} + y\sec^2 u\,\dfrac{\partial u}{\partial y} = 2\tan u$

or $\quad x\dfrac{\partial u}{\partial x} + y\dfrac{\partial u}{\partial y} = 2\dfrac{\tan u}{\sec^2 u}$

or $\quad x\dfrac{\partial u}{\partial x} + y\dfrac{\partial u}{\partial y} = \sin 2u$ $\qquad\qquad$ .....(i)

(ii) Differentiating (i) partially w.r.t x, we have

$$x\frac{\partial^2 u}{\partial x^2} + \frac{\partial u}{\partial x} + y\frac{\partial^2 u}{\partial x\partial y} = 2\cos 2u\frac{\partial u}{\partial x}$$

or $\quad x\dfrac{\partial^2 u}{\partial x^2} + y\dfrac{\partial^2 u}{\partial x\partial y} = (2\cos 2u - 1)\dfrac{\partial u}{\partial x}$ $\qquad\qquad$ .....(ii)

Differentiating (i) partially w.r.t y, we have

$$x\frac{\partial^2 u}{\partial x\partial y} + y\frac{\partial^2 u}{\partial y^2} + \frac{\partial u}{\partial y} = 2\cos 2u\frac{\partial u}{\partial y}$$

or $\quad x\dfrac{\partial^2 u}{\partial x\partial y} + y\dfrac{\partial^2 u}{\partial y^2} = (2\cos 2u - 1)\dfrac{\partial u}{\partial y}$ $\qquad\qquad$ .....(iii)

Multiply (ii) by x and (iii) by y and adding, we have

$$x^2\frac{\partial^2 u}{\partial x^2} + 2xy\frac{\partial^2 u}{\partial x\partial y} + y^2\frac{\partial^2 u}{\partial y^2} = (2\cos 2u - 1)\left(x\frac{\partial u}{\partial x} + y\frac{\partial y}{\partial y}\right)$$

$$= (2\cos 2u - 1)\sin 2u$$

$$= 2\sin 2u\cos 2u - \sin 2u$$

$$= \sin 4u - \sin 2u$$

$$= 2\cos 3u\sin u$$

Hence proved.

***Example* 16:** If $u = \sin^{-1}\left(\dfrac{\sqrt{x}-\sqrt{y}}{\sqrt{x}+\sqrt{y}}\right)$, prove by Euler's theorem that $\quad x\dfrac{\partial u}{\partial x} + y\dfrac{\partial u}{\partial y} = 0$.

***Solution*:** Given

$$u = \sin^{-1}\left(\frac{\sqrt{x}-\sqrt{y}}{\sqrt{x}+\sqrt{y}}\right)$$

$$\therefore \quad \sin u = \frac{\sqrt{x}-\sqrt{y}}{\sqrt{x}+\sqrt{y}} = f\,(\text{say})$$

Hence f is a homogeneous function of x and y of degree 0.

$\therefore$ By Euler's theorem,

$$x\frac{\partial f}{\partial x} + y\frac{\partial f}{\partial y} = 0$$

$$x\frac{\partial}{\partial x}(\sin u) + y\frac{\partial}{\partial y}(\sin u) = 0$$

$$x\cos u\,\frac{\partial u}{\partial x} + y\cos u\frac{\partial u}{\partial y} = 0$$

or $\quad x\dfrac{\partial u}{\partial x} + y\dfrac{\partial u}{\partial y} = 0$

***Example* 17:** If $u = \sin^{-1}\left(\dfrac{x^2 + y^2}{x + y}\right)$, then show that $x\dfrac{\partial u}{\partial x} + y\dfrac{\partial u}{\partial y} = \tan u$.

(RGPV June 2004, June 2009)

Proof: Given $\qquad\qquad u = \sin^{-1}\left(\dfrac{x^2 + y^2}{x + y}\right)$

$\therefore \qquad\qquad\qquad \sin u = \dfrac{x^2 + y^2}{x + y} = f \qquad\qquad$ (say)

Then f is a homogeneous function of x and y of degree 1.

$\therefore$ By Euler's Theorem.

$$x\frac{\partial f}{\partial x} + y\frac{\partial f}{\partial y} = f$$

$$x\frac{\partial}{\partial x}(\sin u) + y\frac{\partial}{\partial y}(\sin u) = \sin u \qquad (\because f = \sin u)$$

$$x\cos u\,\frac{\partial u}{\partial x} + y\cos u\,\frac{\partial u}{\partial y} = \sin u$$

or $\qquad\qquad\qquad x\dfrac{\partial u}{\partial x} + y\dfrac{\partial u}{\partial y} = \dfrac{\sin u}{\cos u}$

or $\qquad\qquad\qquad x\dfrac{\partial u}{\partial x} + y\dfrac{\partial u}{\partial y} = \tan u$

Hence proved.

***Example* 18:** If $u = \log\left(\dfrac{x^4 + y^4}{x + y}\right)$, then find

(i) $\qquad x\dfrac{\partial u}{\partial x} + y\dfrac{\partial u}{\partial y}$

(ii) $\qquad x^2\dfrac{\partial^2 u}{\partial x^2} + 2xy\dfrac{\partial^2 u}{\partial x \partial y} + y^2\dfrac{\partial^2 u}{\partial y^2}$ $\qquad\qquad\qquad\qquad$ (RGPV June 2006)

***Solution*:**

(i) Given $\qquad\qquad u = \log\left(\dfrac{x^4 + y^4}{x + y}\right)$

$\therefore \qquad\qquad\qquad e^u = \dfrac{x^4 + y^4}{x + y} = f \quad$ (say)

Then f is a homogeneous function of x and y of degree 3.

$\therefore$ By Euler's theorem,

$$x\frac{\partial f}{\partial x} + y\frac{\partial f}{\partial y} = 3f$$

or

$$x\frac{\partial}{\partial x}e^u + y\frac{\partial}{\partial y}e^u = 3e^u$$

or

$$xe^u\frac{\partial u}{\partial x} + ye^u\frac{\partial u}{\partial y} = 3e^u$$

or

$$x\frac{\partial u}{\partial x} + y\frac{\partial u}{\partial y} = 3 \qquad\qquad .....(i)$$

(ii) Differentiating (i) partially w.r.t x, we get

$$x\frac{\partial^2 u}{\partial x^2} + \frac{\partial u}{\partial x} + y\frac{\partial^2 u}{\partial x\partial y} = 0$$

Multiply by x, we get, $x^2\dfrac{\partial^2 u}{\partial x^2} + xy\dfrac{\partial^2 u}{\partial x\partial y} = -x\dfrac{\partial u}{\partial x} \qquad .....(ii)$

Similarly, $\qquad xy\dfrac{\partial^2 u}{\partial x\partial y} + y^2\dfrac{\partial^2 u}{\partial y} = -y\dfrac{\partial u}{\partial y} \qquad .....(iii)$

Adding (ii) and (iii), we get

$$x^2\frac{\partial^2 u}{\partial x^2} + 2xy\frac{\partial^2 u}{\partial x\partial y} + y^2\frac{\partial^2 u}{\partial y^2} = -\left(x\frac{\partial u}{\partial x} + y\frac{\partial u}{\partial y}\right) = -3, \quad \text{from (i)}$$

***Example* 19:** If $u = \sin^{-1}\left(\dfrac{x+y}{\sqrt{x}+\sqrt{y}}\right)$, show that $x\dfrac{\partial u}{\partial x} + y\dfrac{\partial u}{\partial y} = \dfrac{1}{2}\tan u$. (RGPV June 2003)

***Solution*:** Given

$$u = \sin^{-1}\left(\frac{x+y}{\sqrt{x}+\sqrt{y}}\right)$$

$$\therefore \qquad \sin u = \frac{x+y}{\sqrt{x}+\sqrt{y}} = f\,(\text{say})$$

Then f is a homogeneous function of x and y of degree $\frac{1}{2}$.

$\therefore$ By Euler's theorem,

$$x\frac{\partial f}{\partial x} + y\frac{\partial f}{\partial y} = \frac{1}{2}f$$

or

$$x\frac{\partial}{\partial x}(\sin u) + y\frac{\partial}{\partial y}(\sin u) = \frac{1}{2}\sin u$$

or $\qquad x \cos u \dfrac{\partial u}{\partial x} + y \cos u \dfrac{\partial u}{\partial y} = \dfrac{1}{2} \sin u$

or $\qquad x \dfrac{\partial u}{\partial x} + y \dfrac{\partial u}{\partial y} = \dfrac{1}{2} \tan u$

Hence proved.

***Example* 20:** If $\sin u = \dfrac{x^2 y^2}{x + y}$, show that $x \dfrac{\partial u}{\partial x} + y \dfrac{\partial u}{\partial y} = 3 \tan u$.   (RGPV June 2008)

***Solution*:** Given $\sin u = \dfrac{x^2 y^2}{x + y} = f$ (say)

Then f is a homogeneous function of x and y of degree 3.

$\therefore$ By Euler's Theorem

$$x \dfrac{\partial f}{\partial x} + y \dfrac{\partial f}{\partial y} = 3f$$

or $\qquad x \dfrac{\partial}{\partial x}\left(\sin u\right) + y \dfrac{\partial}{\partial y}\left(\sin u\right) = 3 \sin u$

or $\qquad x \cos u \dfrac{\partial u}{\partial x} + y \cos u \dfrac{\partial u}{\partial y} = 3 \sin u$

or $\qquad x \dfrac{\partial u}{\partial x} + y \dfrac{\partial u}{\partial y} = 3 \tan u$

Hence proved.

***Example* 21:** Verify Euler's theorem for the function $u = \dfrac{\left(x^{\frac{1}{4}} + y^{\frac{1}{4}}\right)}{\left(x^{\frac{1}{5}} + y^{\frac{1}{5}}\right)}$.

***Solution*:** Here u is a homogeneous function of x and y of degree $\left(\dfrac{1}{4} - \dfrac{1}{5}\right) = \dfrac{1}{20}$.

$\therefore$ By Euler's theorem.

$$x \dfrac{\partial u}{\partial x} + y \dfrac{\partial u}{\partial y} = \dfrac{1}{20} u$$

Now, $\qquad u = \dfrac{x^{\frac{1}{4}} + y^{\frac{1}{4}}}{x^{\frac{1}{5}} + y^{\frac{1}{5}}}$ .....(i)

Differentiating (i) partially w.r.t x, we get

$$\frac{\partial u}{\partial x} = \frac{\left(x^{\frac{1}{5}} + y^{\frac{1}{5}}\right)\left(\frac{1}{4}x^{\frac{-3}{4}}\right) - \left(x^{\frac{1}{4}} + y^{\frac{1}{4}}\right)\left(\frac{1}{5}x^{\frac{-4}{5}}\right)}{\left(x^{\frac{1}{5}} + y^{\frac{1}{5}}\right)^2}$$

or $$x\frac{\partial u}{\partial x} = \frac{\left(x^{\frac{1}{5}} + y^{\frac{1}{5}}\right)\left(\frac{1}{4}x^{\frac{-1}{4}}\right) - \left(x^{\frac{1}{4}} + y^{\frac{1}{4}}\right)\left(\frac{1}{5}x^{\frac{1}{5}}\right)}{\left(x^{\frac{1}{5}} + y^{\frac{1}{5}}\right)^2} \qquad \ldots\ldots(ii)$$

again, differentiating (i) partially w.r.t y, we get

$$\frac{\partial u}{\partial y} = \frac{\left(x^{\frac{1}{5}} + y^{\frac{1}{5}}\right)\left(\frac{1}{4}y^{\frac{-3}{4}}\right) - \left(x^{\frac{1}{4}} + y^{\frac{1}{4}}\right)\left(\frac{1}{5}y^{\frac{-4}{5}}\right)}{\left(x^{\frac{1}{5}} + y^{\frac{1}{5}}\right)^2}$$

or $$y\frac{\partial u}{\partial y} = \frac{\left(x^{\frac{1}{5}} + y^{\frac{1}{5}}\right)\left(\frac{1}{4}y^{\frac{1}{4}}\right) - \left(x^{\frac{1}{4}} + y^{\frac{1}{4}}\right)\left(\frac{1}{5}y^{\frac{1}{5}}\right)}{\left(x^{\frac{1}{5}} + y^{\frac{1}{5}}\right)^2} \qquad \ldots\ldots(iii)$$

Adding (ii) and (iii), we get

$$x\frac{\partial u}{\partial x} + y\frac{\partial u}{\partial y} = \frac{\frac{1}{4}\left(x^{\frac{1}{5}} + y^{\frac{1}{5}}\right)\left(x^{\frac{1}{4}} + y^{\frac{1}{4}}\right) - \frac{1}{5}\left(x^{\frac{1}{4}} + y^{\frac{1}{4}}\right)\left(x^{\frac{1}{5}} + y^{\frac{1}{5}}\right)}{\left(x^{\frac{1}{5}} + y^{\frac{1}{5}}\right)^2}$$

$$= \frac{1}{20}\frac{\left(x^{\frac{1}{4}} + y^{\frac{1}{4}}\right)}{\left(x^{\frac{1}{5}} + y^{\frac{1}{5}}\right)} = \frac{1}{20}u, \qquad \text{from (i)}$$

which verify Euler's theorem.

***Example* 22:** If u be a homogeneous function of degree n, show that

(i) $$x\frac{\partial^2 u}{\partial x^2} + y\frac{\partial^2 u}{\partial x \partial y} = (n-1)\frac{\partial u}{\partial x}$$

(ii) $x\dfrac{\partial^2 u}{\partial x \partial y} + y\dfrac{\partial^2 u}{\partial y^2} = (n-1)\dfrac{\partial u}{\partial y}$

(iii) $x\dfrac{\partial^2 u}{\partial x^2} + 2xy\dfrac{\partial^2 u}{\partial x \partial y} + y^2\dfrac{\partial^2 u}{\partial y^2} = n(n-1)u$

Proof: If u be a homogeneous function of degree n,

$\therefore$ By Euler's theorem, we get

$$x\frac{\partial u}{\partial x} + y\frac{\partial u}{\partial y} = nu \qquad\qquad .....(i)$$

(i) Differentiating (i) partially w.r.t x, we get

$$x\frac{\partial^2 u}{\partial x^2} + \frac{\partial u}{\partial x} + y\frac{\partial^2 u}{\partial x \partial y} = n\frac{\partial u}{\partial x}$$

'or' $\qquad x\dfrac{\partial^2 u}{\partial x^2} + y\dfrac{\partial^2 u}{\partial x \partial y} = (n-1)\dfrac{\partial u}{\partial x} \qquad\qquad .....(ii)$

(ii) Differentiating (i) partially w.r.t y, we get

$$x\frac{\partial^2 u}{\partial x \partial y} + y\frac{\partial^2 u}{\partial y^2} + \frac{\partial u}{\partial y} = n\frac{\partial u}{\partial y}$$

'or; $\qquad x\dfrac{\partial^2 u}{\partial x \partial y} + y\dfrac{\partial^2 u}{\partial y^2} = (n-1)\dfrac{\partial u}{\partial y} \qquad\qquad .....(iii)$

(iii) Multiply (ii) by x and (iii) by y and adding, we get

$$x^2\frac{\partial^2 u}{\partial x^2} + 2xy\frac{\partial^2 u}{\partial x \partial y} + y^2\frac{\partial^2 u}{\partial y^2} = (n-1)\left(x\frac{\partial u}{\partial x} + y\frac{\partial u}{\partial y}\right)$$

$$= n(n-1)u, \text{ from (i)}$$

Hence proved.

Practice Problems

1. If $f = a\,\tan^{-1}\left(\dfrac{x}{y}\right)$, verify that $\dfrac{\partial^2 f}{\partial y \partial x} = \dfrac{\partial^2 f}{\partial x \partial y}$.

2. If $u = x^2\,\tan^{-1}\left(\dfrac{y}{x}\right) - y^2\,\tan^{-1}\left(\dfrac{x}{y}\right)$, prove that $\dfrac{\partial^2 u}{\partial y \partial x} = \dfrac{x^2 - y^2}{x^2 + y^2}$.

3. If $u = x^2 + y^2 + z^2$, prove that $x\dfrac{\partial u}{\partial x} + y\dfrac{\partial u}{\partial y} + z\dfrac{\partial u}{\partial z} = 2u$.

4. If $u = e^x (x \cos y - y \sin y)$, prove that $\dfrac{\partial^2 u}{\partial x^2} + \dfrac{\partial^2 u}{\partial y^2} = 0$.

5. If $u = \tan^{-1}\left(\dfrac{x^2 + y^2}{x + y}\right)$, find $\dfrac{\partial u}{\partial x}$ and $\dfrac{\partial u}{\partial y}$.

$$\text{[Ans. } \frac{x^2 + 2xy - y^2}{\left(x + y\right)^2 + \left(x^2 + y^2\right)^2}, \frac{y^2 + 2xy - x^2}{\left(x + y\right)^2 + \left(x^2 + y^2\right)^2} \text{]}$$

6. If $f = \phi (y + ax) + \varphi (y - ax)$, show that $\dfrac{\partial^2 f}{\partial x^2} = a^2 \dfrac{\partial^2 f}{\partial y^2}$.

7. If $u = f(r^2)$ and $r^2 = x^2 + y^2 + z^2$, prove that $\dfrac{\partial^2 u}{\partial x^2} + \dfrac{\partial^2 u}{\partial y^2} + \dfrac{\partial^2 u}{\partial z^2} = 4r^2 f''\left(r^2\right) + 6f'\left(r^2\right)$

8. If $u = x^2 y + y^2 z + z^2 x$, prove that $\dfrac{\partial u}{\partial x} + \dfrac{\partial u}{\partial y} + \dfrac{\partial u}{\partial z} = \left(x + y + z\right)^2$.

9. Verify Euler's Theorem when $f(x, y, z) = 3x^2 yz + 4xy^2 z + 5y^4$.

10. Verify Euler's theorem when $u = \dfrac{x\left(x^3 - y^3\right)}{x^3 + y^3}$

11. Show that if $u = \cos^{-1}\left(\dfrac{x + y}{\sqrt{x} + \sqrt{y}}\right)$, then $x\dfrac{\partial u}{\partial x} + y\dfrac{\partial u}{\partial y} = -\dfrac{1}{2}\cot u$

12. If $u = \sin^{-1}\left(\dfrac{x^{1/3} + y^{1/3}}{x^{1/2} + y^{1/2}}\right)^{1/2}$, prove that $x^2 \dfrac{\partial^2 u}{\partial x^2} + 2xy\dfrac{\partial^2 u}{\partial y \partial y} + y^2 \dfrac{\partial^2 u}{\partial y^2} = \dfrac{1}{144}\tan u$.

13. If $u = \log\sqrt{x^2 + y^2 + z^2}$, then prove that $(x^2 + y^2 + z^2)\left(\dfrac{\partial^2 u}{\partial x^2} + \dfrac{\partial^2 u}{\partial y^2} + \dfrac{\partial^2 u}{\partial z^2}\right) = 1$.

14. If $u = \sec^{-1}\left(\dfrac{x^3 + y^3}{x + y}\right)$, prove that $x\dfrac{\partial u}{\partial x} + y\dfrac{\partial u}{\partial y} = 2\cot u$.

15. If $u = \sin^{-1}\left(x/y\right) + \tan^{-1}\left(y/x\right)$, then prove that $x\dfrac{\partial u}{\partial x} + y\dfrac{\partial u}{\partial y} = 0$.

16.   If $u = \dfrac{x^3 \cdot y^3}{x^3 + y^3}$, then prove that $x \dfrac{\partial u}{\partial x} + y \dfrac{\partial u}{\partial y} = 3u$.

17.   If $u = \log \dfrac{x^3 + y^3}{x^2 + y^2}$, then prove that $x \dfrac{\partial u}{\partial x} + y \dfrac{\partial u}{\partial y} = 1$.

9.4   Total Differential Coefficient

if $u = f(x, y)$, where $x = \phi(t)$ and $y = \psi(t)$, then u can be expressed as a function of t. By substituting the values of x and y in terms of t from the last two equations in the first equation. Then in the usual way, we can find $\dfrac{du}{dt}$, which is called the total differential coefficient of u with respect to t.

But sometimes, it becomes very difficult to express u in terms of t by eliminating x and y. so, we require to find $\dfrac{du}{dt}$ without actually substituting the values of x and y in terms of t in u. Let $\delta x, \delta y$ and δu be the increments in x, y and u corresponding to a small increment δt. Then

$u + \delta u = f(x + \delta x, y + \delta y)$

Where $x + \delta x = \phi(t + \delta t), y + \delta y = \psi(t + \delta t)$

Then by definition

$$\frac{du}{dt} = \lim_{\delta t \to 0} \frac{f(x+\delta x, y+\delta y) - f(x,y)}{\delta t}$$

Adding and subtracting $f(x + \delta x, y)$ on R.H.S, we get

$$\frac{du}{dt} = \lim_{\delta t \to 0} \frac{f(x+\delta x, y+\delta y) - f(x+\delta x, y) + f(x+\delta x, y) - f(x,y)}{\delta t}$$

$$= \lim_{\delta t \to 0} \left[\frac{f(x+\delta x, y+\delta y) - f(x+\delta x, y)}{\delta t} \cdot \frac{\delta y}{\delta t} + \frac{f(x+\delta x, y) - f(x,y)}{\delta x} \cdot \frac{\delta x}{\delta t} \right] \qquad \ldots..(1)$$

Now, proceeding to limit as $\delta t \to 0$, we have

$$\lim_{\delta t \to 0} \frac{\delta x}{\delta t} = \frac{dx}{dt} \text{ and } \lim_{\delta t \to 0} \frac{\delta y}{\delta t} = \frac{dy}{dt}$$

Also as $\delta t \to 0, \delta x \to 0$ and $\delta y \to 0$,

$$\lim_{\delta t \to 0} \frac{f(x+\delta x, y+\delta y) - f(x+\delta x, y)}{\delta y} = \lim_{\delta y \to 0} \frac{f(x+\delta x, y+\delta y) - f(x+\delta x, y)}{\delta y} = \frac{\partial u}{\partial y}$$

(as $x + \delta x$ remains unchanged while y changes to $y + \delta y$)

Similarly, $$\lim_{\delta t \to 0} \frac{f(x+\delta x, y) - f(x,y)}{\delta x} = \frac{\partial u}{\partial x}$$

$\therefore$ from equation (1), we have

$$\frac{du}{dt} = \frac{\partial u}{\partial y} \cdot \frac{dy}{dt} + \frac{\partial u}{\partial x} \frac{dx}{dt}$$

or
$$\frac{du}{dt} = \frac{\partial u}{\partial x} \cdot \frac{dx}{dt} + \frac{\partial u}{\partial y}\frac{dy}{dt} \qquad \qquad \dots(2)$$

In general, if $u = f(x_1, x_2, x_3, \dots \dots x_n)$, where $x_1, x_2, x_3, \dots \dots x_n$ are all functions of t, then

$$\frac{du}{dt} = \frac{\partial u}{\partial x_1} \cdot \frac{dx_1}{dt} + \frac{\partial u}{\partial x_2}\frac{dx_2}{dt} + \dots + \frac{\partial u}{\partial x_n}\frac{dx_n}{dt}$$

9.5   First Differential Coefficient of an Implicit Function

Let the implicit function be $f(x, y) = c$, where c is a constant and y is a function of x.

$\therefore$ from equation (2), we have

$$0 = \frac{\partial f}{\partial x} \cdot \frac{dx}{dx} + \frac{\partial f}{\partial y}\frac{dy}{dx}$$

$$\frac{dy}{dx} = -\frac{\frac{\delta f}{\partial x}}{\frac{\delta f}{\partial y}} = \frac{-p}{q}$$

Where $\qquad p = \frac{\delta f}{\partial x}, \quad q = \frac{\delta f}{\partial y}$

9.6 Second Differential Coefficient of an Implicit Function

Let $f(x, y) = c$ be an implicit function of x and y. then

$$\frac{dy}{dx} = -\frac{\frac{\delta f}{\partial x}}{\frac{\delta f}{\partial y}} \qquad \qquad \dots(1)$$

Let us denote $\frac{\partial f}{\partial x}, \frac{\partial f}{\partial y}, \frac{\partial^2 f}{\partial x^2}, \frac{\partial^2 f}{\partial x \partial y}$ and $\frac{\partial^2 f}{\partial y^2}$ by p, q r, s and t respectively.

Then from equation (1), we have

$$\frac{dy}{dx} = \frac{-p}{q}$$

$$\therefore \quad \frac{d^2 y}{dx^2} = \frac{d}{dx}\left(-\frac{p}{q}\right)$$

$$\frac{d^2 y}{dx^2} = -\left[\frac{q\frac{dp}{dx} - p\frac{dq}{dx}}{q^2}\right] \qquad \qquad \dots(2)$$

But $\qquad \dfrac{dp}{dx} = \dfrac{\partial p}{\partial x} + \dfrac{\partial p}{\partial y}\dfrac{dy}{dx}$

$$= \frac{\partial}{\partial x}\left(\frac{\partial f}{\partial x}\right) + \frac{\partial}{\partial y}\left(\frac{\partial f}{\partial x}\right)\frac{dy}{dx}$$

$$= \frac{\partial^2 f}{\partial x^2} + \frac{\partial^2 f}{\partial x \partial y} \cdot \frac{dy}{dx}$$

$$= r + s\left(-\frac{p}{q}\right)$$

$$= \frac{rq - sp}{q}$$

Similarly, $\dfrac{dq}{dx} = \dfrac{\partial q}{\partial x} + \dfrac{\partial q}{\partial y}\dfrac{dy}{dx}$

$$= \frac{\partial^2 f}{\partial x\,\partial y} + \frac{\partial^2 f}{\partial y^2}\frac{dy}{dx}$$

$$= s + t\left(-\frac{p}{q}\right)$$

$$= \frac{sq - pt}{q}$$

Substituting the values of $\dfrac{dp}{dx}$ and $\dfrac{dq}{dx}$ in equation (2), we get

$$\frac{d^2 y}{dx^2} = -\left(\frac{q^2 r - 2pqs + p^2 t}{q^3}\right)$$

***Example* 23:** Find $\dfrac{dy}{dx}$ if $x^y + y^x = c$.

***Solution*:** Let $f(x, y) = x^y + y^x - c = 0$

$$\frac{\partial f}{\partial x} = yx^{y-1} + y^x.\log y$$

and $\dfrac{\partial f}{\partial y} = x^y.\log x + xy^{x-1}$

$\therefore$ $\dfrac{dy}{dx} = -\dfrac{\frac{\partial f}{\partial x}}{\frac{\partial f}{\partial y}}$

$$= -\left(\frac{yx^{y-1} + y^x \log y}{x^y \log x + xy^{x-1}}\right)$$

***Example* 24:** If $u = \sqrt{x^2 + y^2}$ and $x^3 + y^3 + 3axy = 5a^2$, find the value of $\dfrac{du}{dx}$

when $x = a, y = a$.

***Solution*:** Given $u = \sqrt{x^2 + y^2}$

$\therefore$ $\dfrac{\partial u}{\partial x} = \dfrac{1}{2}(x^2 + y^2)^{-\frac{1}{2}}.2x = \dfrac{x}{\sqrt{x^2+y^2}}$

and $\dfrac{\partial u}{\partial y} = \dfrac{1}{2}(x^2 + y^2)^{-\frac{1}{2}}.2y = \dfrac{y}{\sqrt{x^2+y^2}}$

Now $x^3 + y^3 + 3axy = 5a^2$

Differentiating w. r. t. x, we have

$$3x^2 + 3y^2 \frac{dy}{dx} + 3a\left(x\frac{dy}{dx} + y\right) = 0$$

$\therefore$ $\qquad \dfrac{dy}{dx} = -\dfrac{x^2+ay}{y^2+ax}$

$\therefore$ $\qquad \dfrac{du}{dx} = \dfrac{\partial u}{\partial x} + \dfrac{\partial u}{\partial y}\cdot\dfrac{dy}{dx}$

$$= \frac{x}{\sqrt{x^2+y^2}} + \frac{y}{\sqrt{x^2+y^2}}\left(-\frac{x^2+ay}{y^2+ax}\right)$$

At $\qquad\qquad x = a, y = a,$

$$\frac{du}{dx} = \frac{a}{a\sqrt{2}} + \frac{a}{a\sqrt{2}}\left(-\frac{2a^2}{2a^2}\right) = 0$$

***Example* 25:** If $f(x, y) = 0, \phi(y, z) = 0$, show that

$$\frac{\partial f}{\partial y}\cdot\frac{\partial \phi}{\partial z}\cdot\frac{dz}{dx} = \frac{\partial f}{\partial x}\cdot\frac{\partial \phi}{\partial y}$$

Solution: If $\qquad f(x, y) = 0$

Then $\qquad \dfrac{dy}{dx} = -\dfrac{\frac{\partial f}{\partial x}}{\frac{\partial f}{\partial y}}$ $\qquad\qquad\qquad\qquad\qquad$ .....(i)

If $\qquad\qquad \phi(y, z) = 0$

Then $\qquad \dfrac{dz}{dy} = -\dfrac{\frac{\partial \phi}{\partial y}}{\frac{\partial \phi}{\partial z}}$ $\qquad\qquad\qquad\qquad\qquad$ .....(ii)

Multiplying (i) and (ii), we get

$$\frac{dy}{dx}\cdot\frac{dz}{dy} = \frac{\frac{\partial f}{\partial x}\frac{\partial \phi}{\partial y}}{\frac{\partial f}{\partial y}\frac{\partial \phi}{\partial z}}$$

$$\Rightarrow \frac{\partial f}{\partial y}\frac{\partial \phi}{\partial z}\frac{dz}{dx} = \frac{\partial f}{\partial x}\frac{\partial \phi}{\partial y}$$

***Example* 26:** If $u = x\log xy$, where $x^3 + y^3 + 3xy = 1$, find $\dfrac{du}{dx}$.

Solution: Given $u = x\log xy$

$\therefore$ $\qquad \dfrac{\partial u}{\partial x} = x.\dfrac{1}{xy}.y + \log xy$

$$= 1 + \log xy$$

and $\qquad \dfrac{\partial u}{\partial y} = x.\dfrac{i}{xy}.x = \dfrac{x}{y}$

Now $\qquad x^3 + y^3 + 3xy = 1$

Differentiating, we have

$$3x^2 + 3y^2 \frac{dy}{dx} + 3\left(x\frac{dy}{dx} + y\right) = 0$$

$$\Rightarrow \frac{dy}{dx} = -\frac{x^2+y}{y^2+x}$$

$$\therefore \frac{du}{dx} = \frac{\partial u}{\partial x} + \frac{\partial u}{\partial y}\frac{dy}{dx}$$

$$\Rightarrow \frac{du}{dx} = (1+\log xy) + \frac{x}{y}\left(-\frac{x^2+y}{y^2+x}\right)$$

***Example* 27:** If $u = f(y-z, z-x, x-y)$, prove that $\dfrac{\partial u}{\partial x} + \dfrac{\partial u}{\partial y} + \dfrac{\partial u}{\partial z} = 0$.

***Solution*:** Let $X = y-z, Y = z-x, Z = x-y$ .....(i)

Then $u = f(X, Y, Z)$, where X, Y, Z are functions of x, y, z

Then $\qquad \dfrac{\partial u}{\partial x} = \dfrac{\partial u}{\partial X}\dfrac{\partial X}{\partial x} + \dfrac{\partial u}{\partial Y}\dfrac{\partial Y}{\partial x} + \dfrac{\partial u}{\partial Z}\dfrac{\partial Z}{\partial x}$ .....(ii)

Also from (i), we get $\dfrac{\partial X}{\partial x} = 0, \dfrac{\partial X}{\partial y} = 1, \dfrac{\partial X}{\partial z} = -1$

$$\frac{\partial Y}{\partial x} = -1, \frac{\partial Y}{\partial y} = 0, \frac{\partial Y}{\partial z} = 1$$

And $\qquad \dfrac{\partial Z}{\partial x} = 1, \dfrac{\partial Z}{\partial y} = -1, \dfrac{\partial Z}{\partial z} = 0$

Substituting these values in (ii), we have

$$\frac{\partial u}{\partial x} = -\frac{\partial u}{\partial Y} + \frac{\partial u}{\partial Z} \qquad\qquad .....(iii)$$

Similarly, $\qquad \dfrac{\partial u}{\partial y} = \dfrac{\partial u}{\partial X}\dfrac{\partial X}{\partial y} + \dfrac{\partial u}{\partial Y}\dfrac{\partial Y}{\partial y} + \dfrac{\partial u}{\partial Z}\dfrac{\partial Z}{\partial y}$

$$= \frac{\partial u}{\partial X} - \frac{\partial u}{\partial Z} \qquad\qquad .....(iv)$$

And $\qquad \dfrac{\partial u}{\partial z} = \dfrac{\partial u}{\partial X}\dfrac{\partial X}{\partial z} + \dfrac{\partial u}{\partial Y}\dfrac{\partial y}{\partial z} + \dfrac{\partial u}{\partial Z}\dfrac{\partial Z}{\partial z}$

$$= -\frac{\partial u}{\partial X} + \frac{\partial u}{\partial Y} \qquad\qquad .....(v)$$

Adding (iii) (iv) and (v), we get

$$\frac{\partial u}{\partial x} + \frac{\partial u}{\partial y} + \frac{\partial u}{\partial z} = 0$$

***Example* 28:** If $z = z(u, v), u = x^2 - 2xy - y^2, v = y$, show that $(x+y)\dfrac{\partial z}{\partial x} + (x-y)\dfrac{\partial z}{\partial y} = 0$ is equivalent to $\dfrac{\partial z}{\partial v} = 0$.

***Solution*:** Given that z is a function of u and v and u and v are functions of x and y.

$$\therefore \frac{\partial z}{\partial x} = \frac{\partial z}{\partial u}\frac{\partial u}{\partial x} + \frac{\partial z}{\partial v}\frac{\partial v}{\partial x} \qquad\qquad .....(i)$$

and $$\frac{\partial Z}{\partial y} = \frac{\partial Z}{\partial u}\frac{\partial u}{\partial y} + \frac{\partial Z}{\partial v}\frac{\partial v}{\partial y}$$ .....(ii)

Now, given $u = x^2 - 2xy - y^2$ and $v = y$

$\therefore \qquad \frac{\partial u}{\partial x} = 2x - 2y, \frac{\partial u}{\partial y} = -2x - 2y$

and $\qquad \frac{\partial v}{\partial x} = 0, \frac{\partial v}{\partial y} = 1$

$\therefore$ from equation (i) and (ii), we have

$$\frac{\partial z}{\partial x} = 2(x - y)\frac{\partial z}{\partial u}$$

and $$\frac{\partial z}{\partial y} = -2(x + y)\frac{\partial z}{\partial u} + \frac{\partial z}{\partial v}$$

$\therefore \qquad (x + y)\frac{\partial z}{\partial x} + (x - y)\frac{\partial z}{\partial y} = 0$

$$\Rightarrow (x + y)\left\{2(x - y)\frac{\partial z}{\partial u}\right\} + (x - y)\left\{-2(x + y)\frac{\partial z}{\partial u} + \frac{\partial z}{\partial v}\right\} = 0$$

$$\Rightarrow (x - y)\frac{\partial z}{\partial v} = 0$$

$$\Rightarrow \frac{\partial z}{\partial v} = 0$$

Example 29: Find $\frac{d^2y}{dx^2}$ if $ax^2 + 2hxy + by^2 = 1$.

Solution: Let $\quad f = ax^2 + 2hxy + by^2 - 1 = 0$

$$p = \frac{\partial f}{\partial x} = 2(ax + by)$$

$$q = \frac{\partial f}{\partial y} = 2(hx + by)$$

$$r = \frac{\partial^2 f}{\partial x^2} = 2a, s = \frac{\partial^2 f}{\partial x\,\partial y} = 2h, t = \frac{\partial^2 f}{\partial y^2} = 2b$$

$\therefore \qquad \frac{d^2y}{dx^2} = -\left(\frac{q^2r - 2pqs + p^2t}{q^3}\right)$

$$= -\left[\frac{(hx+by)^2a - 2(ax+hy)(hx+by)h + (ax+by)^2b}{(hx+by)^3}\right]$$

$$= \frac{h^2 - ab}{(hx+by)^2}$$

Example 30: If $u = 3(lx + my + nz)^2 - (x^2 + y^2 + z^2)$ and $l^2 + m^2 + n^2 = 1$, show that
$\frac{\partial^2 u}{\partial x^2} + \frac{\partial^2 u}{\partial y^2} + \frac{\partial^2 u}{\partial z^2} = 0$.

Solution: Given $u = 3(lx + my + nz)^2 - (x^2 + y^2 + z^2)$

$\therefore \qquad \dfrac{\partial u}{\partial x} = 6l(lx + my + nz) - 2x$

$\dfrac{\partial^2 u}{\partial x^2} = 6l^2 - 2$

Similarly $\quad \dfrac{\partial^2 u}{\partial y^2} = 6m^2 - 2$ and $\dfrac{\partial^2 u}{\partial z^2} = 6n^2 - 2$

$\therefore \qquad \dfrac{\partial^2 u}{\partial x^2} + \dfrac{\partial^2 u}{\partial y^2} + \dfrac{\partial^2 u}{\partial z^2} = 6l^2 - 2 + 6m^2 - 2 + 6n^2 - 2$

$$= 6(l^2 + m^2 + n^2) - 6$$

$$= 0 \qquad\qquad (\because l^2 + m^2 + n^2 = 1)$$

***Example* 31**: If $u = u\left(\dfrac{y-x}{xy}, \dfrac{z-x}{xz}\right)$, show that $x^2 \dfrac{\partial u}{\partial x} + y^2 \dfrac{\partial u}{\partial y} + z^2 \dfrac{\partial u}{\partial z} = 0$.

Solution: Let $x_1 = \dfrac{y-x}{xy} = \dfrac{1}{x} - \dfrac{1}{y}, x_2 = \dfrac{z-x}{xz} = \dfrac{1}{x} - \dfrac{1}{z}$

Then $u = u(x_1, x_2)$ and x_1, x_2 be the function of x, y, z

Now $\qquad \dfrac{\partial x_1}{\partial x} = \dfrac{-1}{x^2}, \dfrac{\partial x_1}{\partial y} = \dfrac{1}{y_2}, \dfrac{\partial x_1}{\partial z} = 0$

And $\qquad \dfrac{\partial x_2}{\partial x} = \dfrac{-1}{x^2}, \dfrac{\partial x_2}{\partial y} = 0, \dfrac{\partial x_2}{\partial z} = \dfrac{1}{z^2}$

Also, $\qquad \dfrac{\partial u}{\partial x} = \dfrac{\partial u}{\partial x_1}\dfrac{\partial x_1}{\partial x} + \dfrac{\partial u}{\partial x_2}\dfrac{\partial x_2}{\partial x} = \dfrac{\partial u}{\partial x_1}\left(\dfrac{-1}{x^2}\right) + \dfrac{\partial u}{\partial x_2}\left(\dfrac{-1}{x^2}\right)$

or $\qquad x^2 \dfrac{\partial u}{\partial x} = \dfrac{-\partial u}{\partial x_1} - \dfrac{\partial u}{\partial x_2}$ $\qquad\qquad\qquad$ .....(i)

Similarly $\qquad \dfrac{\partial u}{\partial y} = \dfrac{\partial u}{\partial x_1}\dfrac{\partial x_1}{\partial y} + \dfrac{\partial u}{\partial x_2}\dfrac{\partial x_2}{\partial y} = \dfrac{\partial u}{\partial x_1}\left(\dfrac{1}{y^2}\right)$

or $\qquad y^2 \dfrac{\partial u}{\partial y} = \dfrac{\partial u}{\partial x_1}$ $\qquad\qquad\qquad\qquad\qquad$ .....(ii)

and $\qquad \dfrac{\partial u}{\partial z} = \dfrac{\partial u}{\partial x_1}\dfrac{\partial x_1}{\partial z} + \dfrac{\partial u}{\partial x_2}\dfrac{\partial x_2}{\partial z} = \dfrac{\partial u}{\partial x_1}(0) + \dfrac{\partial u}{\partial x_2}\left(\dfrac{1}{z^2}\right)$

$z^2 \dfrac{\partial u}{\partial z} = \dfrac{\partial u}{\partial x_2}$ $\qquad\qquad\qquad\qquad\qquad$ .....(iii)

Adding (i) (ii) and (iii), we get

$$x^2 \dfrac{\partial u}{\partial x} + y^2 \dfrac{\partial u}{\partial y} + z^2 \dfrac{\partial u}{\partial z} = 0$$

9.7 Change of Independent Variable into Dependent Variable

Let $y = f(x)$, then

$$\frac{dy}{dx} = \frac{1}{\frac{dx}{dy}} = \left(\frac{dx}{dy}\right)^{-1} \qquad \qquad \text{.....(1)}$$

and $\qquad \dfrac{d^2y}{dx^2} = \dfrac{d}{dx}\left(\dfrac{dy}{dx}\right)$

$$= \frac{d}{dx}\left(\frac{dx}{dy}\right)^{-1}$$

$$= \frac{d}{dy}\left(\frac{dx}{dy}\right)^{-1}\frac{dy}{dx}$$

$$= -\left(\frac{dx}{dy}\right)^{-2}\cdot\frac{d^2x}{dy^2}\left(\frac{dx}{dy}\right)^{-1}$$

$$= -\left(\frac{dx}{dy}\right)^{-3}\frac{d^2x}{dy^2} \qquad \qquad \text{.....(2)}$$

and so on.

9.8 To Change the Independent Variable x into another Variable t

Let $y = f(x)$, where $x = f(t)$

Then $\qquad \dfrac{dy}{dx} = \dfrac{dy}{dt}\cdot\dfrac{dt}{dx}$

$$= \frac{dy}{dt}\left(\frac{dx}{dt}\right)^{-1} \qquad \qquad \text{.....(1)}$$

and $\qquad \dfrac{d^2y}{dx^2} = \dfrac{d}{dx}\left(\dfrac{dy}{dx}\right)$

$$= \left(\frac{dx}{dt}\right)^{-1}\frac{d}{dt}\left[\left(\frac{dx}{dt}\right)^{-1}\frac{dy}{dt}\right] \qquad \left[\because \frac{d}{dx} = \left(\frac{dx}{dt}\right)^{-1}\frac{d}{dt}\right]$$

$$= \left(\frac{dx}{dt}\right)^{-1}\left[\left(\frac{dx}{dt}\right)^{-1}\frac{d^2y}{dt^2} + \frac{dy}{dt}(-1)\left(\frac{dx}{dt}\right)^{-2}\frac{d^2x}{dt^2}\right]$$

$$= \left(\frac{dx}{dt}\right)^{-3}\left[\frac{dx}{dt}\frac{d^2y}{dt^2} - \frac{d^2x}{dt^2}\frac{dy}{dt}\right] \qquad \qquad \text{....(2)}$$

and so on.

***Example* 32:** Transform the equation

$$\sin^2 2z\frac{d^2y}{dz^2} + \sin 4z\frac{dy}{dz} + 4y = 0 \ \text{ by putting } \tan z = e^x.$$

Solution: Given $\tan z = e^x$

$$\therefore \sec^2 z \frac{dz}{dx} = e^x = \tan z$$

$$\therefore \frac{dz}{dx} = \frac{\tan z}{\sec^2 z} = \sin z \cos z$$

or

$$\frac{dz}{dx} = \frac{1}{2} \sin 2z$$

Now,

$$\frac{dy}{dz} = \frac{dy}{dx} \cdot \frac{dx}{dz}$$

$$= \frac{dy}{dx} \cdot \frac{2}{\sin 2z}$$

$$\Rightarrow \frac{d}{dz} = \frac{2}{\sin 2z} \frac{d}{dx}$$

$$\therefore \frac{d^2y}{dz^2} = \frac{d}{dz}\left(\frac{dy}{dz}\right)$$

$$= \frac{2}{\sin 2z} \frac{d}{dx}\left(\frac{2}{\sin 2z} \frac{dy}{dx}\right)$$

$$= \frac{2}{\sin 2z}\left[\frac{2}{\sin 2z} \frac{d^2y}{dx^2} - \frac{4\cos 2z}{\sin^2 2z} \frac{dz}{dx}\frac{dy}{dx}\right]$$

$$= \frac{2}{\sin 2z}\left[\frac{2}{\sin 2z} \frac{d^2y}{dx^2} - \frac{4\cos 2z}{\sin^2 2z} \cdot \frac{\sin 2z}{2}\frac{dy}{dx}\right]$$

$$\frac{4}{\sin^2 2z} \frac{d^2y}{dx^2} - \frac{4\cos 2z}{\sin^2 2z}\frac{dy}{dx}$$

Putting the values of $\dfrac{dy}{dz}$ and $\dfrac{d^2y}{dz^2}$ in the given equation, we have

$$\sin^2 2z\left[\frac{4}{\sin^2 2z} \frac{d^2y}{dx^2} - \frac{4\cos 2z}{\sin^2 2z}\frac{dy}{dx}\right] + \sin 4z \cdot \frac{2}{\sin 2z}\frac{dy}{dx} + 4y = 0$$

$$\Rightarrow 4\frac{d^2y}{dx^2} - 4\cos 2z\frac{dy}{dx} + 4\cos 2z\frac{dy}{dx} + 4y = 0$$

$$\Rightarrow \frac{d^2y}{dx^2} + y = 0$$

Which is the required transformed equation.

***Example* 33**: Transform the equation

$$\frac{d^2y}{dx^2} + \cot x\frac{dy}{dx} + 4y\,\text{cosec}^2 x = 0 \text{ by putting } z = 2\log\tan\frac{x}{2}.$$

Solution: Given $z = 2\log\tan\dfrac{x}{2}$

$$\therefore \qquad \frac{dz}{dx} = \frac{2}{\tan\frac{x}{2}}\sec^2\frac{x}{2} \cdot \frac{1}{2} = \frac{2}{\sin x}$$

Now,
$$\frac{dy}{dx} = \frac{dy}{dz} \cdot \frac{dz}{dx}$$

$$= \frac{2}{\sin x} \frac{dy}{dz}$$

$$\Rightarrow \frac{d}{dx} = \frac{2}{\sin x} \frac{d}{dz}$$

Again
$$\frac{d^2y}{dx^2} = \frac{d}{dx}\left(\frac{dy}{dx}\right)$$

$$= \frac{2}{\sin x}\frac{d}{dz}\left(\frac{2}{\sin x}\frac{dy}{dz}\right)$$

$$= \frac{2}{\sin x}\left[\frac{2}{\sin x}\frac{d^2y}{dz^2} - \frac{2\cos x}{\sin^2 x}\frac{dx}{dz}\frac{dy}{dz}\right]$$

$$= \frac{2}{\sin x}\left[\frac{2}{\sin x}\frac{d^2y}{dz^2} - \frac{2\cos x}{\sin^2 x}\cdot\frac{\sin x}{2}\frac{dy}{dz}\right]$$

$$= \frac{4}{\sin^2 x}\frac{d^2y}{dz^2} - \frac{2\cos x}{\sin^2 x}\frac{dy}{dz}$$

Putting the values of $\frac{dy}{dx}$ and $\frac{d^2y}{dx^2}$ in the given equation, we get

$$\frac{4}{\sin^2 x}\frac{d^2y}{dz^2} - \frac{2\cos x}{\sin^2 x}\frac{dy}{dz} + \cot x \cdot \frac{2}{\sin x}\frac{dy}{dz} + 4y\,\text{cosec}^2 x = 0$$

$$\Rightarrow 4\,\text{cosec}^2 x\,\frac{d^2y}{dz^2} + 4y\,\text{cosec}^2 x = 0$$

$$\Rightarrow \frac{d^2y}{dz^2} + y = 0$$

Which is the required transformed equation.

9.9   Change of Two Independent Variables

If $u = f(x, y)$, where $x = \emptyset(t_1, t_2)$ and $y = \psi(t_1, t_2)$

ie. if u is a function of x and y where x and y be the functions of two other variables t_1 and t_2 then

$$\frac{\partial u}{\partial t_1} = \frac{\partial u}{\partial x}\frac{\partial x}{\partial t_1} + \frac{\partial u}{\partial y}\frac{\partial y}{\partial t_1}$$

And
$$\frac{\partial u}{\partial t_2} = \frac{\partial u}{\partial x}\frac{\partial x}{\partial t_2} + \frac{\partial u}{\partial y}\frac{\partial y}{\partial t_2}$$

Now, if $t_1 = F_1(x, y)$ and $t_2 = F_2(x, y)$

Then
$$\frac{\partial u}{\partial x} = \frac{\partial u}{\partial t_1}\frac{\partial t_1}{\partial x} + \frac{\partial u}{\partial t_2}\frac{\partial t_2}{\partial x}$$

and $$\frac{\partial u}{\partial y} = \frac{\partial u}{\partial t_1}\frac{\partial t_1}{\partial y} + \frac{\partial u}{\partial t_2}\frac{\partial t_2}{\partial y}$$

***Example* 34:** If $x = r\cos\theta, y = r\sin\theta$ show that

$$\frac{\partial r}{\partial x} = \frac{\partial x}{\partial r},\ \ \frac{\partial x}{r\,\partial\theta} = \frac{r\,\partial\theta}{\partial x}\ \text{and find the value of }\ \frac{\partial^2\theta}{\partial x^2} + \frac{\partial^2\theta}{\partial y^2}$$

Solution: If $x = r\cos\theta, y = r\sin\theta$

Then we have $\quad r^2 = x^2 + y^2$ $\hfill .....(i)$

and $\quad\quad\quad\quad \theta = \tan^{-1}(y/x)$ $\hfill .....(ii)$

From (i), we get $2r\dfrac{\partial r}{\partial x} = 2x$

or $\quad\quad\quad \dfrac{\partial r}{\partial x} = \dfrac{x}{r} = \dfrac{r\cos\theta}{r} = \cos\theta$ $\hfill ....(iii)$

Also from $\quad\quad x = r\cos\theta$

$$\therefore \frac{\partial x}{\partial r} = \cos\theta \hspace{4cm} .....(iv)$$

From equation (iii) and (iv), we have

$$\frac{\partial r}{\partial x} = \frac{\partial x}{\partial r}$$

Again from $x = r\cos\theta$

$$\therefore \frac{\partial x}{\partial\theta} = -r\sin\theta$$

$$\Rightarrow \frac{\partial x}{r\,\partial\theta} = -\sin\theta \hspace{3cm} .....(v)$$

and from (ii) $\quad \dfrac{\partial\theta}{\partial x} = \dfrac{1}{1+\left(\frac{y}{x}\right)^2}\cdot\left(\dfrac{-y}{x^2}\right)$

$$= \frac{-y}{x^2+y^2} = -\frac{r\sin\theta}{r^2}$$

$$\therefore \frac{r\,\partial\theta}{\partial x} = -\sin\theta \hspace{3cm} .....(vi)$$

From (v) and (vi), we have

$$\frac{\partial x}{r\,\partial\theta} = \frac{r\,\partial\theta}{\partial x}$$

Now, $\quad\quad\quad \dfrac{\partial\theta}{\partial x} = -\dfrac{y}{x^2+y^2}$

$$\therefore \frac{\partial^2\theta}{\partial x^2} = y\frac{1}{(x^2+y^2)^2}\cdot 2x = \frac{2xy}{(x^2+y^2)^2} \hspace{1.5cm} .....(vii)$$

Also from (ii) $\dfrac{\partial\theta}{\partial y} = \dfrac{1}{1+\left(\frac{y}{x}\right)^2}\cdot\dfrac{1}{x} = \dfrac{x}{x^2+y^2}$

$$\therefore \frac{\partial^2\theta}{\partial y^2} = \frac{-2xy}{(x^2+y^2)^2} \qquad\qquad .....(viii)$$

Adding equation (vii) and (viii), we have

$$\therefore \frac{\partial^2\theta}{\partial x^2} + \frac{\partial\theta^2}{\partial y^2} = 0$$

***Example* 35:** If $x = r\cos\theta, y = r\sin\theta,$ prove that

$$\frac{\partial^2 r}{\partial x^2} + \frac{\partial^2 r}{\partial y^2} = \frac{1}{r}\left\{\left(\frac{\partial r}{\partial x}\right)^2 + \left(\frac{\partial r}{\partial y}\right)^2\right\}$$

Solution: If $x = r\cos\theta, y = r\sin\theta$

Then $\dfrac{\partial r}{\partial x} = \dfrac{x}{r}, \dfrac{\partial r}{\partial y} = \dfrac{y}{r}, \dfrac{\partial^2 r}{\partial x^2} = \dfrac{r^2-x^2}{r^3}, \dfrac{\partial^2 r}{\partial y^2} = \dfrac{r^2-y^2}{r^3}$

$$\therefore \frac{\partial^2 r}{\partial x^2} + \frac{\partial^2 r}{\partial y^2} = \frac{r^2-x^2}{r^3} + \frac{r^2-y^2}{r^3}$$

$$= \frac{2r^2-(x^2+y^2)}{r^3}$$

$$= \frac{2r^2-r^2}{r^3} = \frac{1}{r} \qquad\qquad [\because (x^2 + y^2) = r^2]$$

Also $\dfrac{1}{r}\left\{\left(\dfrac{\partial r}{\partial x}\right)^2 + \left(\dfrac{\partial r}{\partial y}\right)^2\right\} = \dfrac{1}{r}\left\{\dfrac{x^2}{r^2} + \dfrac{y^2}{r^2}\right\} = \dfrac{1}{r}$

Hence $\dfrac{\partial^2 r}{\partial x^2} + \dfrac{\partial^2 r}{\partial y^2} = \dfrac{1}{r}\left\{\left(\dfrac{\partial r}{\partial x}\right)^2 + \left(\dfrac{\partial r}{\partial y}\right)^2\right\}$

***Example* 36:** If $x = e^\theta, y = e^\phi$, transform the equation

$$e^{2\theta} \frac{\partial^2 v}{\partial x^2} + e^{2\phi} \frac{\partial^2 v}{\partial y^2} + e^\theta \frac{\partial v}{\partial x} + e^\phi \frac{\partial v}{\partial y} = 0$$

so that θ and ϕ are the independent variables.

Solution: we have $x = e^\theta \Rightarrow \theta = \log x$

And $y = e^\phi \Rightarrow \phi = \log y$

$$\therefore \frac{\partial v}{\partial x} = \frac{\partial v}{\partial\theta}\cdot\frac{\partial\theta}{\partial x} + \frac{\partial v}{\partial\phi}\frac{\partial\phi}{\partial x}$$

$$= \frac{\partial v}{\partial\theta}\cdot\frac{1}{x} + \frac{\partial v}{\partial\phi}\cdot 0 = \frac{1}{x}\frac{\partial v}{\partial\theta}$$

$$\Rightarrow x\frac{\partial v}{\partial x} = \frac{\partial v}{\partial\theta}$$

$$\Rightarrow x \frac{\partial}{\partial x} = \frac{\partial}{\partial \theta}$$

Now
$$x \frac{\partial}{\partial x}\left(x \frac{\partial v}{\partial x}\right) = \frac{\partial}{\partial \theta}\left(\frac{\partial v}{\partial \theta}\right)$$

$$\Rightarrow x\left[x \frac{\partial^2 v}{\partial x^2} + \frac{\partial v}{\partial x}\right] = \frac{\partial^2 v}{\partial \theta^2}$$

$$\Rightarrow x^2 \frac{\partial^2 v}{\partial x^2} + x \frac{\partial v}{\partial x} = \frac{\partial^2 v}{\partial \theta^2}$$

$$\Rightarrow e^{2\theta} \frac{\partial^2 v}{\partial x^2} + e^{\theta} \frac{\partial v}{\partial x} = \frac{\partial^2 v}{\partial \theta^2} \qquad \qquad .....(i)$$

Similarly, $\quad e^{2\phi} \dfrac{\partial^2 v}{\partial y^2} + e^{\phi} \dfrac{\partial v}{\partial y} = \dfrac{\partial^2 v}{\partial \phi^2} \qquad \qquad .....(ii)$

Adding (i) and (ii), we get

$$e^{2\theta} \frac{\partial^2 v}{\partial x^2} + e^{\theta} \frac{\partial v}{\partial x} + e^{2\phi} \frac{\partial^2 v}{\partial y^2} + e^{\phi} \frac{\partial v}{\partial y} = \frac{\partial^2 v}{\partial \theta^2} + \frac{\partial^2 v}{\partial \phi^2}$$

***Example* 37:** Transform the equation

$$\frac{\partial^2 z}{\partial x^2} + 2xy^2 \frac{\partial z}{\partial x} + 2(y - y^3) \frac{\partial z}{\partial y} + x^2 y^2 z = 0$$

by the substitution $x = uv, y = \dfrac{1}{v}$ and hence, show that z is the same function of u and v as of x and y .

***Solution*:** Given $x = uv, y = \dfrac{1}{v}$

$$\Rightarrow u = xy, v = \frac{1}{y}$$

$$\therefore \frac{\partial z}{\partial x} = \frac{\partial z}{\partial u}\frac{\partial u}{\partial x} + \frac{\partial z}{\partial v}\frac{\partial v}{\partial x}$$

$$= \frac{\partial z}{\partial u}\cdot y + \frac{\partial z}{\partial v}\cdot 0 = \frac{1}{v}\frac{\partial z}{\partial u} \Rightarrow \frac{\partial}{\partial x} = \frac{1}{v}\frac{\partial}{\partial u}$$

and
$$\frac{\partial^2 z}{\partial x^2} = \frac{\partial}{\partial x}\left(\frac{\partial z}{\partial x}\right)$$

$$= \frac{1}{v}\frac{\partial}{\partial u}\left(\frac{1}{v}\frac{\partial z}{\partial u}\right)$$

$$= \frac{1}{v^2}\frac{\partial^2 z}{\partial u^2}$$

Also
$$\frac{\partial z}{\partial y} = \frac{\partial z}{\partial u}\frac{\partial u}{\partial y} + \frac{\partial z}{\partial v}\frac{\partial v}{\partial y}$$

$$= \frac{\partial z}{\partial u}\cdot x + \frac{\partial z}{\partial v}\left(-\frac{1}{y^2}\right)$$

$$\therefore y\frac{\partial z}{\partial y} = xy\frac{\partial z}{\partial u} - \frac{1}{y}\frac{\partial z}{\partial v}$$

$$= u\frac{\partial z}{\partial u} - v\frac{\partial z}{\partial v}$$

Putting these values in the given equation, we have

$$\frac{1}{v^2}\frac{\partial^2 z}{\partial u^2} + \frac{2u}{v}\left(\frac{1}{v}\frac{\partial z}{\partial u}\right) + 2\left(1 - \frac{1}{v^2}\right)\left(u\frac{\partial z}{\partial u} - v\frac{\partial z}{\partial v}\right) + u^2 z = 0$$

$$\Rightarrow \frac{\partial^2 z}{\partial u^2} + \frac{\partial z}{\partial u}(2u + 2u(v^2 - 1)) - 2v(v^2 - 1)\frac{\partial z}{\partial v} + u^2 v^2 z = 0$$

$$\Rightarrow \frac{\partial^2 z}{\partial u^2} + 2uv^2\frac{\partial z}{\partial u} + 2(v - v^3)\frac{\partial z}{\partial v} + u^2 v^2 z = 0$$

This equation is exactly similar to the given equation which shows that z is the same function of u and v as of x and y

***Example* 38:** Prove that $\dfrac{\partial^2 u}{\partial x^2} + \dfrac{\partial^2 u}{\partial y^2} = \dfrac{\partial^2 u}{\partial \zeta^2} + \dfrac{\partial^2 u}{\partial \eta^2}$,

Where $x = \zeta\cos\alpha - \eta\sin\alpha, y = \sin\alpha + \eta\cos\alpha$.

Solution: Given $x = \zeta\cos\alpha - \eta\sin\alpha, y = \zeta\sin\alpha + \eta\cos\alpha$

$$\therefore \frac{\partial x}{\partial \zeta} = \cos\alpha, \frac{\partial x}{\partial \eta} = -\sin\alpha, \frac{\partial y}{\partial \zeta} = \sin\alpha, \frac{\partial y}{\partial \eta} = \cos\alpha$$

$$\therefore \frac{\partial u}{\partial \zeta} = \frac{\partial u}{\partial x}\frac{\partial x}{\partial \zeta} + \frac{\partial u}{\partial y}\frac{\partial y}{\partial \zeta} = \frac{\partial u}{\partial x}\cos\alpha + \frac{\partial u}{\partial y}\sin\alpha$$

$$\therefore \frac{\partial^2 u}{\partial \zeta^2} = \frac{\partial}{\partial \zeta}\left(\frac{\partial u}{\partial \zeta}\right) = \cos\alpha\frac{\partial}{\partial x}\left(\frac{\partial u}{\partial \zeta}\right) + \sin\alpha\frac{\partial}{\partial y}\left(\frac{\partial u}{\partial \zeta}\right)$$

$$= \cos\alpha\frac{\partial}{\partial x}\left[\cos\alpha\frac{\partial u}{\partial x} + \sin\alpha\frac{\partial u}{\partial y}\right] + \sin\alpha\frac{\partial}{\partial y}\left[\cos\alpha\frac{\partial u}{\partial x} + \sin\alpha\frac{\partial u}{\partial y}\right]$$

$$\frac{\partial^2 u}{\partial \zeta^2} = \cos^2\alpha\frac{\partial^2 u}{\partial x^2} + 2\sin\alpha\cos\alpha\frac{\partial^2 u}{\partial x\,\partial y} + \sin\alpha\frac{\partial^2 u}{\partial y^2}$$

Similarly,
$$\frac{\partial^2 u}{\partial \eta^2} = \sin^2\alpha\frac{\partial^2 u}{\partial x^2} - 2\sin\alpha\cos\alpha\frac{\partial^2 u}{\partial x\,\partial y} + \cos^2\alpha\frac{\partial^2 u}{\partial y^2}$$

$$\therefore \frac{\partial^2 u}{\partial \zeta^2} + \frac{\partial^2 u}{\partial \eta^2} = \frac{\partial^2 u}{\partial x^2} + \frac{\partial^2 u}{\partial y^2}$$

***Example* 39:** If $\theta = t^n e^{-\frac{r^2}{4t}}$, find what value of n will make $\dfrac{1}{r^2}\dfrac{\partial}{\partial r}\left(r^2\dfrac{\partial\theta}{\partial r}\right) = \dfrac{\partial\theta}{\partial t}$

Solution: Given $\theta = t^n e^{-\frac{r^2}{4t}}$ .....(i)

Differentiating partially w. r. t. t, we get

$$\frac{\partial\theta}{\partial t} = t^n\, e^{-\frac{r^2}{4t}}.\left(\frac{r^2}{4t^2}\right) + e^{-\frac{r^2}{4t}}.nt^{n-1}$$

$$= \frac{1}{4}\, t^{n-2}\, e^{-\frac{r^2}{4t}}(r^2 + 4nt)$$

Again from (1), $\dfrac{\partial\theta}{\partial r} = t^n\, e^{-\frac{r^2}{4t}}\left(\dfrac{-2r}{4t}\right)$

$$= -\frac{1}{2}\, t^{n-1}r\, e^{-\frac{r^2}{4t}}$$

Now, $\dfrac{\partial}{\partial r}\left(r^2\dfrac{\partial\theta}{\partial r}\right) = \dfrac{\partial}{\partial r}\left[r^2.\left(-\dfrac{1}{2}\right)t^{n-1}\, r\, e^{-\frac{r^2}{4t}}\right]$

$$= \frac{\partial}{\partial r}\left[-\frac{1}{2}r^3 t^{n-1}\, e^{-\frac{r^2}{4t}}\right]$$

$$= -\frac{1}{2}t^{n-1}\left[r^3\, e^{-\frac{r^2}{4t}}\left(\frac{-2r}{4t}\right) + 3r^2\, e^{-\frac{r^2}{4t}}\right]$$

or $\dfrac{1}{r^2}\dfrac{\partial}{\partial r}\left(r^2\dfrac{\partial\theta}{\partial r}\right) = \dfrac{1}{4}t^{n-2}(r^2 - 6t)e^{-\frac{r^2}{4t}}$

If $\dfrac{1}{r^2}\dfrac{\partial}{\partial r}\left(r^2\dfrac{\partial\theta}{\partial r}\right) = \dfrac{\partial\theta}{\partial t}$

Then $r^2 - 6t = r^2 + 4nt$

or $4n = -6$

$$\Rightarrow n = -\frac{3}{2}$$

Practice Problems

1. Find $\dfrac{du}{dx}$ if $u = \sin(x^2 + y^2)$, where $a^2x^2 + b^2y^2 = c^2$

$$\left[\textbf{Ans. } 2x\cos(x^2 + y^2)\left(1 - \frac{a^2}{b^2}\right)\right]$$

2. If $u = x^2y$, where $x^2 + xy + y^2 = 1$, find $\dfrac{dy}{dx}$ $\left[\textbf{Ans. } 2xy - x^2\left(\dfrac{2x+y}{x+2y}\right)\right]$

3. If the curves $f(x, y) = 0$ and $\phi(x, y) = 0$ touch, show that at the point of contact
$$\frac{\partial f}{\partial x}\frac{\partial\phi}{\partial y} - \frac{\partial f}{\partial y}\frac{\partial\phi}{\partial x} = 0$$

4. If $ax^2 + 2hxy + by^2 + 2gx + 2fy + c = 0$ Prove that $\dfrac{d^2y}{dx^2} = \dfrac{abc+2fgh-af^2-bg^2-ch^2}{(hx+by+f)^3}$

5. If $z = f(u, v), u = x^2 - 2xy - y^2, v = y$, show that $(x + y)\dfrac{\partial z}{\partial x} + (x - y)\dfrac{\partial z}{\partial y} = 0$

6. If z is a function of x & y, prove that, if $x = e^u + e^{-v}, y = e^{-u} - e^v$,

$$\frac{\partial z}{\partial u} - \frac{\partial z}{\partial v} = x\frac{\partial z}{\partial x} - y\frac{\partial z}{\partial y}$$

7. If x, y, z are connected by the equations $\phi(x, y, z) = 0$ and $\psi(x, y, z) = 0$, find $\frac{dy}{dx}$

$$\left[\textbf{Ans. } \frac{\dfrac{\partial\phi}{\partial z}\dfrac{\partial\psi}{\partial x} - \dfrac{\partial\psi}{\partial z}\dfrac{\partial\phi}{\partial x}}{\dfrac{\partial\psi}{\partial z}\dfrac{\partial\phi}{\partial y} - \dfrac{\partial\psi}{\partial y}\dfrac{\partial\phi}{\partial z}}\right]$$

8. If $u = f(y - x, z - y)$ prove that $\frac{\partial u}{\partial x} + \frac{\partial u}{\partial y} + \frac{\partial u}{\partial z} = 0$

9. Change the independent variable from x to z in the equation $x^2\frac{d^2y}{dx^2} + x\frac{dy}{dx} + y = 0$, where $x = \frac{1}{z}$

$$\left[\textbf{Ans. } z^2\frac{d^2y}{dz^2} + z\frac{dy}{dz} + y = 0\right]$$

10. Transform the equation $(1 - x^2)\frac{d^2y}{dx^2} - x\frac{dy}{dx} = 0$, where $x = \cos\theta$

$$\left[\textbf{Ans. } \frac{d^2y}{d\theta^2} = 0\right]$$

11. Transform the equation $\frac{d^2y}{dx^2} + xy\frac{dy}{dx} + \sec^2 x = 0$, where $x = \tan^{-1} z$

$$\left[\textbf{Ans. } \frac{d^2y}{dz^2} + y = 0\right]$$

12. Transform the equation $\cos x\left(\frac{d^2y}{dx^2}\right) + \sin x\frac{dy}{dx} - 2y\cos^3 x = 2\cos^5 x$ into one in which z is independent variable given by $z = \sin x$
$$\left[\textbf{Ans. } \frac{d^2y}{dz^2} - 2y = 2(1 - z^2)\right]$$

13. If $x = r\cos\theta, y = r\sin\theta$, prove that $\left(\frac{\partial r}{\partial x}\right)^2 + \left(\frac{\partial r}{\partial y}\right)^2 = 1$.

14. If $x = r\cos\theta, y = r\sin\theta$, prove that $\frac{\partial^2 r}{\partial x^2} \cdot \frac{\partial^2 r}{\partial y^2} = \left(\frac{\partial^2 r}{\partial y\,\partial x}\right)^2$

15. If $v = At^{-1/2}e^{-x^2/4a^2 t}$, prove that $\frac{\partial v}{\partial t} = a^2\frac{\partial^2 v}{\partial x^2}$

16. If $x = e^v\sec u, y = e^v\tan u$ and ϕ is a function of x and y, show that

$$\cos u\left(\frac{\partial^2\phi}{\partial u\,\partial v} - \frac{\partial\phi}{\partial u}\right) = xy\left(\frac{\partial^2\phi}{\partial x^2} + \frac{\partial^2\phi}{\partial y^2}\right) + (x^2 + y^2)\frac{\partial^2\phi}{\partial x\,\partial y}$$

17. If $u = f(x, y), x^2 = \xi n, y^2 = \frac{\xi}{n}$, change the independent variable to ξ, η in the equation

$$x^2\frac{\partial^2 u}{\partial x^2} - 2xy\frac{\partial^2 u}{\partial x\,\partial y} + y^2\frac{\partial^2 u}{\partial y^2} + 2y\frac{\partial u}{\partial y} = 0 \quad \left[\textbf{Ans. } 2\eta\frac{\partial^2 u}{\partial\eta^2} + \frac{\partial u}{\partial\eta} = 0\right]$$

18. If $x + y = 2e^\theta\cos\phi, x - y = 2ie^\theta\sin\phi$, prove that $\frac{\partial^2 v}{\partial\theta^2} + \frac{\partial^2 v}{\partial\phi^2} = 4xy\frac{\partial^2 v}{\partial x\,\partial y}$

CHAPTER – 10

Jacobian

10.1 Jacobian

If u and v are functions of two independent variables x and y then the determinant

$$\begin{vmatrix} \dfrac{\partial u}{\partial x} & \dfrac{\partial u}{\partial y} \\[2mm] \dfrac{\partial v}{\partial x} & \dfrac{\partial v}{\partial y} \end{vmatrix} \quad i.e., \quad \begin{vmatrix} u_x & u_y \\ v_x & v_y \end{vmatrix}$$ is called the Jacobian of u, v with respect to x and y and is denoted by

$$\frac{\partial(u,v)}{\partial(x,y)} \quad \text{or} \quad J\left(\frac{u,v}{x,y}\right)$$

If u, v, w are functions of three independent variables x, y, z then the determinant

$$\begin{vmatrix} u_x & u_y & u_z \\ v_x & v_y & v_z \\ w_x & w_y & w_z \end{vmatrix}$$

is called the Jacobian of u, v, w with respect to x, y, z and is denoted by

$$\frac{\partial(u,v,w)}{\partial(x,y,z)} \quad \text{or} \quad J\left(\frac{u,v,w}{x,y,z}\right)$$

Similarly

$$\frac{\partial(u_1,u_2.....u_n)}{\partial(x_1,x_2.....x_n)} = \begin{vmatrix} \dfrac{\partial u_1}{\partial x_1} & \dfrac{\partial u_1}{\partial x_2} & \cdots & \dfrac{\partial u_1}{\partial x_n} \\[2mm] \vdots & \vdots & \vdots & \vdots \\[2mm] \dfrac{\partial u_n}{\partial x_1} & \dfrac{\partial u_n}{\partial x_2} & \cdots & \dfrac{\partial u_n}{\partial x_n} \end{vmatrix}$$

is called the Jacobian of u_1, u_2 ..... u_n w.r.t x_1, x_2.....x_n and is denoted by

$$\frac{\partial(u_1,u_2,...,u_n)}{\partial(x_1,x_2,...,x_n)} \quad \text{or} \quad J\left(\frac{u_1,u_2,...,u_n}{x_1,x_2,...,x_n}\right)$$

10.2   Some Properties of Jacobian

(1)
$$\frac{\partial(u,v)}{\partial(x,y)} \cdot \frac{\partial(x,y)}{\partial(u,v)} = 1 \text{ or } JJ' = 1$$

Where
$$J = \frac{\partial(u,v)}{\partial(x,y)} \text{ and } J' = \frac{\partial(x,y)}{\partial(u,v)}$$

Proof: Let u = u(x, y) and v = u(x, y) so that u and v are functions of x and y. Then differentiating these partially each with respect to u and v, we get

$$1 = \frac{\partial u}{\partial x} \cdot \frac{\partial x}{\partial u} + \frac{\partial u}{\partial y} \cdot \frac{\partial y}{\partial u}$$

$$0 = \frac{\partial u}{\partial x} \cdot \frac{\partial x}{\partial v} + \frac{\partial u}{\partial y} \cdot \frac{\partial y}{\partial v}$$

$$0 = \frac{\partial v}{\partial x} \cdot \frac{\partial x}{\partial u} + \frac{\partial v}{\partial y} \cdot \frac{\partial y}{\partial u}$$

$$1 = \frac{\partial v}{\partial x} \cdot \frac{\partial x}{\partial v} + \frac{\partial v}{\partial y} \cdot \frac{\partial y}{\partial v}$$

Now
$$\frac{\partial(u,v)}{\partial(x,y)} \cdot \frac{\partial(x,y)}{\partial(u,v)} = \begin{vmatrix} \dfrac{\partial u}{\partial x} & \dfrac{\partial u}{\partial y} \\[2mm] \dfrac{\partial v}{\partial x} & \dfrac{\partial v}{\partial y} \end{vmatrix} \times \begin{vmatrix} \dfrac{\partial x}{\partial u} & \dfrac{\partial x}{\partial v} \\[2mm] \dfrac{\partial y}{\partial u} & \dfrac{\partial y}{\partial v} \end{vmatrix}$$

$$= \begin{vmatrix} \dfrac{\partial u}{\partial x} & \dfrac{\partial u}{\partial y} \\[2mm] \dfrac{\partial v}{\partial x} & \dfrac{\partial v}{\partial y} \end{vmatrix} \times \begin{vmatrix} \dfrac{\partial x}{\partial u} & \dfrac{\partial y}{\partial u} \\[2mm] \dfrac{\partial x}{\partial v} & \dfrac{\partial y}{\partial v} \end{vmatrix}$$

(Interchanging rows and columns in 2nd determinant)

$$= \begin{vmatrix} \dfrac{\partial u}{\partial x} \cdot \dfrac{\partial x}{\partial u} + \dfrac{\partial u}{\partial y} \cdot \dfrac{\partial y}{\partial u} & \dfrac{\partial u}{\partial x} \cdot \dfrac{\partial x}{\partial v} + \dfrac{\partial u}{\partial y} \cdot \dfrac{\partial y}{\partial u} \\[3mm] \dfrac{\partial v}{\partial u} \cdot \dfrac{\partial x}{\partial u} + \dfrac{\partial v}{\partial y} \cdot \dfrac{\partial y}{\partial u} & \dfrac{\partial v}{\partial x} \cdot \dfrac{\partial x}{\partial v} + \dfrac{\partial v}{\partial y} \cdot \dfrac{\partial y}{\partial v} \end{vmatrix}$$

$$= \begin{vmatrix} 1 & 0 \\ 0 & 1 \end{vmatrix} = 1 \text{ Proved}$$

(2)
$$\frac{\partial(u,v)}{\partial(x,y)} = \frac{\partial(u,v)}{\partial(r,s)} \cdot \frac{\partial(r,s)}{\partial(x,y)}$$

Where u, v are functions of r, s and r, s are functions of x, y.

Proof: By differentiation of function of a functions, partially; we get

$$\frac{\partial u}{\partial x} = \frac{\partial u}{\partial r} \cdot \frac{\partial r}{\partial x} + \frac{\partial u}{\partial s} \cdot \frac{\partial s}{\partial x};$$

$$\frac{\partial u}{\partial y} = \frac{\partial u}{\partial r} \cdot \frac{\partial r}{\partial y} + \frac{\partial y}{\partial s} \cdot \frac{\partial s}{\partial y};$$

$$\frac{\partial v}{\partial x} = \frac{\partial v}{\partial r} \cdot \frac{\partial r}{\partial x} + \frac{\partial v}{\partial s} \cdot \frac{\partial s}{\partial x};$$

and
$$\frac{\partial v}{\partial y} = \frac{\partial v}{\partial r} \cdot \frac{\partial r}{\partial y} + \frac{\partial v}{\partial s} \cdot \frac{\partial s}{\partial y};$$

..... (1)

$$\Rightarrow \frac{\partial(u,v)}{\partial(r,s)} \cdot \frac{\partial(r,s)}{\partial(x,y)} = \begin{vmatrix} \dfrac{\partial u}{\partial r} & \dfrac{\partial u}{\partial s} \\ \dfrac{\partial v}{\partial r} & \dfrac{\partial v}{\partial s} \end{vmatrix} \times \begin{vmatrix} \dfrac{\partial r}{\partial x} & \dfrac{\partial s}{\partial y} \\ \dfrac{\partial s}{\partial y} & \dfrac{\partial s}{\partial y} \end{vmatrix}$$

$$= \begin{vmatrix} \dfrac{\partial u}{\partial r} & \dfrac{\partial u}{\partial s} \\ \dfrac{\partial v}{\partial r} & \dfrac{\partial v}{\partial s} \end{vmatrix} \times \begin{vmatrix} \dfrac{\partial r}{\partial x} & \dfrac{\partial s}{\partial x} \\ \dfrac{\partial r}{\partial y} & \dfrac{\partial s}{\partial y} \end{vmatrix}$$

(Interchanging rows and columns of the second determinant)

$$= \begin{vmatrix} \dfrac{\partial u}{\partial r} \cdot \dfrac{\partial r}{\partial x} + \dfrac{\partial u}{\partial s} \cdot \dfrac{\partial s}{\partial x} & \dfrac{\partial u}{\partial r} \cdot \dfrac{\partial r}{\partial y} + \dfrac{\partial u}{\partial s} \cdot \dfrac{\partial s}{\partial y} \\ \dfrac{\partial v}{\partial r} \cdot \dfrac{\partial r}{\partial x} + \dfrac{\partial v}{\partial s} \cdot \dfrac{\partial s}{\partial x} & \dfrac{\partial v}{\partial r} \cdot \dfrac{\partial r}{\partial y} + \dfrac{\partial v}{\partial s} \cdot \dfrac{\partial s}{\partial y} \end{vmatrix}$$

$$= \begin{vmatrix} \dfrac{\partial u}{\partial x} & \dfrac{\partial u}{\partial y} \\ \dfrac{\partial v}{\partial x} & \dfrac{\partial v}{\partial y} \end{vmatrix} = \frac{\partial(u,v)}{\partial(x,y)} \quad \text{using (i)}$$

(3) If u, v are functions of two independent variables x and y *then* u, v, are independent if $\dfrac{\partial(u,v)}{\partial(x,y)} \neq 0$. Otherwise they are dependent. Similarly if u, v, w (functions of x, v, z) are independent if and only if $\dfrac{\partial(u,v,w)}{\partial(x,y,z)} \neq 0$.

Theorem: If the functions $u_1, u_2 \ldots u_n$ of the variables $x_1, x_2 \ldots x_n$ be defined by the relations $u_1 = f_1(x_1)$, $u_2 = f_2(x_1, x_2)$, $u_3 = f_3(x_1, x_2, x_3) \ldots u_n = f_n(x_1, x_2 \ldots x_n)$.

Then
$$\frac{\partial(u_1, u_2 \ldots u_n)}{\partial(x_1, x_2 \ldots x_n)} = \frac{\partial u_1}{\partial x_1} \cdot \frac{\partial u_2}{\partial x_2} \cdot \frac{\partial u_3}{\partial x_3} \ldots \frac{\partial u_n}{\partial x_n}$$

We know that
$$\frac{\partial u_1}{\partial x_2} = 0 = \frac{\partial u_2}{\partial x_4} = \frac{\partial u_3}{\partial x_5} \ldots$$

Similarly for $u_3, u_4 \ldots u_n$

Then
$$\frac{\partial(u_1, u_2 \ldots u_n)}{\partial(x_1, x_2 \ldots x_n)} = \begin{vmatrix} \dfrac{\partial u_1}{\partial x_1} & \dfrac{\partial u_2}{\partial x_1} & \cdots & \dfrac{\partial u_n}{\partial x_1} \\[2ex] \dfrac{\partial u_1}{\partial x_2} & \dfrac{\partial u_2}{\partial x_2} & \cdots & \dfrac{\partial u_n}{\partial x_2} \\[2ex] \cdots & \cdots & \cdots & \cdots \\[1ex] \cdots & \cdots & \cdots & \cdots \\[1ex] \dfrac{\partial u_1}{\partial x_n} & \dfrac{\partial u_2}{\partial x_n} & \cdots & \dfrac{\partial u_n}{\partial x_n} \end{vmatrix}$$

$$= \begin{vmatrix} \dfrac{\partial u_1}{\partial x_1} & \dfrac{\partial u_2}{\partial x_1} & \cdots & \dfrac{\partial u_n}{\partial x_1} \\[2ex] 0 & \dfrac{\partial u_2}{\partial x_2} & \cdots & \dfrac{\partial u_n}{\partial x_2} \\[2ex] \cdots & \cdots & \cdots & \cdots \\[1ex] \cdots & \cdots & \cdots & \cdots \\[1ex] 0 & 0 & \cdots & \dfrac{\partial u_n}{\partial x_n} \end{vmatrix} = \frac{\partial u_1}{\partial x_1} \cdot \frac{\partial u_2}{\partial x_2} \ldots \frac{\partial u_n}{\partial x_n}$$

Illustrative Examples

***Example* 1:**   If $x = u^2 - v^2$; $y = 2uv$; find $\dfrac{\partial(x,y)}{\partial(u,v)}$.

***Solution*:**   We have $x = u^2 - v^2 \Rightarrow \dfrac{\partial x}{\partial u} = 2u, \quad \dfrac{\partial x}{\partial v} = -2v$

$$y = 2uv \Rightarrow \frac{\partial y}{\partial u} = 2v, \frac{\partial y}{\partial v} = 2u$$

$$\frac{\partial(x,y)}{\partial(u,v)} = \begin{vmatrix} \dfrac{\partial x}{\partial r} & \dfrac{\partial x}{\partial v} \\ \dfrac{\partial y}{\partial u} & \dfrac{\partial y}{\partial v} \end{vmatrix} = \begin{vmatrix} 2u & -2v \\ 2v & 2u \end{vmatrix} = 4(u^2 + v^2)$$

***Example* 2:** If $x = r \cos q$, $y = r \sin q$ find $\dfrac{\partial(x,y)}{\partial(r,\theta)}$ and $\dfrac{\partial(r,\theta)}{\partial(x,y)}$.

***Solution*:** $x = r \cos q, \ y = r \sin q$

$$\Rightarrow \qquad \frac{\partial x}{\partial r} = \cos\theta, \qquad \frac{\partial y}{\partial r} = \sin\theta$$

Then $\dfrac{\partial(x,y)}{\partial(r,\theta)} = \begin{vmatrix} \dfrac{\partial x}{\partial r} & \dfrac{\partial x}{\partial \theta} \\ \dfrac{\partial y}{\partial r} & \dfrac{\partial y}{\partial \theta} \end{vmatrix} = \begin{vmatrix} \cos\theta & -r\sin\theta \\ \sin\theta & r\sin\theta \end{vmatrix} = r(\cos^2 q + \sin^2 q) = r$

Again $r^2 = x^2 + y^2$; $q = \tan^{-1}\dfrac{y}{x}$

$$\Rightarrow \qquad 2r\frac{\partial r}{\partial x} = 2x$$

$$\frac{\partial \theta}{\partial x} = \frac{-y}{x^2 + y^2} = \frac{-y}{r^2}$$

$$\therefore \qquad \frac{\partial r}{\partial x} = \frac{+x}{r}$$

$$\frac{\partial r}{\partial y} = \frac{y}{r},$$

$$\frac{\partial \theta}{\partial y} = \frac{x}{x^2 + y^2} = \frac{x}{r^2}$$

$$\Rightarrow \quad \frac{\partial (r, \theta)}{\partial (x, y)} = \begin{vmatrix} \dfrac{\partial r}{\partial x} & \dfrac{\partial r}{\partial y} \\ \dfrac{\partial \theta}{\partial x} & \dfrac{\partial \theta}{\partial y} \end{vmatrix} = \begin{vmatrix} \dfrac{x}{r} & \dfrac{y}{r} \\ \dfrac{-y}{r^2} & \dfrac{x}{r^2} \end{vmatrix} = \frac{x^2}{r^3} + \frac{y^2}{r^3} = \frac{r^2}{r^3} = \frac{1}{r}$$

Example 3: If $x = uv$, $y = \dfrac{u}{v}$ then show that $JJ' = 1$.

Solution: We have $\quad \dfrac{\partial x}{\partial u} = v, \quad \dfrac{\partial x}{\partial v} = u$

$$\frac{\partial y}{\partial u} = \frac{1}{v}, \quad \frac{\partial y}{\partial v} = -\frac{u}{v^2}$$

$$J = \begin{vmatrix} x_u & x_v \\ y_u & y_v \end{vmatrix} = \begin{vmatrix} v & u \\ \dfrac{1}{v} & \dfrac{-u}{v^2} \end{vmatrix} = -\frac{2u}{v}$$

But $\qquad u^2 = xy \qquad \therefore \qquad 2u \dfrac{\partial u}{\partial x} = y, \qquad 2u \dfrac{\partial u}{\partial y} = x$

and $\qquad v^2 = \dfrac{x}{y}, \quad 2v \dfrac{\partial v}{\partial x} = \dfrac{1}{y}, \quad 2v \dfrac{\partial v}{\partial y} = -\dfrac{x}{y^2}$

$$J' = \begin{vmatrix} u_x & u_y \\ v_x & v_y \end{vmatrix} = \begin{vmatrix} \dfrac{y}{2u} & \dfrac{x}{2u} \\ \dfrac{1}{2vy} & -\dfrac{x}{2vy^2} \end{vmatrix}$$

$$= \left(-\frac{x}{4uvy}\right)\left(-\frac{x}{4uvy}\right) - \frac{1}{2uv} \cdot \frac{x}{y}$$

$$= -\frac{1}{2uv} v^2 = -\frac{v}{2u}$$

$$JJ' = \left(-\frac{2u}{v}\right)\left(-\frac{v}{2u}\right) = 1$$

***Example* 4:** Prove that $= 1$ for $x = e^v \sec u$, $y = e^v \tan u$

Solution:
$$\frac{\partial x}{\partial u} = e^v \sec u \tan u, \qquad \frac{\partial x}{\partial v} = e^v \sec u$$

and
$$\frac{\partial y}{\partial u} = e^v \sec^2 u; \qquad \frac{\partial y}{\partial v} = e^v \tan u$$

$$\therefore \quad J = \begin{vmatrix} x_u & x_v \\ y_u & y_v \end{vmatrix} = \begin{vmatrix} e^v \sec u \tan u & e^v \sec u \\ e^v \sec^2 u & e^v \tan u \end{vmatrix} = -e^{2v} \sec u = -xe^v$$

Now $\sec^2 u - \tan^2 u = 1$

$$\therefore \quad x^2 e^{-2v} - y^2 e^{-2v} = 1$$

$$e^{2v} = x^2 - y^2, \text{ also } \frac{e^v \tan u}{e^v \sec u} = \frac{y}{x}$$

$$\therefore \quad \sin u = \frac{y}{x}$$

$$\therefore \quad u = \sin^{-1}\left(\frac{y}{x}\right) \text{ and } v = \frac{1}{2}\log(x^2 - y^2)$$

$$\therefore \quad \frac{\partial u}{\partial x} = -\frac{y}{x\sqrt{x^2 - y^2}}, \qquad \frac{\partial u}{\partial y} = \frac{1}{\sqrt{x^2 - y^2}}$$

$$\frac{\partial v}{\partial x} = \frac{x}{x^2 - y^2}, \qquad \frac{\partial v}{\partial y} = -\frac{y}{x^2 - y^2}$$

$$\therefore \quad J' = \begin{vmatrix} u_x & u_y \\ v_x & v_y \end{vmatrix} = \begin{vmatrix} -\dfrac{y}{x\sqrt{x^2 - y^2}} & \dfrac{1}{\sqrt{x^2 - y^2}} \\ \dfrac{x}{x^2 - y^2} & \dfrac{y}{x^2 - y^2} \end{vmatrix}$$

$$= \frac{y^2}{x(x^2 - y^2)^{3/2}} - \frac{x}{(x^2 - y^2)^{3/2}} = \frac{y^2 - x^2}{x(x^2 - y^2)^{3/2}}$$

$$= -\frac{1}{x\sqrt{(x^2 - y^2)}} = \frac{-1}{xe^v}$$

$$\therefore \quad JJ' = xe^v \cdot \frac{1}{xe^v} = 1$$

Example 5: If $x = \dfrac{vw}{u}, y = \dfrac{wu}{v}, z = \dfrac{uv}{w}$ then show that $\dfrac{\partial(x,y,z)}{\partial(u,v,w)} = 4$

Solution:

$$\frac{\partial(x,y,z)}{\partial(u,v,w)} = \begin{vmatrix} \dfrac{\partial x}{\partial u} & \dfrac{\partial x}{\partial v} & \dfrac{\partial x}{\partial w} \\ \dfrac{\partial y}{\partial u} & \dfrac{\partial y}{\partial v} & \dfrac{\partial y}{\partial w} \\ \dfrac{\partial z}{\partial u} & \dfrac{\partial z}{\partial v} & \dfrac{\partial z}{\partial w} \end{vmatrix} = \begin{vmatrix} -\dfrac{vw}{u^2} & \dfrac{w}{u} & \dfrac{v}{u} \\ \dfrac{w}{v} & -\dfrac{wu}{v^2} & \dfrac{u}{v} \\ \dfrac{v}{w} & \dfrac{u}{w} & -\dfrac{uv}{w^2} \end{vmatrix}$$

$$= \begin{vmatrix} -vw & wu & uv \\ vw & -wu & uv \\ vw & uw & -uv \end{vmatrix} \cdot \frac{1}{u^2} \cdot \frac{1}{v^2} \cdot \frac{1}{w^2}$$

$$= \begin{vmatrix} -1 & 1 & 1 \\ 1 & -1 & 1 \\ 1 & 1 & -1 \end{vmatrix}$$

Example 6: If $x = r \sin\theta \cos\phi,\ y = \phi \sin\theta \sin r,\ z = r \cos\theta$ show that $\dfrac{\partial(x,y,z)}{\partial(r,\theta,\phi)} = r^2 \sin\theta$

Solution:

$$\frac{\partial(x,y,z)}{\partial(r,\theta,\phi)} = \begin{vmatrix} \dfrac{\partial x}{\partial r} & \dfrac{\partial x}{\partial\theta} & \dfrac{\partial y}{\partial\phi} \\ \dfrac{\partial y}{\partial r} & \dfrac{\partial y}{\partial\theta} & \dfrac{\partial y}{\partial\phi} \\ \dfrac{\partial z}{\partial r} & \dfrac{\partial z}{\partial\theta} & \dfrac{\partial z}{\partial\phi} \end{vmatrix} = \begin{vmatrix} \sin\theta\cos\phi & r\cos\theta\cos\phi & -r\sin\theta\sin\phi \\ \sin\theta\sin\phi & r\cos\theta\sin\phi & r\sin\theta\cos\phi \\ \cos\theta & -r\sin\theta & 0 \end{vmatrix}$$

$$= \sin q \cos f\,[0 + r^2 \sin^2\theta \cos\phi] + r\cos\theta\cos\phi . r\sin\theta\cos\theta\cos\phi +$$

$$+ (-r\sin\theta\sin\phi)\,[-r\sin^2\theta\sin\phi - r\cos^2\theta\sin\phi]$$

$$= r^2 \sin^3\theta \cos\phi + r^2 \sin\theta \cos^2\theta \cos^2\phi + r^2 \sin^3\theta \sin^2\phi + r^2 \cos^2\theta$$
$$\sin\theta \sin^2\phi$$

$$= r^2 \sin^3\theta + r^2 \sin\theta \cos^2\theta = r^2 \sin\theta.$$

Hence proved.

Practice Problems

1. If $y + x + z = u$, $z + y = uv$, $z = uvw$, find $\dfrac{\partial(x,y,z)}{\partial(u,v,w)}$ **[Ans : $(u^2\,v)$]**

(JNTU 2006, 2008)

2. If $u = x = 2y^2 - z^2$, $v = x^2yz$, $w = 2x^2 - xy$ find $\dfrac{\partial(u,v,w)}{\partial(x,y,z)}$ at the point $(1, -1.0)$

[Ans:19]

3. if $x = r\cos q$, $y = r\sin q$, find $\dfrac{\partial(x,y)}{\partial(r,\theta)}$ and $\dfrac{\partial(r,\theta)}{\partial(x,y)}$ and show that $\dfrac{\partial(x,y)}{\partial(r,\theta)}\cdot\dfrac{\partial(r,\theta)}{\partial(x,y)}=1$

4. If $y_1 = \dfrac{x_2 x_3}{x_1}$, $y_2 = \dfrac{x_1 x_3}{x_2}$, $y_3 = \dfrac{x_1 x_2}{x_3}$, show that $\dfrac{\partial(y_1,y_2,y_3)}{\partial(x_1,x_2,x_3)} = 4$

5. If $y_1 = 1 - x_1$, $y_2 = x_1(1 - x_2)$, $y_3 = x_1 x_3(1 - x_3)$, prove that $\dfrac{\partial(y_1,y_2,y_3)}{\partial(x_1,x_2,x_3)} = (-1)^3\, x_1^2 x_3$.

6. If $x = a(\cosh u)\cos v$, $y = a(\sin h\, u)\sin v$, then show that $\dfrac{\partial(x,y)}{\partial(u,v)} = \dfrac{a^2}{2}\left(\cosh 2u - \cosh 2v\right)$

7. If $u = f(x_1)$, $v = \phi(x_1, x_2)$, $w = \psi(x_1, x_2, x_3)$ then show that $\dfrac{\partial(u,v,w)}{\partial(x_1,x_2,x_3)} = \dfrac{\partial u}{\partial x_1},\dfrac{\partial v}{\partial x_2},\dfrac{\partial w}{\partial x_3}$

(JNTU 2006 S 2007)

10.3   Jacobian of Composite Functions

Let (u, v) be function of x, y themselves are functions of r, θ. Then u, v are composite functions of r, θ and $\dfrac{\partial(u,v)}{\partial(r,\theta)} = \dfrac{\partial(u,v)}{\partial(x,y)},\dfrac{\partial(x,y)}{\partial(r,\theta)}$ (from 3.6.3)

Note 1: If u, v are functions of x, y we may regard x, y as functions of u, v.

Taking $r = u$, $\theta = v$ in the above result, we have

$$\frac{\partial(u,v)}{\partial(x,y)} = \frac{\partial(x,y)}{\partial(u,v)},\frac{\partial(u,v)}{\partial(u,v)} = \begin{vmatrix} \dfrac{\partial u}{\partial u} & \dfrac{\partial u}{\partial v} \\ \dfrac{\partial v}{\partial u} & \dfrac{\partial v}{\partial v} \end{vmatrix} = \begin{vmatrix} 1 & 0 \\ 0 & 1 \end{vmatrix} = 1$$

$$\therefore \qquad \frac{\partial(u,v)}{\partial(x,y)} = \frac{\partial(x,y)}{\partial(u,v)} = 1 \Rightarrow JJ^1 = 1$$

Where $\qquad J = \dfrac{\partial(u,v)}{\partial(x,y)}, \quad J^1 \dfrac{\partial(x,y)}{\partial(u,v)}$

Note 2: $J = \dfrac{\partial(u,v,w)}{\partial(r,\theta,\phi)} = \dfrac{\partial(u,v,w)}{\partial(x,y,z)} \cdot \dfrac{\partial(x,y,z)}{\partial(r,\theta,\phi)}$ where u, v, w are functions of x, y, z and x, y, z are function of r, θ, ϕ.

Also $JJ^1 = 1$; where $J = \dfrac{\partial(u,v,w)}{\partial(x,y,z)}, J^1 = \dfrac{\partial(x,y,z)}{\partial(u,v,w)}$

Eample 7: If u = xyz, v = $x^2 + y^2 + z^2$, w = x + y + z find $J = \dfrac{\partial(x,y,z)}{\partial(u,v,w)}$,

Solution: we frist evaluate $J^1 = \dfrac{\partial(u,v,w)}{\partial(x,y,z)}$, since u, v, w are explicity given as functions of

x, y, z of x, y, z. $J^1 = \begin{vmatrix} yz & zx & xy \\ 2x & 2y & 2x \\ 1 & 1 & 1 \end{vmatrix} = -2(x-y)\,(y-z)\,(z-x)$

Hence from the relation that $JJ^1 = 1$, we have

$$J = \frac{\partial(x,y,z)}{\partial(u,u,w)} = \frac{-1}{2(x-y)(y-z)(z-x)}$$

Practice Problems

1. If $u = 2xy$, $v = x^2 - y^2$, where $x = r \sin\theta$ show that $J = \dfrac{\partial(u,v)}{\partial(r,\theta)} = -4r^3$.

2. If $x = \sqrt{vw}$, $y = \sqrt{wu}$, $z = \sqrt{uv}$

    where $u = r \sin\theta \cos\phi$, $v = r \sin\theta \sin\phi$, $w = r \cos\theta$,

    Show that $\dfrac{\partial(x,y,z)}{\partial(r,\theta,\phi)} = \dfrac{1}{4}r^2 \sin\theta$

3. If $x = uv$, $y = \dfrac{u+v}{u-v}$, find $\dfrac{\partial(u,v)}{\partial(x,y)}$ $\qquad$ **[Ans:** $\dfrac{(u-v)^2}{4uv}$]

4. If $u = \dfrac{yz}{x}$, $v = \dfrac{zx}{y}$, $w = \dfrac{xy}{z}$ prove that $\dfrac{\partial(x,y,z)}{\partial(u,v,w)} = \dfrac{1}{4}$.

5. If $u = \dfrac{x+y}{1-xy}$, $\theta = \tan^{-1}(x) - \tan^{-1}(y)$ show that $\dfrac{\partial(u,\theta)}{\partial(x,y)} = 0$.

6. If $u = x^2 - 2y$, $v = x + y + z$, $w = x - 2y + 3z$ show that $\dfrac{\partial(u,v,w)}{\partial(x,y,z)} = 10x + 4$

10.4 Jacobian of Implicit Functions

Let u_1, u_2, u_3, be implicit functions of x_1, x_2, x_3 so that

$$f_r(u_1, u_2, u_3, x_1, x_2, x_3) = 0, \ r = 1, 2, 3$$

Then

$$\frac{\partial f_1}{\partial x_1} + \frac{\partial f_1}{\partial u_1} \cdot \frac{\partial u_1}{\partial x_1} + \frac{\partial f_1}{\partial u_2} \cdot \frac{\partial u_2}{\partial x_1} + \frac{\partial f_1}{\partial u_3} \cdot \frac{\partial u_3}{\partial x_1} = 0 \text{ and so on}$$

then

$$\frac{\partial(f_1, f_2, f_3)}{\partial(u_1, u_2, u_3)} \cdot \frac{\partial(u_1, u_2, u_3)}{\partial(x_1, x_2, x_3)}$$

$$= \begin{vmatrix} \sum \dfrac{\partial f_1}{\partial u_r} \cdot \dfrac{\partial u_r}{\partial x_1} & \sum \dfrac{\partial f_1}{\partial u_r} \cdot \dfrac{\partial u_r}{\partial x_2} & \sum \dfrac{\partial f_1}{\partial u_r} \cdot \dfrac{\partial u_r}{\partial x_3} \\[3mm] \sum \dfrac{\partial f_2}{\partial u_r} \cdot \dfrac{\partial u_r}{\partial x_1} & \sum \dfrac{\partial f_2}{\partial u_r} \cdot \dfrac{\partial u_r}{\partial x_2} & \sum \dfrac{\partial f_2}{\partial u_r} \cdot \dfrac{\partial u_r}{\partial x_3} \\[3mm] \sum \dfrac{\partial f_3}{\partial u_r} \cdot \dfrac{\partial u_r}{\partial x_1} & \sum \dfrac{\partial f_3}{\partial u_r} \cdot \dfrac{\partial u_r}{\partial x_2} & \sum \dfrac{\partial f_3}{\partial u_r} \cdot \dfrac{\partial u_r}{\partial x_3} \end{vmatrix}$$

(by the rule of product of determinants)

$$= \begin{vmatrix} -\dfrac{\partial f_1}{\partial x_1} & -\dfrac{\partial f_1}{\partial x_2} & -\dfrac{\partial f_1}{\partial x_3} \\[3mm] -\dfrac{\partial f_2}{\partial u_1} & -\dfrac{\partial f_2}{\partial u_2} & -\dfrac{\partial f_2}{\partial u_3} \\[3mm] -\dfrac{\partial f_3}{\partial x_1} & -\dfrac{\partial f_3}{\partial x_2} & -\dfrac{\partial f_3}{\partial x_3} \end{vmatrix}$$

$$= (-1)^3 \frac{\partial(f_1, f_2, f_3)}{\partial(x_1, x_2, x_3)}$$

Hence
$$\frac{\partial(u_1, u_2, u_3)}{\partial(x_1, x_2, x_3)} = (-1)^3 \frac{\dfrac{\partial(f_1, f_2, f_3)}{\partial(x_1, x_2, x_3)}}{\dfrac{\partial(f_1, f_2, f_3)}{\partial(u_1, u_2, u_3)}}$$

(The result if obvious)

***Example* 8:**

If $u = \dfrac{x}{\sqrt{1-r^2}}, v = \dfrac{y}{\sqrt{1-r^2}}, w = \dfrac{z}{\sqrt{1-r^2}}$ prove that

$$\frac{\partial(u, v, w)}{\partial(x, y, z)} = \frac{1}{\left(1-r^2\right)^{5/2}}$$

where $r^2 = x^2 + y^2 + z$

***Solution*:** we have $u^2 (1 - r^2) = x^2$

$$\Rightarrow \qquad f_1(u, v, w, x, y, z) = u^2(1 - x^2 - y^2 - z^2) - x^2 = 0$$

Similarly
$$f_2\ v^2(1 - x^2 - y^2 - z^2) - y^2 = 0$$
$$f_3\ w^2(1 - x^2 - y^2 - z^2) - z^2 = 0$$

$$\frac{\partial(u, v, w)}{\partial(x, y, z)} = (-1)^3 \frac{\dfrac{\partial(f_1, f_2 f_3)}{\partial(x, y, z)}}{\dfrac{\partial(f_1, f_2 f_3)}{\partial(u, v, w)}} \qquad\qquad \text{.... (i)}$$

$$\therefore \qquad = \frac{\partial(f_1, f_2, f_3)}{\partial(x, y, z)} = \begin{vmatrix} -2u^2x - 2x & -2u^2y & -2u^2z \\ -2v^2x & -2v^2y - 2y & -2v^2z \\ -2w^2x & -2w^2y & -2w^2z - 2z \end{vmatrix}$$

$$= (-2x)(-2y)(-2z) \begin{vmatrix} u^2 + 1 & u^2 & u^2 \\ v^2 & v^2 + 1 & v^2 \\ w^2 & w^2 & w^2 + 1 \end{vmatrix}$$

$$-8xyz \begin{vmatrix} 1 & 0 & u^2 \\ -1 & 1 & v^2 \\ 0 & -1 & w^2 + 1 \end{vmatrix}$$

By $C_1 - C_2$

$C_2 - C_3$

$= -8\ xyz\ (u^2 + v^2 + w^2 + 1)$

$$= -8xyz\left[\frac{x^2}{1-r^2} + \frac{y^2}{1-r^2} + \frac{z^2}{1-r^2} + 1\right]$$

$$= -8xyz\,\frac{x^2 + y^2 + z^2 + 1 - r^2}{1-r^2}$$

$$= \frac{-8xyz}{1-r^2};\ \text{also.}$$

$$\frac{\partial(f_1, f_2, f_3)}{\partial(u_1, v_2, w_3)} = \begin{vmatrix} 2u\left(1 - x^2 - y^2 - z^2\right) & 0 & 0 \\ 0 & 2u\left(1 - x^2 - y_2 - z^2\right) & 0 \\ 0 & 0 & 2u\left(1 - x^2 - y^2 - z^2\right) \end{vmatrix}$$

$= 8\ uvw\ (1 - x^2 - y^2 - z^2)^3$

$$= 8\frac{x}{\sqrt{1-x^2}} \cdot \frac{y}{\sqrt{1-r^2}} \cdot \frac{z}{\sqrt{1-r^2}}\left(1-r^2\right)^3 = 8xyz\left(1-r^2\right)^{1/2}$$

Hence from (i), we have $\dfrac{\partial(u, v, w)}{\partial(x, y, z)} = (-1)^3\,\dfrac{-8xyz}{1-r^2}\Big/ 8xyz\left(1-r^2\right)^{3/2} = \dfrac{1}{\sqrt{1-r^2}}$

Cor: Given $u = \dfrac{x}{k}, v = \dfrac{y}{k}, w = \dfrac{z}{k}$

Where $k = \sqrt{1 - x^2 - y^2 - z^2}$, find J $\dfrac{(x, y, z)}{(u, v, w)}$ in terms of k.

[**Ans:** k^5]

Practice Problems

1. If $x + y + z = u$, $y + z = uvw$, then prove that $\dfrac{\partial(x, y, z)}{\partial(u, v, w)} = u^2 v$

2. If $x^2 + y^2 + u^2 - v^2 = 0$, $uv + xy = 0$, prove that $\dfrac{\partial(u, v)}{\partial(x, y)} = \dfrac{x^2 - y^2}{u^2 + v^2}$

3. If $u_1 = x_1 + x_2 + x_3 + x_4$, $u_1 u_2 = x_2 + x_3 + x_4$, $u_1 u_2 u_3 = x_4$ then prove that

$$\frac{\partial(x_1, x_2, x_3, x_4)}{\partial(u_1, u_2, u_3, u_4)} = u_1^3 . u_2^2 . u_3$$

4. If $u^3 + v^3 + w^3 = x + y + z$, $u^2 + v^2 + w^2 = x^3 + y^3 + z^3$, $u + v + w = x^2 + y^2 + z^2$, then show

that $\dfrac{\partial(u,v,w)}{\partial(x,y,z)} = \dfrac{(y-z)(z-x)(x-y)}{(v-w)(w-u)(u-v)}$

10.5   Functional Dependence and Independence of Functions

Let u and v be two functions of x and y connected by the relation $v = f(u)$, then we say that u and v are functionally dependent. We shall prove that the condition for functional denpendence is

$$\frac{\partial(u,v)}{\partial(x,y)} = 0$$

Consider $w = v - f(u) = 0$

Now w is a function of u and v where u, v are functions of x and y.

$\therefore$ w is a composite function of x and y

$$\therefore \qquad \frac{\partial w}{\partial x} = \frac{\partial w}{\partial x} \cdot \frac{\partial u}{\partial x} + \frac{\partial w}{\partial v} \cdot \frac{\partial v}{\partial x} \qquad \qquad \text{..... (1)}$$

$$\therefore \qquad \frac{\partial w}{\partial x} = \frac{\partial w}{\partial u} \cdot \frac{\partial u}{\partial x} + \frac{\partial w}{\partial v} \cdot \frac{\partial v}{\partial x} \qquad \qquad \text{..... (2)}$$

But w considered as a function of x, y is identically zero.

i.e., $\dfrac{\partial w}{\partial x} = 0, \qquad \dfrac{\partial w}{\partial y} = 0$

$\therefore$ From (1) and (2) we get

$$\frac{\partial w}{\partial u} \cdot \frac{\partial u}{\partial x} + \frac{\partial w}{\partial v} \cdot \frac{\partial v}{\partial x} = 0 \qquad \qquad \text{..... (3)}$$

and $$\frac{\partial w}{\partial u} \cdot \frac{\partial u}{\partial y} + \frac{\partial w}{\partial v} \cdot \frac{\partial v}{\partial y} = 0 \qquad \qquad \text{..... (4)}$$

Eliminating $\dfrac{\partial w}{\partial u} \cdot \dfrac{\partial w}{\partial v}$ from (3) and (4) we have

$$\begin{vmatrix} \dfrac{\partial u}{\partial x} & \dfrac{\partial v}{\partial x} \\[2mm] \dfrac{\partial u}{\partial y} & \dfrac{\partial v}{\partial y} \end{vmatrix} = 0$$

$$\Rightarrow \quad \begin{vmatrix} \dfrac{\partial u}{\partial x} & \dfrac{\partial u}{\partial y} \\[2mm] \dfrac{\partial u}{\partial x} & \dfrac{\partial u}{\partial y} \end{vmatrix} = 0$$

$$\Rightarrow \quad \frac{\partial(u,v)}{\partial(x,y)} = 0$$

The concept of functional dependence can be extended to any number of variables.

Thus if u, v, w are functions of x, y, z then u, v, w will be functionally dependent (i.e., there will exist a relation between u, v, w) if

$$\frac{\partial(u,v,w)}{\partial(x,y,z)} = 0$$

Illustrative Problems

Example 9: Are $x + y - z$, $x - y + z$, $x^2 + y^2 + z^2 - 2yz$ functionally dependent? If so, find a relation between them.

Solution: Let $\quad u = x + y - z \hspace{5cm}$ ..... (i)

$$v = x - y + z \hspace{5cm} \text{..... (ii)}$$

$$w = x^2 + y^2 + z^2 - 2yz \hspace{3cm} \text{..... (iii)}$$

Then

$$\frac{\partial(u,v,w)}{\partial(x,y,z)} = \begin{vmatrix} 1 & 1 & -1 \\ 1 & -1 & +1 \\ 2x & 2y-2z & 2z-2y \end{vmatrix}$$

$$= -\begin{vmatrix} 1 & 1 & 1 \\ 1 & -1 & -1 \\ 2x & 2y-2z & 2y-2z \end{vmatrix} = 0$$

(since 2nd and 3rd columns are identical)

$\therefore$ u, v, w are functionally dependent

To find the relation between *u, v, w* we notice that, from (i) and (ii)

$$u + v = 2x, \quad u - v = 2(y - z)$$

$\therefore$ From (iii)

$$w = x^2 + (y - z)^2$$

$$= \left(\frac{(u+v)}{2}\right)^2 + \left(\frac{(u-v)}{2}\right)^2$$

$$= \frac{1}{2}\left(u^2 + v^2\right)$$

$$\Rightarrow \quad u^2 + v^2 = 2w, \text{ the required relation}$$

***Example* 10:** If $u = x + y + z$, $v = x^2 + y^2 + z^2$, $w = x^3 + y^3 + z^3 - 3\,xyz$ then prove that u, v, w are not independent and find the relation between them.

***Solution*:** Condition for functional dependence is $\dfrac{\partial(u,v,w)}{\partial(x,y,z)} = 0$

Now
$$\frac{\partial(u,v,w)}{\partial(x,y,z)} = \begin{vmatrix} 1 & 1 & 1 \\ 2x & 2y & 2z \\ 3x^2 - 3yz & 3y^2 - 3zx & 3z^2 - 3xy \end{vmatrix}$$

$$= 6 \begin{vmatrix} 1 & 1 & 1 \\ x & y & z \\ x^2 - yz & y^2 - zx & z^2 - xy \end{vmatrix} \quad C_1 - C_2, C_2 - C_3$$

$$= 6 \begin{vmatrix} 0 & 0 & 1 \\ x - y & y - z & z \\ (x-y)(x+y+z) & (y-z)(x+y+z) & z^2 - xy \end{vmatrix}$$

$$= 6(x-y)(y-z) \begin{vmatrix} 0 & 0 & 1 \\ 1 & 1 & z \\ x+y+z & x+y+z & z^2 - xy \end{vmatrix}$$

$$= 0 \ (\because C_1 = C_2)$$

$\therefore$ u, v, w are functionally dependent. We shall now find the relation between them

we have $\qquad w = (x + y + z)(x^2 + y^2 + z^2 - xy - yz - zx)$ ..... (i)

Now $\qquad u^2 - v = (x + y + z)^2 - (x^2 + y^2 + z^2)$

$$= 2(xy + yz + zx)$$

$\therefore$ $\qquad$ We write (i) as

$$w = u\left[v - \frac{u^2 - v}{2}\right]$$

Thus $\quad 2w = u(3v - u^2)$

***Example* 11**: Test whether $u = \dfrac{x+y}{x-y}, v = \dfrac{xy}{(x-y)^2}$ are functionally dependent and if so, find the

relation between them.

Solution: Treating u, v as functions of x, y the condition for functional dependence is

$$\frac{\partial(u,v)}{\partial(x,y)} = 0$$

$$u = \frac{x+y}{x-y}, \qquad v = \frac{xy}{(x-y)^2}$$

We have
$$u_x = \frac{(x-y).1-(x+y).1}{(x-y)^2} = \frac{-2y}{(x-y)^2}$$

$$u_y = \frac{(x-y).(1)-(x+y)(1)}{(x-y)^2} = \frac{2x}{(x-y)^2}$$

$$v_x = \frac{(x-y)^2\, y - 2(x-y).1.xy}{(x-y)^4}$$

$$= \frac{\big((x-y)y - 2xy\big)}{(x-y)^3}$$

$$= \frac{xy-y^2-2xy}{(x-y)^3} = \frac{-y(y+x)}{(x-y)^3}$$

Similarly
$$v_y = \frac{x(x+y)}{(x-y)^3}$$

$\therefore$
$$\frac{\partial(u,v)}{\partial(x,y)} = \begin{vmatrix} u_x & u_v \\ v_x & v_y \end{vmatrix}$$

$$= u_x v_y - v_x u_y$$

$$= \frac{-2y}{(x-y)^2}.\frac{x(x+y)}{(x-y)^3} + \frac{y(x+y)}{(x-y)^3}.\frac{2x}{(x-y)^2} = 0$$

$\therefore$ u, v are functionally dependent.

We shall now find the relation between them :

We have $u = \dfrac{x+y}{x-y}$

$\therefore \qquad \dfrac{x-y}{x+y} = \dfrac{1}{u}$

By componendo and dividend

$$\dfrac{x}{y} = \dfrac{1+u}{1-u} \qquad\qquad ......(i)$$

Now $v = \dfrac{xy}{(x-y)^2} = \dfrac{y^2\dfrac{x}{y}}{y^2\left(\dfrac{x}{y}-1\right)^2} = \dfrac{\dfrac{x}{y}}{\left(\dfrac{x}{y}-1\right)^2} = \dfrac{\dfrac{1+u}{1-u}}{\left(\dfrac{(1+u)}{(1-u)}-1\right)^2}$ by (i)

$$v = \left(\dfrac{1-u^2}{4u^2}\right)$$

$\therefore \qquad 2(4v+1) = 1$

The required relation of functional depedence between u, v.

***Example* 12:** Examine for functional dependence of

$u = \sin^{-1} x + \sin^{-1} y$

$$v = x\sqrt{1-y^2} + y\sqrt{1-x^2}$$

and find the relation between them, if it exists.

***Solution*:** We have $\qquad u_x = \dfrac{1}{\sqrt{1-x^2}}$

$$u_y = \dfrac{1}{\sqrt{1-y^2}}$$

$$v_x = \sqrt{1-y^2} + y\dfrac{1}{2\sqrt{1-x^2}}(-2x)$$

$$= \sqrt{1-y^2} - \dfrac{xy}{\sqrt{1-x^2}}$$

Similarly $\qquad v_y = \sqrt{1-x^2} - \dfrac{xy}{\sqrt{1-y^2}}$

$\therefore \qquad \dfrac{\partial(u,v)}{\partial(x,y)} = \begin{vmatrix} u_x & u_y \\ v_x & v_y \end{vmatrix}$

$$= u_x y_y - u_y v_x$$

$$= \left(1 - \dfrac{xy}{\sqrt{1-x^2}\,\sqrt{1-y^2}}\right) - \left(1 - \dfrac{xy}{\sqrt{1-x^2}\,\sqrt{1-y^2}}\right) = 0$$

$\therefore$ u, v are functionally dependent .

We shall now determine the relation between u and v:

Let $A = \sin^{-1} x$ and $B = \sin^{-1} y$

The given function can then be written as $u = A + B$

$$V = \sin A.\sqrt{1 - \sin^2 B} + \sin B\sqrt{1 - \sin^2 A} = \sin(A + B) = \sin u$$

Hence the required relation of functional dependence between u and v is $v = \sin u$

***Example* 13:** It $u = e^x \sin y$, $v = e^x \cos y$, show that u, v are functionally independent

Solution: The Jacobian $\dfrac{\partial(u,v)}{\partial(x,y)} = \begin{vmatrix} \dfrac{\partial u}{\partial x} & \dfrac{\partial u}{\partial y} \\ \dfrac{\partial v}{\partial x} & \dfrac{\partial v}{\partial y} \end{vmatrix} = \begin{vmatrix} e^x \sin y & e^x \cos y \\ e^x \cos y & -e^x \sin y \end{vmatrix} = -e^x \neq 0.$

$\therefore$ x, v are functionally independent.

***Example* 14:** Verify whether $u = \dfrac{x-y}{x+z}$ $v = \dfrac{x+y}{y+z}$ are functionally dependent, and if so find the relation.

Solution: Treating u, v as function of x, y regarding z as an absolute constant, the condition for functional dependent in

$$\dfrac{\partial(u,v)}{\partial(x,y)} = 0,$$

Now
$$\frac{\partial u}{\partial x} = \frac{y+z}{(x+z)^2}, \frac{\partial u}{\partial y} = -\frac{1}{x+z} \text{ and } \frac{\partial v}{\partial x} = \frac{1}{y+z}, \frac{\partial v}{\partial y} = -\frac{x+z}{(y+z)^3}$$

$$\frac{\partial(u,v)}{\partial(x,y)} = \begin{vmatrix} \dfrac{y+z}{(x+z)^2} & \dfrac{1}{x+z} \end{vmatrix} = 0$$

$\therefore$ u, v are functionally dependent. To find the relation between u and v, we eliminate 'x' between the given relation viz:

$$u = \frac{x-y}{x+z} \qquad\qquad \ldots\ldots(i),$$

$$v = \frac{x+z}{y+z} \qquad\qquad \ldots\ldots(ii)$$

From (i) $ux + vy = x - y, \Rightarrow (1-u)x = uz + y$

$$\Rightarrow \qquad\qquad x = x = \frac{uz+y}{1-u}$$

Substituting for 'x' in 'ii' $v = \dfrac{\dfrac{uz+y}{1-u}}{y+z} = \dfrac{1}{1-u}, \dfrac{(y+z)}{y+z} = \dfrac{1}{1-u}$

Hence the required functional relation between 'u' and 'v' is $v(1-u) = 1$.

Practice problems

1. Verify whether the following are functionally dependent and if so, find the relation between them:

    (a) $u = \dfrac{x-y}{x+y}, v = \dfrac{x+y}{x}$ [**Ans:** $u(1+v) = 2$]

    (b) $u = \dfrac{x-y}{1-xy}, v = \tan^{-1}x + \tan^{-1}y$ [**Ans:** $v = \tan^{-1}u$]

    (c) $u = \dfrac{x-y}{x+y}, v = \dfrac{xy}{(x+y)^2}$ [**Ans:** $4y + u^2 = 1$]

2. Prove that the following functions are not independent:

    (a) $u = x + 2y + z, v = x - 2y + 3z, w = 2xy - 2z + 4yz - 2z^2$

    (b) $u = x + y - z, v = x - y + z, w = x^2 + y^2 + z^2 - 2yz$

    (c) $u = x + y + z, v = xy + yz + zx, w = x^3 + y^3 + z^3 - 3xyz$

3. If $u = \dfrac{x+y}{z}, v = \dfrac{y+z}{x}, w = \dfrac{y(x+y+z)}{xz}$, show that u, v are connected by a functional relation.

4. Examine for function dependence between $u = \dfrac{x-y}{1-xy}$, $v = \tan^{-1} x - \tan^{-1} y$.

   If dependent. find the relation. **[Ans:** $u = \tan v$**]**

5. Prove that functions $u = y + z$, $v = x + 2z^2$, $w = x - 4yz - 2y^2$ are functionally dependent and find the relation between them. **[Ans:** $v = w + 2u^2$**]**

CHAPTER – 11

Approximations and Errors

11.1 Approximations and Errors

Let $y = f(x)$ be a function of x.

If δx and δy be the increments of x and y, then

$$y + \delta y = f(x + \delta x)$$

$$\therefore \quad \delta y = f(x + \delta x) - y$$

$$\therefore \quad \delta y = f(x + \delta x) - f(x) \qquad \qquad \dots\dots(1)$$

Now $\dfrac{dy}{dx} = \lim\limits_{\delta x \to 0} \dfrac{f(x + \delta x) - f(x)}{\delta x} = f'(x)$

or $\quad \dfrac{f(x + \delta x) - f(x)}{\delta x} = f'(x) + \eta$, where $\eta \to 0$ as $\delta x \to 0$

or $\quad f(x + \delta x) - f(x) = f'(x) . \delta x + \eta \delta x$,

or $\quad \delta y = f'(x) . \delta x$, from (1)

or $\quad \dfrac{\delta y}{y} = \dfrac{f'(x)}{f(x)} \delta x \qquad \qquad \dots\dots(2)$

Hence, if δx is an error in x, then

(a) $\quad |\delta x|$ in x is called an absolute error

(b) $\quad \dfrac{\delta x}{x}$ in x is called the relative error (or proportional error)

(c) $\quad \dfrac{\delta x}{x} \times 100$ in x is called the percentage error

Illustrative Examples

***Example* 1:** If the the radius of a spherical balloon increases by 0.1 percent, find approximately the percentage increase in the volume.

***Solution*:** Let V be the volume of the balloon and r be the radius. Then

$$V = \frac{4}{3}\pi r^3$$

Taking log on both sides, we get

$$\log V = \log \frac{4\pi}{3} + 3 \log r$$

Differentiating, we get

$$\frac{1}{V}\delta v = 0 + \frac{3}{r}\delta r$$

or $\qquad \dfrac{\delta v}{v} \times 100 = 3\dfrac{\delta r}{r} \times 100$

or $\qquad \dfrac{\delta v}{v} \times 100 = 3 \times 0.1 \quad$ (Given $\dfrac{\delta r}{r} \times 100 = 0.1$)

$$= 0.3$$

Hence percentage increase in V = 0.3

***Example* 2:** The period T of a simple pendulum is given by $T = 2\pi\sqrt{\dfrac{l}{g}}$. Find maximum error in T due to possible errors upto 1% in l and 2.5% in g. $\hfill$ (RGPV Dec 2006)

***Solution*:** Here $\qquad T = 2\pi\sqrt{\dfrac{l}{g}}$

Taking log on both sides, we get

$$\log T = \log 2\pi + \frac{1}{2}(\log l - \log g)$$

Differentiating, we get

$$\frac{\delta T}{T} = 0 + \frac{1}{2}\left(\frac{\delta l}{l} - \frac{\delta g}{g}\right)$$

or $\quad \dfrac{\delta T}{T} \times 100 = \dfrac{1}{2}\left(\dfrac{\delta l}{l} \times 100 - \dfrac{\delta g}{g} \times 100\right)$

$$= \dfrac{1}{2}\ (1 \pm 2.5) \quad \left(\text{Given } \dfrac{\delta l}{l} \times 100 = 1,\ \dfrac{\delta g}{g} \times 100 = 2.5\right)$$

$\therefore$ Maximum error in T = 1.75%

***Example* 3:** Find the percentage error in the area of an ellipse if one percentage error is made in measuring the major and minor axes.

(RGPV June 2002, June 2004)

***Solution*:** Let the equation of ellipse be

$$\dfrac{x^2}{a^2} + \dfrac{y^2}{b^2} = 1$$

Where major axis = 2a, and minor axis = 2b

The area of ellipse be $A = \pi\, ab$

Taking log on both sides, we get

$$\log A = \log \pi + \log a + \log b$$

Differentiating, we get

$$\dfrac{\delta A}{A} = 0 + \dfrac{\delta a}{a} + \dfrac{\delta b}{b}$$

or $\quad \dfrac{\delta A}{A} \times 100 = \dfrac{\delta a}{a} \times 100 + \dfrac{\delta b}{b} \times 100$

$$= 1 + 1$$

$$\left(\text{Given } \dfrac{\delta a}{a} \times 100 = 1,\ \dfrac{\delta b}{b} \times 100 = 1\right)$$

$$= 2$$

$\therefore$ Percentage error in area = 2%

***Example* 4:** Find the percentage error in the area is made of a rectangle when an error of +1 percent is made in measuring its length and breadth. $\qquad$ (RGPV 2004, Dec 2005)

***Solution*:** Let x be the length and y be the breadth of a rectangle. The area of rectangle

$$A = x \,.\, y$$

Taking log on both sides

$$\log A = \log x + \log y$$

Differentiating, we get

$$\dfrac{\delta A}{A} = \dfrac{\delta x}{x} + \dfrac{\delta y}{y}$$

or $\quad \dfrac{\delta A}{A} \times 100 = \dfrac{\delta x}{x} \times 100 + \dfrac{\delta y}{y} \times 100$

$$= 1 + 1 \qquad \left(\text{Given } \dfrac{\delta x}{x} \times 100 = 1, \dfrac{\delta y}{y} \times 100 = 1\right)$$

$$= 2$$

$\therefore$ Percentage error in A = 2%

Example 5: Find the approximate value of cube root of 127.

Solution: Let $\qquad f(x) = x^{1/3}, f'(x) = \dfrac{1}{3x^{\frac{2}{3}}}$

Now, $\quad f(x + \delta x) - f(x) = f'(x).\delta x$

$$f(x + \delta x) - f(x) = \dfrac{1}{3x^{\frac{2}{3}}}.\delta x \qquad \qquad .....(i)$$

Since $\qquad x + \delta x = 127$, then $x = 125$, $\delta x = 2$

$\therefore$ from (i), we get

$$f(127) - f(125) = \dfrac{1}{3(125)^{2/3}}.2$$

$$f(127) = f(125) + \dfrac{2}{3(25)}$$

$$f(127) = (125)^{1/3} + \dfrac{2}{75}$$

$$(127)^{1/3} = 5 + \dfrac{2}{75}$$

$$= \dfrac{377}{75} = 5.027 \text{ (approx)}$$

Example 6: If H.P required to propel a steamer is proportional to cube of its velocity and square of its length, prove that 2% increase in velocity and 3% increase in length will require an approximately 12% increase in H.P.

(RGPV Feb 2005)

Solution: Let v be the velocity an l be the length of a steamer

If H be the H.P of the steamer then

$$H \alpha v^3 l^2$$

or $\qquad H = k v^3 l^2,$ where k is a constant

Taking log on both the sides, we get

$$\log H = \log k + 3 \log v + 2 \log l$$

Differentiating, we get

$$\frac{\delta H}{H} = 0 + 3\frac{\delta v}{v} + 2\frac{\delta l}{l}$$

or

$$\frac{\delta H}{H} \times 100 = 3\frac{\delta v}{v} \times 100 + 2\frac{\delta l}{l} \times 100$$

$$= 3 \times 2 + 2 \times 3 \qquad \left(\text{Given } \frac{\delta v}{v} \times 100 = 2, \ \frac{\delta l}{l} \times 100 = 3\right)$$

$$= 12$$

$\therefore$ Percentage error in H.P $= 12\%$

***Example* 7:** The deflection at the centre of a rod of length l and diameter d supported at its ends and loaded at the centre with a weight w varies as $wl^3 \, d^{-4}$. What is the percentage increase in the deflection corresponding to the percentage increase in w, l and d of 3, 2 and 1 respectively.

***Solution*:** Let the deflection of the rod at the centre be D

Then, we have

$$D = k \cdot wl^3 \, d^{-4}$$

$$D = k \cdot \frac{wl^3}{d^4}$$

Taking log on both sides, we get

$$\log D = \log k + \log w + 3 \log l - 4 \log d$$

Differentiating, we get

$$\frac{\delta D}{D} = 0 + \frac{\delta w}{w} + 3\frac{\delta l}{l} - 4\frac{\delta d}{d}$$

'or'

$$\frac{\delta D}{D} \times 100 = \frac{\delta w}{w} \times 100 + 3\frac{\delta l}{l} \times 100 - 4\frac{\delta d}{d} \times 100$$

$$= 3 + 3 \times 2 - 4 \times 1$$

$$= 5$$

$\therefore$ Percentage error in deflection D is 5%.

***Example* 8:** What error in the common logarithm of a number will be produced by an error of 1% in the number.

***Solution*:** Let a be the number and let $b = \log_{10} a$

Differentiating, we get

$$\delta b = \frac{1}{a} \log_{10} . e . \delta a \qquad \left(\because \log_{10} a = \frac{\log_e a}{\log_e 10} = (\log_e a).(\log_{10} e) \right)$$

or $\qquad \delta b = \frac{\delta a}{a} \log_{10} e$

$$= \left(\frac{\delta a}{a} \times 100 \right) \log_{10} e \times \frac{1}{100}$$

$$= 1 \times \frac{1}{100} \times \log_{10} e \qquad\qquad \left(\text{given } \frac{\delta a}{a} \times 100 = 1 \right)$$

$$= \frac{0.4343}{100} \qquad\qquad \left(\because \log_{10} e = 0.4343 \right)$$

Percentage error $= 0.004343$

Practice Problems

1. The period T of a simple pendulum of length l is given by $\quad T = 2\pi \sqrt{\frac{l}{g}}$. Find

   (i) The error and

   (ii) Percentage error made in computing T by using $l = 2$ ft and $g = 32$ ft/sec^2, when the true values are $l = 1.95$ ft and $g = 32.2$ ft/sec^2.

   [Ans. (i) 0.0246, (ii) 1.5926]

2. Find the approximate value of $(1.04)^{3.01}$.

   [Ans. 1.12]

3. Evaluate $\sqrt{99}$ approximately.

   [Ans. 9.95]

4. A balloon in the form of right circular cylinder of radius 1.5 m and length 4 m in surmounted by hemispherical ends. If the radius is increased by 0.01 m and the length by 0.05 m, find the percentage change in the volume of the balloon.

   [*Hint*: volume $v = \pi r^2 h + \frac{4}{3} \pi r^3$]   **[Ans. 2.3888]**

5. The radius of a sphere is found to be 20 cm with a possible error of 0.02 cm. Find the relative error in calculating the volume.

   [*Hint*: $r = 20$ cm, $\delta r = 0.02$ cm, $v = \frac{4}{3} \pi r^3$]

   [Ans. 0.003 cubic cm]

6   If the kinetic energy $T = \dfrac{mv^2}{2}$. Find approximately the change in T as m changes from 49 to 49.5 and v changes from 1600 to 1590.

[*Hint*: m = 49 m, m + δm = 49.5, v = 1600, δ + δv = 1590]

[**Ans.** δT = 14400]

Unit – IV

Integral Calculus

- Definite Integral as a Limit of sum
- Beta and Gamma Functions

CHAPTER – 12

Definite Integral as a Limit of Sum

12.1 Definite Integral as Limit of a Sum

Let f(x) be a single valued continuous function defined in the interval (a, b) where a and b both finite quantities and b>a and let the interval (a, b) be divided into n equal parts each of length h, so that nh = b – a, then

$$\int_a^b f(x)dx = \lim_{h \to 0} h\left[f(a)+f(a+h)+f(a+2h)+....+f(a+(n-1)h)\right]$$

or

$$\int_a^b f(x)dx = \lim_{h \to 0} h \sum_{r-0}^{n-1} f(a+rh)$$

The integral $\int_a^b f(x)dx$ is said to be the definite integral of f(x) w.r.t. x between the limits a and b i.e., the lower limit or interior limit and upper or superior limit.

***Example* 1:** Evaluate the following integral as limit of sums $\int_a^b x^2dx$.

(RGPV June 2002, June 2003)

***Solution*:** By definition, we have

$$\int_a^b f(x)dx = \lim_{h \to 0} h\left[f(a)+f(a+h)+f(a+2h)+....+f(a+n-1)h\right]$$

where $h = \dfrac{b-a}{n}$,

Given that f(x) = x², so that f(a) = a², f(a + h) = (a + h)² .... and so on

$$\therefore \quad \int_a^b x^2dx = \lim_{h \to 0} h\left[a^2 +(a+h)^2 +(a+2h)^2 +...+(a+(n-1)h)^2\right]$$

$$\int_a^b x^2 dx = \lim_{h \to 0} h \left[\begin{array}{l} a^2 + \left(a^2 + 2ah + h^2\right) + \left(a^2 + 4ah + 4h^2\right) + \\ \ldots + \left(a^2 + 2ah(n-1) + (n-1)^2 h^2\right) \end{array} \right]$$

$$= \lim_{h \to 0} h \left[na^2 + ah\left(1 + 2 + \ldots + (n-1)\right) + h^2\left(1^2 + 2^2 + \ldots + (n-1)^2\right) \right]$$

$$= \lim_{h \to 0} h \left[na^2 + 2ah\frac{n(n-1)}{2} + h^2 \frac{n(n-1)(2n-1)}{6} \right]$$

$$\left(\because \sum_{n=1}^{n-1} n = \frac{n(n-1)}{2}, \sum_{n=1}^{n-1} n^2 = \frac{n(n-1)(2n-1)}{6} \right)$$

$$= \lim_{h \to 0} \left[nha^2 + 2anh(nh - h) + \frac{1}{6} nh(nh - h)(2nh - h) \right]$$

$$= \lim_{h \to 0} \left[\begin{array}{l} (b-a)a^2 + a(b-a)(b-a-h) \\ + \frac{1}{6}(b-a)(b-a-h)(2b-2a-h) \end{array} \right]$$

$$= (b-a) a^2 + a(b-a)^2 + \frac{2}{6} (b-a)^3$$

$$= \frac{b-a}{3} [3a^2 + 3a(b-a) + (b-a)^2]$$

$$= \frac{b-a}{3} [3a^2 + 3ab - 3a^2 + b^2 + a^2 - 2ab]$$

$$= \frac{b-a}{3} (a^2 + b^2 + ab)$$

$$= \frac{1}{3} (b^3 - a^3)$$

***Example* 2:** Evaluate $\int_a^b \sin x \, dx$ as limit of sums .   (RGPV June 2002, June 2003)

***Solution*:**

By definition, we get

$$\int_a^b f(x) \, dx = \lim_{h \to 0} h[f(a) + f(a+h) + f(a+2h) + \ldots + f(a + (n-1)h)]$$

Here, f(x) = sin x, so that f(a) = sin a, f(a + h) = sin (a + h) … so on.

$$\therefore \int_a^b \sin x \, dx = \lim_{h \to 0} h[\sin a + \sin(a+h) + \sin(a+2h) + \ldots + \sin(a + (n-1)h)]$$

$$= \lim_{h \to 0} h \left[\frac{\sin\left(a + \dfrac{(n-1)h}{2}\right) \sin \dfrac{nh}{2}}{\sin h\big/2} \right]$$

$$\left[\because \sin a + \sin (a + h) + \ldots + \sin (a + (n-1)h) = \left(\sin \frac{a + (n-1)h}{2} . \sin \frac{nh}{2}\right)\Big/ \sin h\big/2 \right]$$

$$= \lim_{h \to 0} h \left[\frac{\sin\left(a + \dfrac{nh}{2} - \dfrac{h}{2}\right) . \sin \dfrac{nh}{2}}{\sin h\big/2} \right]$$

$$= \lim_{h \to 0} h \left[\frac{\sin\left(a + \dfrac{b-a}{2} - \dfrac{h}{2}\right) . \sin \dfrac{b-a}{2}}{\sin h\big/2} \right]$$

$$= \lim_{h \to 0} 2 . \frac{h\big/2}{\sin h\big/2} . \sin\left(\frac{a+b}{2} - h\big/2\right) \sin \frac{(b-a)}{2}$$

$$= 2 \lim_{h \to 0} \sin\left(\frac{a+b}{2} - h\big/2\right) . \sin \frac{b-a}{2} \qquad\qquad (\because \lim_{h \to 0} \frac{h}{\sinh} = 1)$$

$$= 2 \sin \frac{a+b}{2} . \sin \frac{b\ \ a}{2}$$

$$= \cos a - \cos b$$

***Example* 3:** Evaluate $\displaystyle\int_a^b e^x \, dx$ as limit of a sum. $\qquad\qquad$ (RGPV Dec 2003)

***Solution*:** By definition, we have

$$\int_a^b f(x)dx = \lim_{h \to 0} h[f(a) + f(a + h) + f(a + 2h) + \ldots + f(a + (n-1)h)]$$

Here. $f(x) = e^x$, so that $f(a) = e^a$, $f(a + h) = e^{a+h}, \ldots$ so on

$$\therefore \int_a^b e^x dx = \lim_{h \to 0} h\left[e^a + e^{a+h} + e^{a+2h} + \ldots + e^{a+(n-1)h} \right]$$

$$= \lim_{h \to 0} h e^a \left[1 + e^h + e^{2h} + \ldots + e^{(n-1)h} \right]$$

$$= \lim_{h\to 0} h.e^a \left[\frac{\left(e^h\right)^n - 1}{e^h - 1}\right] \qquad (\because \text{ sum of n terms in G.P})$$

$$= \lim_{h\to 0} h\, e^a \left[\frac{e^{nh} - 1}{e^h - 1}\right]$$

$$= e^a \lim_{h\to 0}\left(e^{nh} - 1\right).\lim_{h\to 0}\frac{h}{e^h - 1}$$

$$= e^a . \lim_{h\to 0}\left(e^{b-a} - 1\right).1 \qquad \left(\because \lim_{h\to 0}\frac{h}{e^h - 1} = 1\right)$$

$$= e^a . (e^{b-a} - 1)$$

$$= e^b - e^a$$

***Example* 4:** Evaluate $\displaystyle\int_{1}^{2}\left(x^2 + x\right)\,dx$ as the limit of a sum.

***Solution*:** By definition, we get

$$\int_{a}^{b} f(x)\,dx = \lim_{h\to 0}\ h[f(a + b) + f(a + 2h) + \ldots + f(a + nh)],$$

where $\quad h = \dfrac{b-a}{n} = \dfrac{2-1}{n} = \dfrac{1}{n}$

$\therefore \qquad nh = 1$

$$\therefore \qquad \int_{1}^{1}\left(x + x^2\right)dx = \lim_{h\to 0}\ h\sum_{r=1}^{n} f(a + rh)$$

$$= \lim_{h\to 0}\ h\sum_{r=1}^{n}\left[1 + rh + \left(1 + rh\right)^2\right]$$

$$(\because \text{ given } f(a + rh) = (a + rh) + (a + rh)^2 \text{ and } a = 1)$$

$$= \lim_{h\to 0}\ h\sum_{r=1}^{n}\left(2 + 3rh + r^2h^2\right)$$

$$= \lim_{h\to 0}\left[2h\sum_{r=1}^{n}.1 + 3h^2\sum_{r=1}^{n}r + h^3\sum_{r=1}^{n}r^2\right]$$

$$= \lim_{h\to 0}\left[2h.h + 3h^2\frac{n(n+1)}{2} + h^3\frac{n(n+1)(2n+1)}{6}\right]$$

$$= \lim_{h\to 0}\left[2 + \frac{3}{2}(1+h) + \frac{1}{6}(1+h)(2+h)\right]$$

$$= 2 + \frac{3}{2} + \frac{1}{3}$$

$$= \frac{23}{6}$$

12.2   The Sum as a Definite Integral

Procedure

(i)   First find the rth term of the series and put it in the form $\frac{1}{n} f\left(\frac{r}{n}\right)$. Then the series can be written as $\lim_{n \to 0} \sum \frac{1}{n} f\left(\frac{r}{n}\right)$.

(ii)   Replace $\lim_{n \to 0} \sum$ by $\int$, $\frac{r}{n}$ by x, $\frac{1}{n}$ by dx.

(iii)   Obtain lower and upper limits by computing $\lim_{n \to 0} \frac{r}{n}$ for the first term and for the last term.

Example **5:** Show that the limit of the sum $\frac{1}{n} + \frac{1}{n+1} + \frac{1}{n+2} + ... + \frac{1}{3n}$, when n is indefinitely increased is log 3. (RGPV 2001)

Solution: Here, r^{th} term $= \frac{1}{n+r}$

$$= \frac{1}{n}\left[\frac{1}{1 + \frac{r}{n}}\right]$$

$$\therefore \quad \lim_{n \to 0}\left[\frac{1}{n} + \frac{1}{n+1} + \frac{1}{n+2} + ... + \frac{1}{n+2n}\right]$$

$$= \lim_{n \to 0} \sum_{r=0}^{2n} \frac{1}{n}\cdot\left[\frac{1}{1 + \frac{r}{n}}\right]$$

Now, the lower limit $= \lim_{n \to \infty} \frac{r}{n}$ for the first term $(r = 0) = \lim_{n \to \infty} \frac{0}{n} = 0$

The upper limit $= \lim\limits_{n\to\infty} \dfrac{r}{n}$ for the last term $(r = 2n) = \lim\limits_{n\to\infty} \dfrac{2n}{n} = 2$

$\therefore$ limit of the sum of the given series

$$= \int_0^2 \frac{1}{1+x}\,dx = \Big[\log(1+x)\Big]_0^2 = \log 3 - \log 1 = \log 3$$

Example 6: Evaluate $\lim\limits_{n\to\infty}\left[\dfrac{n}{n^2} + \dfrac{n}{n^2+1^2} + ... + \dfrac{n}{n^2+(n-1)^2}\right]$ as a definite integral.

$$\text{(RGPV June 2004, Dec 2008, April 2009)}$$

Solution: Here, r^{th} term $= \dfrac{n}{n^2+r^2}$

$$= \frac{1}{n}\cdot\frac{1}{1+r^2/n^2}$$

$\therefore \qquad \lim\limits_{n\to\infty}\left[\dfrac{n}{n^2} + \dfrac{n}{n^2+1^2} + ... + \dfrac{n}{n^2+(n-1)^2}\right]$

$$= \lim\limits_{n\to\infty}\sum_{r=0}^{n-1}\frac{1}{n}\cdot\frac{1}{1+r^2/n^2}$$

Now, the lower limit $= \lim\limits_{n\to\infty}\dfrac{r}{n}$ for the first term $(r = 0) = \lim\limits_{n\to\infty}\dfrac{0}{n} = 0$

The upper limit $= \lim\limits_{n\to\infty}\dfrac{r}{n}$ for the last term $(r = n-1) = \lim\limits_{n\to\infty}\dfrac{n-1}{n} = 1$

$\therefore$ limit of the sum of the given series

$$= \int_0^1 \frac{dx}{1+x^2} = \Big[\tan^{-1}x\Big]_0^1 = \frac{\pi}{4}$$

Example 7: Evaluate $\lim\limits_{n\to\infty}\left[\dfrac{1}{n} + \dfrac{1}{\sqrt{n^2-1^2}} + \dfrac{1}{\sqrt{n^2-2^2}} + ... + \dfrac{1}{\sqrt{n^2-(n-1)^2}}\right]$ (RGPV Dec 2004)

Solution: Here, r^{th} term $= \dfrac{1}{\sqrt{n^2-r^2}}$

$$= \frac{1}{n}\cdot\frac{1}{\sqrt{1-(r/n)^2}}$$

$$\therefore \quad \lim_{n \to \infty}\left[\frac{1}{n}+\frac{1}{\sqrt{n^2-1^2}}+\frac{1}{\sqrt{n^2-2^2}}+....+\frac{1}{\sqrt{n^2-(n-1)^2}}\right]$$

$$= \lim_{n \to \infty}\sum_{r=0}^{n-1}\frac{1}{n}\cdot\frac{1}{\sqrt{1-\left(\frac{r}{n}\right)^2}}$$

Now, the lower limit $= \lim_{n \to \infty}\frac{r}{n}$ for the first term $(r = 0) = \lim_{n \to \infty}\frac{0}{n}=0$

The upper limit $= \lim_{n \to \infty}\frac{r}{n}$ for the last term $(r = n - 1) = \lim_{n \to \infty}\frac{n-1}{n}=1$

$\therefore$ limit of the sum of the given series

$$= \int_{0}^{1}\frac{1}{\sqrt{1-x^2}}dx = \left[\sin^{-1}x\right]_{0}^{1} = \frac{\pi}{2}$$

Example 8: Evaluate the limit when $n \to \infty$ $\displaystyle\sum_{r=1}^{n-1}\frac{1}{\sqrt{n^2-r^2}}$.   (RGPV Jan 2006, Dec 2007)

Solution: Same as above example.

Example 9: Evaluate $\displaystyle\lim_{n \to \infty}\sum_{r=1}^{n-1}\frac{1}{n}\sqrt{\frac{n+r}{n-r}}$ (RGPV June 2005)

Solution: Here, r^{th} term $= \dfrac{1}{n}\sqrt{\dfrac{1+\frac{r}{n}}{1-\frac{r}{n}}}$

$$\therefore \quad \lim_{n \to 0}\sum_{r=1}^{n-1}\frac{1}{n}\sqrt{\frac{1+\frac{r}{n}}{1-\frac{r}{n}}}$$

Now, the lower limit $= \lim_{n \to \infty}\frac{r}{n}$ for the first term $(r = 1) = \lim_{n \to \infty}\frac{1}{n}=0$

The upper limit $= \lim_{n \to \infty}\frac{r}{n}$ for the last term $(r = n - 1) =$

$$\lim_{n \to \infty}\frac{n-1}{n}=1$$

$\therefore$ limit of the sum of the given series

$$= \int_0^1 \sqrt{\frac{1+x}{1-x}} \, dx$$

$$= \int_0^1 \sqrt{\frac{1+x}{1-x^2}} \, dx$$

$$= \int_0^1 \sqrt{\frac{1}{1-x^2}} \, dx + \int_0^1 \frac{x}{\sqrt{1-x^2}} \, dx$$

$$= \left[\sin^{-1} x \right]_0^1 + \left[-\sqrt{1-x^2} \right]_0^1 = \frac{\pi}{2} - (-1) = \frac{\pi}{2} + 1$$

***Example* 10:**

Evaluate $\displaystyle \lim_{n \to \infty} \left[\frac{1}{n} + \frac{n^2}{(n+1)^3} + \frac{n^2}{(n+2)^3} + \ldots + \frac{1}{8n} \right]$ (RGPV Dec 2005)

***Solution*:**

Here, $\displaystyle \lim_{n \to \infty} \left[\frac{1}{n} + \frac{n^2}{(n+1)^3} + \frac{n^2}{(n+2)^3} + \ldots + \frac{1}{8n} \right]$

$$= \lim_{n \to \infty} \left[\frac{n^2}{n^3} + \frac{n^2}{(n+1)^3} + \frac{n^2}{(n+2)^3} + \ldots + \frac{n^2}{(n+n)^3} \right]$$

$\therefore$ r^{th} term $= \dfrac{n^2}{(n+r)^3} = \dfrac{1}{n} \dfrac{1}{\left(1 + \frac{r}{n}\right)^3}$

$\therefore$ $\displaystyle \lim_{n \to \infty} \sum_{r=0}^{n} \frac{1}{n} \cdot \frac{1}{\left(1 + \frac{r}{n}\right)^3}$

Now, the lower limit $= \displaystyle \lim_{n \to \infty} \frac{r}{n}$ for the first term $(r = 0) = \displaystyle \lim_{n \to \infty} \frac{0}{n} = 0$

The upper limit $= \displaystyle \lim_{n \to \infty} \frac{r}{n}$ for the last term $(r = n) = \displaystyle \lim_{n \to \infty} \frac{n}{n} = 1$

$\therefore$ limit of sum of the given series

$$= \int_0^1 \frac{1}{(1+x)^3} \, dx$$

$$= -\frac{1}{2} \left[\frac{1}{(1+x)^2} \right]_0^1 = -\frac{1}{2} \left[\frac{1}{4} - 1 \right] = \frac{3}{8}$$

***Example* 11:** Evaluate $\lim\limits_{n\to\infty}\dfrac{1}{\sqrt{n}}\left[1+\dfrac{1}{\sqrt{2}}+\dfrac{1}{\sqrt{3}}+...+\dfrac{1}{\sqrt{n}}\right]$ (RGPV June 2006)

***Solution*:** Here, r^{th} term $= \dfrac{1}{\sqrt{n}}.\dfrac{1}{\sqrt{r}}$

$$= \dfrac{1}{n}.\dfrac{1}{\sqrt{r/n}}$$

$\therefore\qquad \lim\limits_{n\to\infty}\dfrac{1}{\sqrt{n}}\left[1+\dfrac{1}{\sqrt{2}}+\dfrac{1}{\sqrt{3}}+...+\dfrac{1}{\sqrt{n}}\right]$

$$= \lim\limits_{n\to\infty}\sum_{r=1}^{n}\dfrac{1}{n}.\dfrac{1}{\sqrt{r/n}}$$

Now, the lower limit $= \lim\limits_{n\to\infty}\dfrac{r}{n}$ for the first term $(r=1) = \lim\limits_{n\to\infty}\dfrac{1}{n} = 0$

The upper limit $= \lim\limits_{n\to\infty}\dfrac{r}{n}$ for the last term $(r=n) = \lim\limits_{n\to\infty}\dfrac{n}{n} = 1$

$\therefore$ limit of sum of the given series

$$\int_{0}^{1}\dfrac{1}{\sqrt{x}}dx = 2\left[\sqrt{x}\right]_{0}^{1} = 2$$

***Example* 12:** Evaluate $\lim\limits_{n\to\infty}\left[\dfrac{1}{n+1}+\dfrac{1}{n+2}+...+\dfrac{1}{2n}\right]$ (RGPV June 2008, 2009)

***Solution*:** Here, r^{th} term $= \dfrac{1}{n+r}$

$$= \dfrac{1}{n}.\dfrac{1}{1+r/n}$$

$\therefore\qquad \lim\limits_{n\to\infty}\left[\dfrac{1}{n+1}+\dfrac{1}{n+2}+...+\dfrac{1}{n+n}\right]$

$$= \lim\limits_{n\to\infty}\sum_{r=1}^{n}\dfrac{1}{n}.\dfrac{1}{1+r/n}$$

Now, the lower limit $= \lim\limits_{n\to\infty}\dfrac{r}{n}$ for the first term $(r=1) = \lim\limits_{n\to 0}\dfrac{1}{n} = 0$

The upper limit $\lim\limits_{n\to\infty}\dfrac{r}{n}$ for the last term $(r=n) = \lim\limits_{n\to\infty}\dfrac{n}{n} = 1$

∴ limit of the sum of the given series

$$= \int_0^1 \frac{1}{1+x}\,dx = \Big[\log\,(1+x)\Big]_0^1 = \log\,2$$

***Example* 13:** Evaluate $\lim\limits_{n\to\infty}\left[\dfrac{1}{n^3}+\dfrac{4}{n^3}+\dfrac{9}{n^3}+...+\dfrac{1}{n}\right]$

***Solution*:** Here, $\lim\limits_{n\to\infty}\left[\dfrac{1}{n^3}+\dfrac{4}{n^3}+\dfrac{9}{n^3}+...+\dfrac{1}{n}\right]$

$$= \lim_{n\to\infty}\left[\frac{1^2}{n^3}+\frac{2^2}{n^3}+\frac{3^3}{n^3}+...+\frac{n^2}{n^3}\right]$$

∴ r^{th} term $= \dfrac{r^2}{n^3} = \dfrac{1}{n}\cdot\left(\dfrac{r}{n}\right)^2$

∴ series be $\lim\limits_{n\to\infty}\sum\limits_{r=1}^{n}\dfrac{1}{n}\left(\dfrac{r}{n}\right)^2$

Now, the lower limit $= \lim\limits_{n\to\infty}\dfrac{r}{n}$ for the first term $(r=1) = \lim\limits_{n\to\infty}\dfrac{1}{n} = 0$

The upper limit $= \lim\limits_{n\to\infty}\dfrac{r}{n}$ for the last term $(r=n) = \lim\limits_{n\to\infty}\dfrac{n}{n} = 1$

∴ limit of sum of the given series

$$= \int_0^1 x^2\,dx = \left[\frac{x^3}{3}\right]_0^1 = \frac{1}{3}$$

***Example* 14:** Evaluate $\lim\limits_{n\to\infty}\left\{\left(1+\dfrac{1}{n}\right)\left(1+\dfrac{2}{n}\right)\left(1+\dfrac{3}{n}\right)....\left(1+\dfrac{n}{n}\right)\right\}^{1/n}$ (RGPV Jan 2008)

***Solution*:** Let A $= \lim\limits_{n\to\infty}\left\{\left(1+\dfrac{1}{n}\right)\left(1+\dfrac{2}{n}\right)....\left(1+\dfrac{n}{n}\right)\right\}^{1/n}$

Taking logarithms on both sides, we get

$$\log A = \lim_{n\to\infty}\frac{1}{n}\left[\log\left(1+\frac{1}{n}\right)+\log\left(1+\frac{2}{n}\right)+...+\log\left(1+\frac{n}{n}\right)\right]$$

$$= \lim_{n\to\infty}\sum_{r=1}^{n}\frac{1}{n}\log\left(1+\frac{r}{n}\right)$$

Now, the lower limit $= \lim\limits_{n \to 0} \dfrac{r}{n}$ for the first term $= \lim\limits_{n \to \infty} \dfrac{1}{n} = 0$

The upper limit $= \lim\limits_{n \to \infty} \dfrac{r}{n}$ for the last term $= \lim\limits_{n \to \infty} \dfrac{n}{n} = 1$

$\therefore$ limit of sum of the given series

$$\log A = \int_0^1 \log(1+x)\,dx$$

$$= \left[\log(1+x)\int dx\right]_0^1 - \int_0^1 \left\{\frac{d}{dx}\log(1+x)\int dx\right\}dx$$

$$= \left[x \log(1+x)\right]_0^1 - \int_0^1 \frac{x}{1+x}\,dx$$

$$= \log 2 - \left[x - \log(1+x)\right]_0^1$$

$$= \log 2 - [1 - \log 2 - 0 + \log 1]$$

$$= \log 2 - 1 + \log 2$$

$$= 2 \log 2 - 1$$

$$= \log 2^2 - \log e$$

$$\log A = \log \frac{4}{e}$$

Hence $A = \dfrac{4}{e}$

***Example* 15:** Evaluate $\lim\limits_{n \to \infty}\left\{\left(1+\dfrac{1^2}{n^2}\right)\left(1+\dfrac{2^2}{n^2}\right)\left(1+\dfrac{3^2}{n^2}\right)\ldots\left(1+\dfrac{n^2}{n^2}\right)\right\}^{1/n}$

(RGPV Dec 2002, Jan 2007)

***Solution*:** Let $A = \lim\limits_{n \to \infty}\left\{\left(1+\dfrac{1^2}{n^2}\right)\left(1+\dfrac{2^2}{n^2}\right)\ldots\left(1+\dfrac{n^2}{n^2}\right)\right\}^{1/n}$

Taking logarithms on both sides, we get

$$\log A = \lim\limits_{n \to \infty}\frac{1}{n}\left[\log\left(1+\frac{1^2}{n^2}\right) + \log\left(1+\frac{2^2}{n^2}\right) + \ldots + \log\left(1+\frac{n^2}{n^2}\right)\right]^{1/n}$$

$$= \lim\limits_{n \to \infty}\sum_{r=1}^{n}\frac{1}{n}\log\left(1+\frac{r^2}{n^2}\right)$$

Now, the lower limit $= \lim_{n \to \infty} \dfrac{r}{n}$ for the first time $(r = 1) = \lim_{n \to \infty} \dfrac{1}{n} = 0$

The upper limit $= \lim_{n \to \infty} \dfrac{r}{n}$ for the last term $(r = n) = \lim_{n \to \infty} \dfrac{n}{n} = 1$

$\therefore$ limit of sum of the given series

$$\log A = \int_0^1 \log\left(1 + x^2\right) dx$$

$$= \left[x \log\left(1 + x^2\right) - \int \frac{2x}{1 + x^2}.x \, dx \right]_0^1$$

$$= \left[x \log\left(1 + x^2\right) - 2\int \frac{x^2 + 1 - 1}{1 + x^2} dx \right]_0^1$$

$$= \left[x \log\left(1 + x^2\right) - 2\int dx + 2\int \frac{1}{1 + x^2} dx \right]_0^1$$

$$= \left[x \log\left(1 + x^2\right) - 2x + 2\tan^{-1} x \right]_0^1$$

$$= \log 2 - 2\,(1 - 0) + 2\tan^{-1} 1$$

$$= \log 2 - 2 + 2.\, {}^{\pi}\!\!\diagup\!\!_{4}$$

$$\log A = \log 2 - 2 + {}^{\pi}\!\!\diagup\!\!_{2}$$

$$\log A - \log 2 = {}^{\pi}\!\!\diagup\!\!_{2} - 2$$

$$\log {}^{A}\!\!\diagup\!\!_{2} = {}^{\pi}\!\!\diagup\!\!_{2} - 2$$

or $\qquad {}^{A}\!\!\diagup\!\!_{2} = e^{\left({}^{\pi}\!/_{2} - 2\right)}$

$\therefore \qquad A = 2\,.\,e^{\left({}^{\pi}\!/_{2} - 2\right)}$

***Example* 16:** Evaluate $\displaystyle \lim_{n \to \infty} \left[\left(\frac{1}{n}\right).\left(\frac{2}{n}\right).\left(\frac{3}{n}\right)....\left(\frac{n}{n}\right) \right]^{1/n}$

***Solution*:** Let $A = \displaystyle \lim_{n \to \infty} \left[\left(\frac{1}{n}\right)\left(\frac{2}{n}\right)\left(\frac{3}{n}\right)....\left(\frac{n}{n}\right) \right]^{1/n}$

Taking logarithm on both sides, we get

$$\log A = \lim_{n\to\infty} \frac{1}{n}\left[\log\frac{1}{n} + \log\frac{2}{n} + \log\frac{3}{n} + ... + \log\frac{n}{n}\right]$$

$$= \lim_{n\to\infty} \sum_{r=1}^{n} \frac{1}{n}.\log\frac{r}{n}$$

Now, the lower limit $= \lim_{n\to\infty}\frac{r}{n}$ for the first term $(r = 1) = \lim_{n\to\infty}\frac{1}{n} = 0$

The upper limit $= \lim_{n\to\infty}\frac{r}{n}$ for the last term $(r = n) = \lim_{n\to\infty}\frac{n}{n} = 1$

$\therefore$ limit of sum of the given series

$$\log A = \int_{0}^{1} \log x \ dx$$

$$= \left[x \log x - x\right]_{0}^{1}$$

$$= \log 1 - 1 = -1$$

$$\therefore \qquad A = e^{-1} \Rightarrow A = \frac{1}{e}$$

***Example* 17:** Evaluate $\lim_{n\to\infty}\left\{\dfrac{n!}{n^n}\right\}^{1/n}$

***Solution*:** Let $\qquad A = \lim_{n\to\infty}\left\{\dfrac{n!}{n^n}\right\}^{1/n}$

$$A = \lim_{n\to\infty}\left(\frac{1.2.3....n}{n^n}\right)^{1/n}$$

$$= \lim_{n\to\infty}\left[\left(\frac{1}{n}\right)\left(\frac{2}{n}\right)\left(\frac{3}{n}\right)....\left(\frac{n}{n}\right)\right]^{1/n}$$

Now, proceeding as above example 20. We get the result.

Practice Problems

1. Evaluate each of the following integrals as the limit of a sum.

(a) $\displaystyle\int_{0}^{2} x \ dx$ (b) $\displaystyle\int_{0}^{2}\left(3x^2 + 5\right)dx$ (c) $\displaystyle\int_{a}^{b} x^3 dx$

[**Ans.** (a) 2 (b) 18 (c) $\dfrac{1}{4}(b^4 - a^4)$]

2. Evaluate each of the following integrals as the limit of a sum

   (a) $\displaystyle\int_{0}^{2} e^{x}\, dx$ (b) $\displaystyle\int_{a}^{b} e^{-x}\, dx$

   (c) $\displaystyle\int_{0}^{\pi/2} \sin x\, dx$ (d) $\displaystyle\int_{a}^{b} \cos x\, dx$

$$[\textbf{Ans.} \ (a)\ e^{2}-1 \ (b)\ e^{-a}-e^{-b}\ (c)\ 1\ (d)\ \sin b - \sin a]$$

3. Evaluate $\displaystyle\lim_{n\to\infty}\left[\frac{1}{n+1}+\frac{1}{n+2}+....+\frac{1}{4n}\right]$

$$[\textbf{Ans. Log 4}]$$

4. Evaluate $\displaystyle\lim_{n\to\infty}\left[\frac{1}{n+m}+\frac{1}{n+2m}+....+\frac{1}{n+nm}\right]$

$$\left[\textbf{Ans.}\ \frac{1}{m}\log(m+1)\right]$$

5. Evaluate $\displaystyle\lim_{n\to\infty}\left[\frac{n}{n^{2}+1^{2}}+\frac{n}{n^{2}+2^{2}}+...+\frac{1}{2n}\right]$

$$\left[\textbf{Ans.}\ \frac{\pi}{4}\right]$$

6. Evaluate $\displaystyle\lim_{n\to\infty}\left[\frac{1^{2}}{1^{3}+n^{3}}+\frac{2^{2}}{2^{3}+n^{3}}+\frac{3^{2}}{3^{3}+n^{3}}+...+\frac{n^{2}}{n^{3}+n^{3}}\right]$

$$\left[\textbf{Ans.}\ \frac{1}{3}\log 2\right]$$

7. Evaluate $\displaystyle\lim_{n\to\infty}\sum_{r=1}^{n}\frac{n}{n^{2}+r^{2}}$

$$\left[\textbf{Ans.}\ \frac{\pi}{4}\right]$$

8. Evaluate $\displaystyle\lim_{n\to\infty}\left[\frac{(n+1)(n+2)....(n+2)}{n^{n}}\right]^{1/n}$

$$\left[\textbf{Ans.}\ \frac{4}{e}\right]$$

9. Evaluate $\displaystyle\lim_{n\to\infty}\frac{1}{2}\left[\tan\left(\frac{\pi}{4n}\right)+\tan\left(\frac{2\pi}{4n}\right)+\tan\left(\frac{3\pi}{4n}\right)+..+\tan\left(\frac{n\pi}{4n}\right)\right]$

$$\left[\textbf{Ans.}\ \frac{2}{\pi}\log 2\right]$$

10. Find the limit, when $n \to \infty$, of the product

$$\left(1+\frac{1}{n}\right)\left(1+\frac{2}{n}\right)^{1/n}\left(1+\frac{3}{n}\right)^{1/3}.....\left(1+\frac{n}{n}\right)^{1/n}$$

[**Ans.** $e^{\pi^2/12}$]

11. Prove that $\displaystyle\lim_{n\to\infty}\left[\frac{1}{3n+1}+\frac{1}{3n+2}+..+\frac{1}{3n+n}\right]=\log \frac{4}{3}$

12. Find the limit, when $n \to \infty$, of the series

$$\frac{1}{n^2}\sec^2\frac{1}{n^2}+\frac{2}{n^2}\sec^2\frac{4}{n^2}+\frac{3}{n^2}\sec^2\frac{9}{n^2}+...+\frac{1}{n}\sec^2 1$$

13. Evaluate $\displaystyle\lim_{n\to\infty}\left[\frac{n+1}{n^2+1^2}+\frac{n+2}{n^2+2^2}+...+\frac{1}{n}\right]$

[**Ans.** $\frac{\pi}{4}-\frac{1}{2}\log 2$]

14. Evaluate $\displaystyle\lim_{n\to\infty}\left[\frac{\sqrt{n+1}+\sqrt{n+2}+...+\sqrt{2n}}{n\sqrt{n}}\right]$

[**Ans.** $\frac{2}{3}\left(2\sqrt{2}-1\right)$]

15. Evaluate $\displaystyle\lim_{n\to\infty}\left[\sin\frac{\pi}{2n}.\sin\frac{2\pi}{2n}.\sin\frac{3\pi}{2n}......\sin\frac{n\pi}{2n}\right]^{1/n}$

[**Ans.** $\frac{1}{2}$]

CHAPTER – 13

Beta and Gamma Functions

13.1 Introduction

A great mathematician, Euler investigated two definite integrals. These first and second Eulerian integrals are Beta and Gamma functions respectively. Beta and Gamma functions are improper integrals. They have their wide applications in science and engineering such as physics, statistics and applied science etc.

13.2 Beta Function

The Beta function denoted by $\beta(l, m)$ is defined as follows:

$$\beta(l, m) = \int_0^1 x^{l-1}(1-x)^{m-1}\, dx, \text{ where } l > 0, m > 0$$

Symmetrical Property

The Beta function is symmetric in l and m, i.e.,

$$\beta(l, m) = \beta(m, l)$$

***Proof*:**   By the definition

$$\beta(l, m) = \int_0^1 x^{l-1}(1-x)^{m-1}\, dx$$

Putting $1 - x = y \Rightarrow dx = -\, dy$. Therefore we get,

$$\beta(l, m) = \int_1^0 (1-y)^{l-1}\, y^{m-1}(-dy)$$

$$= \int_0^1 (1-y)^{l-1}\, y^{m-1}\, dy$$

$$= \beta(m, l)$$

Some other Identities of Beta Functions

(i) $\beta(l,m) = \int\limits_{0}^{\infty} \dfrac{x^{m-1}}{(1+x)^{l+m}}\,dx$

Proof: we have $\beta(l,m) = \int\limits_{0}^{1} x^{l-1}(1-x)^{m-1}\,dx$

Putting $x = \dfrac{1}{1+y} \Rightarrow dx = -\dfrac{1}{(1+y)^2}\,dy$, we get

$$\beta(l,m) = \int\limits_{\infty}^{0} \frac{1}{(1+y)^{l-1}}\cdot\left(1-\frac{1}{1+y}\right)^{m-1}\cdot\left(-\frac{1}{(1+y)^2}\right)dy$$

$$= \int\limits_{0}^{\infty} \frac{y^{m-1}}{(1+y)^{l+m}}\,dy$$

$$= \int\limits_{0}^{\infty} \frac{x^{m-1}}{(1+x)^{l+m}}\,dx \qquad \left[\because \int\limits_{a}^{b} f(x)\,dx = \int\limits_{a}^{b} f(y)\,dy\right]$$

(ii) $\beta(l,m) = 2\int\limits_{0}^{\pi/2} \sin^{2l-1}\theta\cos^{2m-1}\theta\,d\theta$

Proof: we have $\beta(l,m) = \int\limits_{0}^{1} x^{l-1}(1-x)^{m-1}\,dx$

Putting $x = \sin^2\theta \Rightarrow dx = 2\sin\theta\cos\theta\,d\theta$, we get

$$\beta(l,m) = \int\limits_{0}^{\pi/2} (\sin^2\theta)^{l-1}(1-\sin^2\theta)^{m-1}\,2\sin\cos\theta\,d\theta$$

$$= 2\int\limits_{0}^{\pi/2} \sin^{2l-1}\theta\cos^{2m-1}\theta\,d\theta$$

13.3   Gamma Function

The gamma function is denoted by $\overline{|n}$ for n is positive and the relation be

$$\overline{|n} = \int\limits_{0}^{\infty} e^{-x}\cdot x^{n-1}\,dx$$

Properties of Gamma Function

I. $\overline{|1} = 1$

Proof: We have

$$\overline{|n} = \int_0^\infty e^{-x}\, x^{n-1} dx$$

when n = 1, we get

$$\overline{|1} = \int_0^\infty e^{-x}\, dx$$

$$= \lim_{h \to \infty} \int_0^h e^{-x}\, dx$$

$$= \lim_{h \to \infty} \left[-e^{-x} \right]_0^h$$

$$= \lim_{h \to \infty} \left[-e^{-h} + e^0 \right]$$

$$= \lim_{h \to \infty} \left[1 - e^{-h} \right] = 1$$

$$\therefore \quad \overline{|1} = 1$$

II. $\overline{|n+1} = n\overline{|n}, \ n > 0$

Proof: We have $\overline{|n} = \int_0^\infty e^{-x}\, x^{n-1}\, dx$

Replacing n by n +1, we get

$$\overline{|n+1} = \int_0^\infty e^{-x} . \, x^n\, dx$$

$$= \left[x^n \left(-e^{-x} \right) \right]_0^\infty - \int_0^\infty n x^{n-1} \left(-e^{-x} \right) dx$$

$$= \left[-x^n e^{-x} \right]_0^\infty + n \int_0^\infty x^{n-1} \left(e^{-x} \right) dx$$

$$= \lim_{h \to \infty} \left[-x^n e^{-x} \right]_0^h + n\overline{|n}$$

$$= \lim_{h \to \infty} \left[-h^n e^{-h} + 0 \right] + n\overline{|n}$$

$$= \lim_{h \to \infty} \left(\frac{-h^n}{e^h} \right) + n \overline{\lfloor n}$$

$$= \lim_{h \to \infty} \left(\frac{-h^n}{1 + h + \dfrac{h^2}{2!} + \ldots + \dfrac{h^n}{n!} + \ldots} \right) + n \overline{\lfloor n}$$

$$= \lim_{h \to \infty} \left[\frac{-1}{\dfrac{1}{h^n} + \dfrac{1}{h^{n-1}} + \ldots + \dfrac{1}{n!} + \ldots} \right] + n \overline{\lfloor n}$$

$$= -\frac{1}{\infty} + n \overline{\lfloor n}$$

$$= 0 + \ n \overline{\lfloor n}$$

Hence $\overline{\lfloor n+1} = n \overline{\lfloor n}$. It is called recurrence formula for the gamma function

Note: (i) $\overline{\lfloor n} = (n-1) \overline{\lfloor n-1}$ (ii) $\overline{\lfloor n} = (n-1)!$ (iii) $\overline{\lfloor 0} = \infty$

Transformation of Gamma Function

Show that

(a) $\overline{\lfloor n} = \int\limits_0^1 \left(\log \frac{1}{y} \right)^{n-1} dy$

(b) $\overline{\lfloor n} = P^n \int\limits_0^\infty e^{-Px} \, x^{n-1} \, dx$

(c) $\overline{\lfloor n+1} = \int\limits_0^\infty e^{-y^{1/n}} \, dy$

(d) $\overline{\lfloor \frac{1}{2}} = \sqrt{\pi}$ (RGPV Dec 2005, June 2007)

Proof:

(a) we have $\overline{\lfloor n} = \int\limits_6^\infty e^{-x} \cdot x^{n-1} dx$ .....(i)

   Putting $x = \log \dfrac{1}{y} \Rightarrow y = e^{-x} \Rightarrow dy = -e^{-x} dx$, we get

$$\overline{n} = -\int_{1}^{0} \left(\log \frac{1}{y} \right)^{n-1} dy$$

$$= \int_{0}^{1} \left(\log \frac{1}{y} \right)^{n-1} dy$$

(b) Replacing x by px in eqn. (i), we get

$$\overline{n} = \int_{0}^{\infty} e^{-px} (px)^{n-1} \cdot p\,dx$$

$$= P^{n} \int_{0}^{\infty} e^{-px} \cdot x^{n-1} dx$$

(c) Putting $x^{n} = y \Rightarrow nx^{n-1} dx = dy$ in eqn (i), we get

$$\overline{n} = \int_{0}^{\infty} e^{-y^{1/n}} \cdot x^{n-1} \cdot \frac{dy}{nx^{n-1}}$$

$$= \frac{1}{n} \int_{0}^{\infty} e^{-y^{1/n}} dy$$

$$\text{or} \quad n\sqrt{n} = \int_{0}^{\infty} e^{-y^{1/n}} dy$$

$$\text{or} \quad \overline{n+1} = \int_{0}^{\infty} e^{-y^{1/n}} dy$$

(d) Putting $n = \dfrac{1}{2}$ in eqn (i), we get

$$\overline{\frac{1}{2}} = \int_{0}^{\infty} e^{-x} \, x^{-1/2} dx$$

Putting $x = t^{2} \Rightarrow dx = 2t\, dt$, we get

$$\overline{\frac{1}{2}} = 2 \int_{0}^{\infty} e^{-t^{2}} dt$$

Replacing t by x, we have $\overline{\dfrac{1}{2}} = 2 \displaystyle\int_{0}^{\infty} e^{-x^{2}} dx$ .....(ii)

Again substituting y for x, we have

$$\left|\frac{1}{2}\right. = 2\int_0^\infty e^{-y^2}\, dy \qquad\qquad .....(iii)$$

Multiply eqn (ii) and (iii), we get

$$\left(\left|\frac{1}{2}\right.\right)^2 = 4\int_0^\infty e^{-x^2}\, dx \,.\, \int_0^\infty e^{-y^2}\, dy$$

$$= 4\int_0^\infty\int_0^\infty e^{-(x^2+y^2)}\, dx\, dy$$

Transforming into polar coordinates, we put

$$x = r\cos\theta,\ y = r\sin\theta \text{ and } dx\, dy = r\, d\theta\, dr, \text{ we get}$$

$$\left(\left|\frac{1}{2}\right.\right)^2 = 4\int_{\theta=0}^{\pi/2}\int_{r=0}^{\infty} e^{-r^2}\,.\, r\, d\theta\, dr \qquad (\because x^2+y^2 = r^2)$$

$$= 2\int_{\theta=0}^{\pi/2}\left[-e^{-r^2}\right]_{r=0}^{\infty}\, d\theta$$

$$= 2\int_{\theta=0}^{\pi/2}\, d\theta$$

$$= 2\,.\,\frac{\pi}{2} = \pi$$

Hence $\left|\dfrac{1}{2}\right. = \sqrt{\pi}\,.$ Hence proved.

13.4   Relation between Beta and Gamma Function

I.   Prove that $\beta(m, n) = \dfrac{\left|m\right.\left|n\right.}{\left|m+n\right.}$, $m > 0, n > 0$

(RGPV June 2003, Feb 2005, June 2009)

Proof: we have $\left|m\right. = \displaystyle\int_0^\infty e^{-x}\,.\, x^{m-1}\, dx$

Replacing x by x^2, we get

$$\left|m\right. = 2\int_0^\infty e^{-x^2}\,.\, x^{2m-1}\, dx \qquad\qquad .....(i)$$

Similarly $\overline{|n} = 2\int\limits_{0}^{\infty} e^{-y^2} \cdot y^{2n-1}\, dy$ .....(ii)

multiplying eqn. (i) and (ii), we get

$$\overline{|m}\,\overline{|n} = 4\int\limits_{0}^{\infty}\int\limits_{0}^{\infty} e^{-(x^2+y^2)} \cdot x^{2m-1}\, y^{2n-1}\, dxdy \qquad \text{.....(iii)}$$

Changing to polar coordinates, we put

$$x = r\cos\theta,\ y = r\sin\theta,\ dxdy = r\, dr\, d\theta$$

Then eqn. (iii) becomes

$$\overline{|m}\,\overline{|n} = 4\int\limits_{0}^{\pi/2}\int\limits_{0}^{\infty} e^{-r^2} r^{2m+2n-1}\ \cos^{2m-1}\theta \cdot \sin^{2n-1}\theta\, drd\theta$$

$$= 4\int\limits_{0}^{\pi/2} \cos^{2m-1}\theta\ \sin^{2n-1}\theta\, d\theta \int\limits_{0}^{\infty} e^{-r^2} \cdot r^{2(m+n)-1}\, dr \qquad \text{.....(iv)}$$

Since $\beta(m,\, n) = 2\int\limits_{0}^{\pi/2} \cos^{2m-1}\theta \cdot \sin^{2n-1}\theta\, d\theta$

and $\overline{|m+n} = 2\int\limits_{0}^{\infty} e^{-r^2} \cdot r^{2(m+n)-1}\, dr$

Hence, eqn. (iv) may be written in the form

$$\overline{|m}\,\overline{|n} = \beta\,(m,n) \cdot \overline{|m+n}$$

or $\beta(m,\, n) = \dfrac{\overline{|m}\,\overline{|n}}{\overline{|m+n}}$ Hence proved.

II. Prove that $\overline{|n}\ \overline{|1-n} = \dfrac{\pi}{\sin n\,\pi},\quad 0 < n < 1$

Proof: We know that $\beta(m,n) = \dfrac{\overline{|m}\ \overline{|n}}{\overline{|m+n}}$ .....(i)

and $\beta(m,\, n) = \displaystyle\int\limits_{0}^{\infty} \dfrac{x^{n-1}}{(1+x)^{m+n}}\, dx$ .....(ii)

equating eqn. (i) and (ii) we get

$$\int\limits_{0}^{\infty} \dfrac{x^{n-1}}{(1+x)^{m+n}}\, dx = \dfrac{\overline{|m}\,\overline{|n}}{\overline{|m+n}}$$

Putting $m + n = 1$, so that $m = 1 - n$; we get

$$\int_0^\infty \frac{x^{n-1}}{1+x}\, dx = \frac{\overline{|1-n}\,.\,\overline{|n}}{\overline{|1}}$$

$$= \overline{|n}\;\overline{|1-n} \qquad (\because \overline{|1}=1)$$

Since $\qquad \displaystyle\int_0^\infty \frac{x^{n-1}}{1+x}\,dx = \frac{\pi}{\sin n\pi}, \quad 0<n<1$

Hence $\qquad \overline{|n}\,\overline{|1-n} = \dfrac{\pi}{\sin n\pi}, \quad 0<n<1$

III. Prove that $\displaystyle\int_0^{\pi/2} \sin^{2m-1}\theta \cos^{2n-1}\theta\, d\theta = \frac{\overline{|m}\,\overline{|n}}{2\,\overline{|m+n}}$

'or' $\qquad \displaystyle\int_0^{\pi/2} \sin^{P} \cos^{q}\theta\, d\theta = \frac{\overline{\left|\dfrac{P+1}{2}\right.}\;\overline{\left|\dfrac{q+1}{2}\right.}}{2\,\overline{\left|\dfrac{P+q+2}{2}\right.}}$

Proof: By the definition

$$\beta(m,\,n) = \int_0^1 x^{m-1}(1-x)^{n-1}\, dx$$

Substituting $x = \sin^2\theta \Rightarrow dx = 2\sin\theta\cos\theta\, d\theta$, we get

$$\beta(m,n) = \int_0^{\pi/2} \sin^{2m-2}\theta \cos^{2n-2}\theta\,.\,2\sin\theta\cos\theta\, d\theta$$

$$= 2\int_0^{\pi/2} \sin^{2m-1}\theta \cos^{2n-1}\theta\, d\theta$$

$\therefore \qquad \displaystyle\int_0^{\pi/2} \sin^{2m-1}\theta \cos^{2n-1}\theta\, d\theta = \frac{\overline{|m}\,\overline{|n}}{2\,\overline{|m+n}}$

$$\left(\because \beta(m,n) = \frac{\overline{|m}\,\overline{|n}}{\overline{|m+n}} \right)$$

Again, putting $2m - 1 = P$ and $2n - 1 = q$

$$\Rightarrow \quad m = \frac{P+1}{2} \text{ and } n = \frac{q+1}{2}, \text{ we get}$$

$$\int_0^{\pi/2} \sin^P \theta \cos^q \theta \, d\theta = \frac{\left\lfloor \frac{P+1}{2} \right\rfloor \left\lfloor \frac{q+1}{2} \right\rfloor}{2 \left\lfloor \frac{P+q+2}{2} \right\rfloor} \quad \text{Hence proved}$$

13.5   Duplication Formula

Prove that $\lceil m \lceil m + \frac{1}{2} = \frac{\sqrt{\pi} \lceil 2m}{2^{2m-1}}$, where m is positive real number   (RGPV Dec 2003)

Proof: By definition

$$\beta(m, n) = \frac{\lceil m \lceil n}{\lceil m + n}$$

or $$2 \int_0^{\pi/2} \sin^{2m-1} \theta \cos^{2n-1} \theta \, d\theta = \frac{\lceil m \lceil n}{\lceil m + n} \qquad \qquad .....(i)$$

Putting $2n - 1 = 0 \Rightarrow n = \frac{1}{2}$ in eqn (i), we have

$$2 \int_0^{\pi/2} \sin^{2m-1} \theta \, d\theta = \frac{\lceil m \lceil 1/2}{\lceil m + 1/2}$$

$$\Rightarrow \quad 2 \int_0^{\pi/2} \sin^{2m-1} \theta \, d\theta = \frac{\lceil m \sqrt{\pi}}{\lceil m + 1/2} \qquad \qquad .....(ii)$$

Again, putting $n = m$ in eqn. (i), we have

$$2 \int_0^{\pi/2} \sin^{2m-1} \theta \cos^{2m-1} \theta \, d\theta = \frac{\lceil m \lceil m}{\lceil m + m}$$

$$\Rightarrow \quad \frac{2}{2^{2m-1}} \int_0^{\pi/2} 2^{2m-1} \sin^{2m-1} \theta \cos^{2m-1} \theta \, d\theta = \frac{\left(\lceil m\right)^2}{\lceil 2m}$$

$$\Rightarrow \quad \frac{2}{2^{2m-1}} \int_0^{\pi/2} \left(\sin 2\theta\right)^{2m-1} d\theta = \frac{\left(\lceil m\right)^2}{\lceil 2m}$$

Putting $2\theta = \phi \Rightarrow 2\,d\theta = d\phi$, we get

$$\frac{1}{2^{2m-1}} \int_0^{\pi} \sin^{2m-1}\phi \; d\phi = \frac{\left(\overline{|m}\right)^2}{\overline{|2m}}$$

$$\Rightarrow \qquad \frac{2}{2^{2m-1}} \int_0^{\pi/2} \sin^{2m-1}\phi \; d\phi = \frac{\left(\overline{|m}\right)^2}{\overline{|2m}}$$

Replacing ϕ by θ, we get

$$2\int_0^{\pi/2} \sin^{2m-1}\theta \; d\theta = \frac{\left(\overline{|m}\right)^2}{\overline{|2m}} \cdot 2^{2m-1} \qquad\qquad \dots\text{(iii)}$$

Equating eqn. (ii) and (iii), we have

$$\frac{\overline{|m}\;\sqrt{\pi}}{\overline{|m+\tfrac{1}{2}}} = \frac{\left(\overline{|m}\right)^2}{\overline{|2m}} \cdot 2^{2m-1}$$

$$\overline{|m}\;\overline{\left|m+\frac{1}{2}\right.} = \frac{\sqrt{\pi}\;\overline{|2m}}{2^{2m-1}}$$

Hence proved.

***Example* 1:** Prove that

$$\beta\,(m,\,n) = \beta\,(m+1,\,n) + \beta\,(m,\,n+1) \qquad\qquad \text{(RGPV June 2004)}$$

Solution: R.H.S. $= \beta\,(m+1,\,n) + \beta\,(m,\,n+1)$

$$= \frac{\overline{|m+1}\;\overline{|n}}{\overline{|m+1+n}} + \frac{\overline{|m}\;\overline{|n+1}}{\overline{|m+n+1}}$$

$$= \frac{m\,\overline{|m}\;\overline{|n}}{(m+n)\;\overline{|m+n}} + \frac{\overline{|m}\;n\;\overline{|n}}{(m+n)\;\overline{|m+n}} \qquad \left(\because \overline{|n} = (n-1)\,\overline{|n-1}\,\right)$$

$$= \frac{m\,\overline{|m}\;\overline{|n} + n\,\overline{|m}\;\overline{|n}}{(m+n)\;\overline{|m+n}}$$

$$= \frac{\overline{|m}\;\overline{|n}\,(m+n)}{(m+n)\;\overline{|m+n}}$$

$$= \frac{\overline{|m}\;\overline{|n}}{\overline{|m+n}}$$

$$= \beta\,(m,\,n) = \text{L.H.S}$$

***Example* 2**: Prove that $\dfrac{\beta\,(m+1,n)}{m} = \dfrac{\beta(m,n+1)}{n} = \dfrac{\beta(m,n)}{m+n}$

(RGPV June 2002, 2004, 2008 Dec 2006)

***Solution*:**

(i) Let

$$\frac{\beta\,(m+1,n)}{m} = \frac{\overline{|m+1}\,\overline{|n}}{m\,\overline{|m+1+n}}$$

$$= \frac{m\,\overline{|m}\,\overline{|n}}{m(m+n)\,\overline{|m+n}}$$

$$= \frac{\overline{|m}\,\overline{|n}}{(m+n)\,\overline{|m+n}}$$

$$= \frac{\beta\,(m,n)}{m+n} \qquad\qquad \ldots(i)$$

(ii) Let

$$\frac{\beta\,(m,n+1)}{n} = \frac{\overline{|m}\;\overline{|n+1}}{n\,\overline{|m+n+1}}$$

$$= \frac{\overline{|m}\,.\,n\,\overline{|n}}{n(m+n)\,\overline{|m+n}}$$

$$= \frac{\overline{|m}\,\overline{|n}}{(m+n)\,\overline{|m+n}}$$

$$= \frac{\beta\,(m,n)}{m+n} \qquad\qquad \ldots(ii)$$

Hence, from (i) and (ii), we have

$$\frac{\beta\,(m+1,n)}{m} = \frac{\beta(m,n+1)}{n} = \frac{\beta(m,n)}{m+n}$$

***Example* 3**: Prove that

$$\int_{a}^{b} (x-a)^{m-1}\,(b-x)^{n-1}\,dx = (b-a)^{m+n-1}\,\beta\,(m,n) \qquad \text{(RGPV Jan 2007)}$$

***Solution*:** By definition

$$\beta\,(m,n) = \int_{0}^{1} u^{m-1}\,(1-u)^{n-1}\,du$$

Substituting $\;u = \dfrac{x-a}{b-a} \Rightarrow du = \dfrac{1}{b-a}\,dx\;$, we get

$$\beta\,(m,\,n) = \int\limits_{a}^{b} \left[\frac{x-a}{b-a}\right]^{m-1} \left[1 - \frac{x-a}{b-a}\right]^{n-1} \cdot \frac{1}{b-a}\,dx$$

$$= \frac{1}{(b-a)^{m+n-1}} \int\limits_{a}^{b} (x-a)^{m-1}\,(b-x)^{n-1}\;dx$$

$$\therefore\;\; \int\limits_{a}^{b} (x-a)^{m-1}\,(b-x)^{n-1}\;ax = (b-a)^{m+n-1} \times \beta(m,\,n)$$

Hence proved

***Example* 4:** Prove that $\displaystyle\int\limits_{0}^{1} x^{m}\,(\log x)^{n}\;dx = \frac{(-1)^{n}\,n!}{(m+1)^{n+1}}$, where m is an integer and $m > -1$

(RGPV Dec 2002, 2008)

***Solution*:** Let $\;\;I = \displaystyle\int\limits_{0}^{1} x^{m}\,(\log x)^{n} dx$

Putting $\log x = -u \Rightarrow x = e^{-u} \Rightarrow dx = -e^{-u}\,du$

$$\therefore\; I = \int_{-\infty}^{0} \left(e^{-u}\right)^{m} \times (-u)^{n}\left(-e^{-u}\right) du$$

$$= (-1)^{n} \int_{0}^{\infty} \left(e^{-u}\right)^{m+1} u^{n}\,du$$

$$= (-1)^{n} \int_{0}^{\infty} e^{-(m+1)u}\, u^{n}\,du$$

Again, putting $(m+1)\,u = t \Rightarrow du = \dfrac{dt}{m+1}$, we get

$$I = (-1)^{n} \int_{0}^{\infty} e^{-t}\left(\frac{t}{m+1}\right)^{n} \times \frac{dt}{m+1}$$

$$= (-1)^{n} \times \frac{1}{(m+1)^{n+1}} \int_{0}^{\infty} e^{-t} \times t^{n}\;dt$$

$$= \frac{(-1)^{n}}{(m+1)^{n+1}} \overline{\lfloor n+1}$$

$$= \frac{(-1)^{n}\,n!}{(m+1)^{n+1}} \qquad \left(\therefore \overline{\lfloor n+1} = n!\right)$$

Hence proved

***Example* 5:** Show that $\int_0^1 y^{q-1}\left(\log \dfrac{1}{y}\right)^{p-1} dy = \dfrac{\overline{\lvert p}}{q^p}$, p, q > 0 $\hspace{2cm}$ (RGPV April 2010)

***Solution*:** Let $I = \int_0^1 y^{q-1}\left(\log \dfrac{1}{y}\right)^{p-1} dy$

Putting $\log \dfrac{1}{y} = t \Rightarrow y = e^{-t}$ and $dy = -e^{-t} dt$

$\therefore I = \int_\infty^0 \left(e^{-t}\right)^{q-1} (t)^{p-1}\left(-e^{-t}\right) dt$

$\hspace{0.8cm} = \int_0^\infty \left(e^{-t}\right)^{q} t^{p-1} dt$

$\hspace{0.8cm} = \int_0^\infty e^{-tq} \times t^{p-1} dt$

Again, putting $tq = x, \Rightarrow q\, dt = dx$, we get

$I = \int_0^\infty e^{-x}.\left(\dfrac{x}{q}\right)^{P-1} \dfrac{dx}{q}$

$\hspace{0.4cm} = \dfrac{1}{q^p} \int_0^\infty e^{-x}. x^{p-1}\, dx = \dfrac{\overline{\lvert P}}{q^p}$

Hence proved.

***Example* 6:** Prove that $\int_0^\infty \dfrac{x^c}{c^x}\, dx = \dfrac{\overline{\lvert c+1}}{(\log c)^{c+1}}$, c > 1 $\hspace{2cm}$ (RGPV June 2007)

***Solution*:** Let $I = \int_0^\infty \dfrac{x^c}{c^x}\, dx$

$\hspace{1.2cm} = \int_0^\infty x^c. c^{-x}\, dx$

$\hspace{1.2cm} = \int_0^\infty x^c. c^{-x \log c}\, dx \hspace{0.8cm} \left(\because c^{-x} = e^{-x \log c}\right)$

Putting $x \log c = t \Rightarrow dx. \log c = dt$, we get

$I = \int_0^\infty \dfrac{t^c}{(\log c)^c}\, e^{-t}\, \dfrac{dt}{\log c}$

$\hspace{0.4cm} = \dfrac{1}{(\log c)^{c+1}} \int_0^\infty e^{-t}. t^c\, dt$

$$= \frac{1}{(\log c)^{c+1}} \int_0^\infty e^{-t} \, t^{(c+1)-1} \, dt$$

$$= \frac{\overline{|c+1}}{(\log c)^{c+1}}$$

Hence proved.

***Example* 7:** Prove that $\left| n + \dfrac{1}{2} \right. = \dfrac{\sqrt{\pi} \, \overline{|2n+1}}{2^{2n} \, \overline{|n+1}}$ (RGPV June 2005)

***Solution*:** Proceeding same as duplication formula.

$$\overline{|m} \, \left| m + \frac{1}{2} \right. = \frac{\sqrt{\pi} \, \overline{|2m}}{2^{2m-1}} \qquad \qquad \dots (i)$$

Putting $m = n + \dfrac{1}{2}$,   we get

$$\left| n + \frac{1}{2} \right. \, \overline{|n+1} = \frac{\sqrt{\pi} \, \overline{|2n+1}}{2^{2n}}$$

$$\Rightarrow \left| n + \frac{1}{2} \right. = \frac{\sqrt{\pi} \, \overline{|2n+1}}{2^{2n} \, \overline{|n+1}}$$

Hence proved.

***Example* 8:** Prove that $\displaystyle\int_0^\infty x^n \, e^{-k^2 x^2} dx - \frac{1}{2k^{n+1}} \left| \frac{n+1}{2} \right., \; n > -1$ (RGPV April 2009)

***Solution*:** Let $I = \displaystyle\int_0^\infty x^n \, e^{-k^2 x^2} dx$

Putting $k^2 x^2 = t \Rightarrow x = \dfrac{\sqrt{t}}{k} \Rightarrow dx = \dfrac{1}{2k\sqrt{t}} \, dt$

Thus $I = \displaystyle\int_0^\infty \left(\frac{\sqrt{t}}{k} \right)^n e^{-t} \, \frac{dt}{2k\sqrt{t}}$

$$= \frac{1}{2k^{n+1}} \int_0^\infty t^{\frac{n-1}{2}} \, e^{-t} \, dt$$

$$= \frac{1}{2k^{n+1}} \int_0^\infty t^{\left(\frac{n+1}{2}\right)-1} e^{-t} \, dt$$

$$= \frac{1}{2k^{n+1}} \left\lfloor \frac{n+1}{2} \right.$$

Hence proved.

Example 9: Prove that $\beta(m, n) = 2^{1-2m} \, \beta\left(m, \frac{1}{2}\right)$

Solution: R.H.S $= 2^{1-2m} \, \beta\left(m, \frac{1}{2}\right)$

$$= 2^{1-2m} \; \frac{\left\lfloor m \right. \left\lfloor \dfrac{1}{2} \right.}{\left\lfloor m + \dfrac{1}{2} \right.}$$

$$= \frac{\left\lfloor m \right. \sqrt{\pi}}{2^{2m-1} \left\lfloor m + \dfrac{1}{2} \right.}$$

$$= \frac{\left\lfloor m \right. \left\lfloor m \right.}{\left\lfloor 2m \right.}$$

$$= \beta(m, m) \hspace{4cm} \text{(by duplication formula)}$$

$$= \beta(m, n) = \text{L.H.S.}$$

Example 10: Show that $\displaystyle\int_0^{\pi/2} \frac{dx}{\sqrt{\sin x}} \times \int_0^{\pi/2} \sqrt{\sin x} \, dx = \pi$

Solution: Let $\displaystyle\int_0^{\pi/2} \frac{dx}{\sqrt{\sin x}} \times \int_0^{\pi/2} \sqrt{\sin x} \, dx$

$$= \int_0^{\pi/2} \sin^{-\frac{1}{2}} x \, dx \, . \int_0^{\pi/2} \sin^{\frac{1}{2}} x \, dx$$

$$= \int_0^{\pi/2} \sin^{2\left(\frac{1}{4}\right)-1} x \cos^{2\left(\frac{1}{2}\right)-1} x \, dx \, . \int_0^{\pi/2} \sin^{2\left(\frac{3}{4}\right)-1} x \cos^{2\left(\frac{1}{2}\right)-1} x \, dx$$

$$= \frac{1}{2} \, \beta\left(\frac{1}{4}, \frac{1}{2}\right) \, . \, \frac{1}{2} \, \beta\left(\frac{3}{4}, \frac{1}{2}\right)$$

$$= \frac{1}{2} \frac{\overline{\left|\frac{1}{4}\right.} \ \overline{\left|\frac{1}{2}\right.}}{\overline{\left|\frac{1}{4} + \frac{1}{2}\right.}} \cdot \frac{1}{2} \ \frac{\overline{\left|\frac{3}{4}\right.} \ \overline{\left|\frac{1}{2}\right.}}{\overline{\left|\frac{3}{4} + \frac{1}{2}\right.}}$$

$$= \frac{1}{2} \frac{\overline{\left|\frac{1}{4}\right.} \ \sqrt{\pi}}{\overline{\left|\frac{3}{4}\right.}} \cdot \frac{1}{2} \ \frac{\overline{\left|\frac{3}{4}\right.} \ \sqrt{\pi}}{\overline{\left|\frac{5}{4}\right.}}$$

$$= \frac{1}{4} \frac{\overline{\left|\frac{1}{4}\right.} \ \pi}{\frac{1}{4} \overline{\left|\frac{1}{4}\right.}} = \pi$$

Example 11: Prove that $\int_0^1 x^2 (1-x)^3 \ dx = \dfrac{1}{60}$

Solution: Let $\int_0^1 x^2 (1-x)^3 \ dx$

$$= \int_0^1 x^{3-1} (1-x)^{4-1} \ dx$$

$$= \beta \,(3, 4)$$

$$= \frac{\overline{|3} \ \overline{|4}}{\overline{|3+4}} = \frac{(2.1)\,(3.2.1)}{6.5.4.3.2.1} = \frac{1}{60}$$

Example 12: Express $\int_0^1 x^m (1-x^n)^P \ dx$ in terms of the beta function and hence evaluate

(i) $\quad \int_0^1 x^5 \ (1-x^3)^{10} \ dx$ $\hspace{3cm}$ (RGPV Jan 2008)

(ii) $\quad \int_0^1 x^3 \ (1-x^2)^4 \ dx$ $\hspace{3cm}$ (RGPV Dec 2004)

Solution: Let $I = \int_0^1 x^m \ (1-x^n)^P \ dx$ $\hspace{3cm}$ .....(i)

Putting $\quad x^n = y \Rightarrow x = y^{\frac{1}{n}}$ and $dx = \dfrac{1}{n} \, y^{\left(\frac{1}{n}-1\right)} \, dy$

Thus
$$I = \int_0^1 y^{\frac{m}{n}} (1-y)^P \cdot \frac{1}{n} y^{\left(\frac{1}{n}-1\right)} \, dy$$

$$= \frac{1}{n} \int_0^1 y^{\left(\frac{m+1}{n}-1\right)} (1-y)^{(P+1)-1} \, dy$$

$$= \frac{1}{n} \beta \left(\frac{m+1}{n}, P+1\right) \qquad\qquad\qquad .....(ii)$$

(i) Now, putting m = 5, n = 3, P = 10 we get from (ii),

$$\int_0^1 x^5 \left(1-x^3\right)^{10} dx = \frac{1}{3} \beta\left(\frac{5+1}{3}, 10+1\right)$$

$$= \beta\,(2, 11)$$

$$= \frac{1}{3} \frac{\lfloor 2 \; \lfloor 11}{\lfloor 2+11} = \frac{1}{3} \frac{1\lfloor 1 \cdot 10!}{12.11.\, 10!} = \frac{1}{396}$$

(ii) Putting m = 3, n = 2, p = 4, we get

$$\int_0^1 x^3 \left(1-x^2\right)^4 dx = \frac{1}{3} \beta\left(\frac{3+1}{2}, 4+1\right)$$

$$= \frac{1}{2} \beta\,(2, 5)$$

$$= \frac{1}{2} \frac{\lfloor 2 \; \lfloor 5}{\lfloor 2+5} = \frac{1}{2} \cdot \frac{1\lfloor 1 \cdot 4!}{6.5.\, 4!} = \frac{1}{60}$$

***Example* 13:** Evaluate $\displaystyle\int_0^\infty \frac{dx}{1+x^4}$

***Solution*:** Let $I = \displaystyle\int_0^\infty \frac{dx}{1+x^4}$

Putting $x^4 = y \Rightarrow x = y^{\frac{1}{4}}$ and $dx = \frac{1}{4} y^{\frac{-3}{4}} \, dy$

$\therefore$
$$I = \int_0^\infty \frac{1}{1+y} \cdot \frac{1}{4} y^{\frac{-3}{4}} \, dy$$

$$= \frac{1}{4} \int_0^\infty \frac{y^{\frac{1}{4}-1}}{(1+y)^{\frac{1}{4}+\frac{3}{4}}} \, dy$$

$$= \frac{1}{4} \beta \left(\frac{1}{4}, \frac{3}{4} \right)$$

$$= \frac{1}{4} \frac{\left\lfloor \frac{1}{4} \right. \left\lfloor \frac{3}{4} \right.}{\left\lfloor \frac{1}{4} + \frac{3}{4} \right.}$$

$$= \frac{1}{4} \frac{\left\lfloor \frac{1}{4} \right. \left\lfloor 1 - \frac{1}{4} \right.}{\left\lfloor 1 \right.}$$

$$= \frac{1}{4} \frac{\pi}{\sin \frac{\pi}{4}} \left(\because \left\lfloor n \right. \left\lfloor 1 - n \right. = \frac{\pi}{\sin n\pi} \right)$$

$$= \frac{1}{4} \cdot \frac{\pi}{\frac{1}{\sqrt{2}}}$$

$$= \frac{\pi\sqrt{2}}{4}$$

***Example* 14:** Evaluate $\int_0^{\frac{\pi}{2}} \tan^n x \, dx$

***Solution*:** Let $I = \int_0^{\frac{\pi}{2}} \tan^n x \, dx$

$$= \int_0^{\frac{\pi}{2}} \sin^n x \cos^{-n} x \, dx$$

$$= \int_0^{\frac{\pi}{2}} \sin^{(n+1)-1} x \cos^{(1-n)-1} dx$$

$$= \int_0^{\frac{\pi}{2}} \sin^{2\left(\frac{n+1}{2}\right)-1} x \cos^{2\left(\frac{1-n}{2}\right)-1} x \, dx$$

$$= \frac{1}{2} \beta \left(\frac{n+1}{2}, \frac{1-n}{2} \right)$$

$$= \frac{\sqrt{\dfrac{n+1}{2}}\,\sqrt{\dfrac{1-n}{2}}}{2\left|\dfrac{n+1}{2}+\dfrac{1-n}{2}\right.}$$

$$= \frac{1}{2}\cdot\sqrt{\dfrac{n+1}{2}}\,\left|1-\left(\dfrac{n+1}{2}\right)\right.$$

$$= \frac{1}{2}\cdot\frac{\pi}{\sin\left(\dfrac{n+1}{2}\right)\pi} \qquad\qquad \left(\because\ \overline{|n}\ \overline{|1-n} = \frac{\pi}{\sin n\pi}\right)$$

$$= \frac{1}{2}\,\frac{\pi}{\cos\dfrac{n\pi}{2}}$$

$$= \frac{\pi}{2}\,\sec\frac{n\pi}{2}$$

***Example* 15:** Evaluate $\displaystyle\int_0^\infty e^{-4x}\cdot x^{\frac{5}{2}}\,dx$ \hfill (RGPV June 2006, Jan 2007)

***Solution*:** Let $I = \displaystyle\int_0^\infty e^{-4x}\cdot x^{\frac{5}{2}}\,dx$

Putting $4x = y \Rightarrow 4dx = dy,$ then

$$I = \int_0^\infty e^{-y}\left(\frac{y}{4}\right)^{\frac{5}{2}}\frac{dy}{4}$$

$$= \frac{1}{4^{\frac{7}{2}}}\int_0^\infty e^{-y}\, y^{\frac{7}{2}-1}\,dy$$

$$= \frac{1}{2^7}\,\overline{\left|\frac{7}{2}\right.}$$

$$= \frac{\dfrac{5}{2}\cdot\dfrac{3}{2}\cdot\dfrac{1}{2}\,\overline{\left|\dfrac{1}{2}\right.}}{2^7} = \frac{15\sqrt{\pi}}{1024}$$

***Example* 16:** Evaluate $\displaystyle\int_0^\infty \sqrt{x}\; e^{-3\sqrt{x}}\; dx$ (RGPV Jan 2006)

***Solution*:** Let $I = \displaystyle\int_0^\infty \sqrt{x}\; e^{-3\sqrt{x}}\; dx$

Putting $\qquad 3\sqrt{x} = t \;\Rightarrow\; \dfrac{3}{2\sqrt{x}} dx = dt \;\Rightarrow\; dx = \dfrac{2t}{9}\; dt$

$$\therefore \qquad I = \int_0^\infty \frac{t}{3}\cdot e^{-t}\cdot \frac{2t}{9}\; dt$$

$$= \frac{2}{27}\int_0^\infty t^2 e^{-t} dt$$

$$= \frac{2}{27}\int_0^\infty e^{-t}\cdot t^{3-1}\; dt$$

$$= \frac{2}{27}\,\overline{|3}$$

$$= \frac{2}{27}\cdot 2.1\,\overline{|1} = \frac{4}{27}$$

***Example* 17:** Prove that $\displaystyle\int_0^\infty e^{-x^2}\; dx = \frac{\sqrt{\pi}}{2}$

***Solution*:** Let $I = \displaystyle\int_0^\infty e^{-x^2}\; dx$

Putting $\quad x^2 = t \Rightarrow 2x\, dx = dt \Rightarrow dx = \dfrac{dt}{2\sqrt{t}}$

$$\therefore \qquad I = \int_0^\infty e^{-t}\cdot \frac{dt}{2\sqrt{t}}$$

$$= \frac{1}{2}\int_0^\infty e^{-t}\; t^{\frac{-1}{2}}\; dt$$

$$= \frac{1}{2}\int_0^\infty e^{-t}\cdot t^{\left(\frac{1}{2}-1\right)}\; dt$$

$$= \frac{1}{2}\,\overline{\left|\frac{1}{2}\right.} = \frac{\sqrt{\pi}}{2}$$

Practice Problems

1. Prove that $\displaystyle\int_0^2 \left(4-x^2\right)^{\frac{3}{2}} dx = 3\pi$

2. Prove that $\displaystyle\int_0^\infty \frac{x\,dx}{1+x^6} = \frac{\pi}{3\sqrt{3}}$

3. Prove that $\displaystyle\int_0^\infty x\left(8-x^3\right)^{\frac{1}{3}} dx = \frac{16\pi}{9\sqrt{3}}$

4. Prove that $\displaystyle\int_0^\infty x^6\, e^{-2x}\, dx = \frac{45}{8}$

5. Prove that $\displaystyle\overline{|n}\ =\ \int_0^1 \left(\log \frac{1}{x}\right)^{n-1} dx, \quad n > 0$

6. Prove that $\displaystyle\int_0^1 \frac{dx}{\sqrt{1-x^n}} = \frac{\sqrt{\pi}\ \overline{\left|\dfrac{1}{n}\right.}}{n\ \overline{\left|\dfrac{1}{n}+\dfrac{1}{2}\right.}}$

7. Prove that $\displaystyle\int_{-\infty}^\infty e^{-K^2 x^2}\, dx = \frac{\sqrt{\pi}}{k}$

8. Prove that $\beta(m, n).\ \beta(m+n,\ l) = \beta(n,\ l)\,\beta(n+l,\ m) = \beta(l,\ m)\,\beta(l+m,\ n)$

9. Prove that $\displaystyle\beta(l,\ m)\,\beta(l+m,\ n)\,\beta(l+m+n,\ p) = \frac{\overline{|m}\ \overline{|n}\ \overline{|l}\ \overline{|p}}{\overline{|m+n+l+p}}$

10. Prove that $\displaystyle\int_0^a (a-x)^{m-1}\, x^{n-1}\, dx = a^{m+n-1}\,\beta(m, n)$

11. Prove that $\displaystyle\int_0^{\frac{\pi}{2}} \sqrt{\tan\theta}\ d\theta = \int_0^{\frac{\pi}{2}} \sqrt{\cot\theta}\ d\theta = \frac{\pi}{\sqrt{2}}$

12. Prove that $\overline{|y}\ \overline{|1-y} = \pi\,\mathrm{cosec}\ \pi y,\ 0 < y < 1$ and deduce that $\displaystyle\int_0^1 \log\overline{|y}\ dy = \frac{1}{2}\log 2\pi$

[Hint: $\displaystyle I = \int_0^1 \log\overline{|y}\ dy,\ $ then $\displaystyle I = \int_0^1 \log\overline{|1-y}\ dy$

$\displaystyle \therefore\ I = \frac{1}{2}\int_0^1 \left(\log\overline{|y} + \log\overline{|1-y}\right) dy = \frac{1}{2}\int_0^1 \log \frac{\pi}{\sin n\pi}\, dy\ \Big]$

13. Evaluate $\displaystyle\int_{a}^{b} (x-a)^{P}(b-x)^{q}\, dx$, where p and q are positive integer.

$$\left[\textbf{Ans. } (b-a)^{p+q-1}\ \frac{P!\, q!}{(p+q+1)!}\right]$$

14. Prove that $\displaystyle\int_{0}^{\frac{\pi}{2}} \sin^{n} x\, dx \times \int_{0}^{\frac{\pi}{2}} \sin^{n+1} x\, dx = \frac{\pi}{2(n+1)}$

15. Prove that $\displaystyle\int_{0}^{1} \frac{x^{m-1}+x^{n-1}}{(1+x)^{m+n}}\, dx = \beta(m,n)$

16. Prove that $\displaystyle\int_{0}^{1} \frac{x^{m-1}(1-x)^{n-1}}{(a+bx)^{m+n}}\, dx = \frac{\beta(m,n)}{(a+b)^{m}.\, a^{n}}$ (RGPV 2000)

17. Prove that $\displaystyle \beta(m+1, n) = \frac{m}{m+n}\ \beta(m, n)$

18. Prove that $\displaystyle\int_{0}^{\infty} \frac{x^{n-1}}{(1+x)^{m+n}}\, dx = \beta(m,n)$

19. Prove that $\displaystyle\int_{0}^{1}\left(1-x^{n}\right)^{\frac{1}{n}} dx = \frac{1}{n}\ \frac{\left(\overline{\left|\frac{1}{n}\right.}\right)^{2}}{2\left|\frac{2}{n}\right.}$

20. Prove that $\displaystyle\int_{0}^{1} \frac{1}{\sqrt{-\log x}}\, dx = \sqrt{\pi}$

Unit – V

Applications of Integral Calculus

- Multiple Integrals
- Volume and Surface of Solid

CHAPTER – 14

Multiple Integrals

14.1 Double Integrals

Let $f(x, y)$ be a continuous and single valued function of the independent variables x, y defined on the region R. Let R be divided into n sub-intervals of areas $\delta R_1, \delta R_2, \ldots \delta R_n$. Let (x_r, y_r) be any point inside the rth sub-region of area δR_r.

Form the sum

$$f(x_1, y_1)\, \delta R_1 + f(x_2, y_2)\, \delta R_2 + \ldots + f(x_n, y_n)\, \delta R_n$$

i.e., $\displaystyle \sum_{r=1}^{n} f(x_r, y_r)\, \delta R_r$ .....(1)

Now, we take the limit of the sum in eqn (1) as the number of sub-divisions i.e., $n \to \infty$ and the areas $\delta A_r \to 0$. If this limit exists, then it is called the double integral of $f(x, y)$ over the region R and is denoted by

$$\iint_R f(x, y)\, dR \quad \text{or} \quad \iint f(x, y)\, dx\, dy$$

14.2 Evaluation of Double Integrals

I. Area

(i) If R is a region bounded by the curves $y = f_1(x)$ and $y = f_2(x)$ and two straight lines $x = a$, $x = b$. Then area of the region be

$$A = \iint_R f(x, y)\, dx\, dy = \int_a^b \left[\int_{f_1(x)}^{f_2(x)} f(x, y)\, dy \right] dx$$

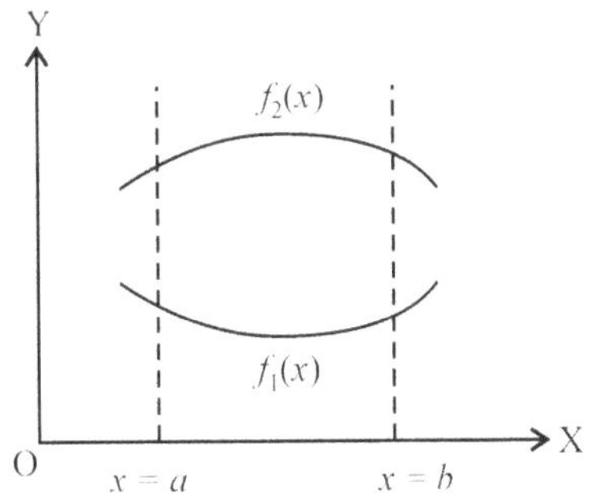

(ii) If R be a region bounded by the curves $x = \phi_1(y)$ and $x = \phi_2(y)$ and two straight lines $y = c$, $y = d$. Then area of the region.

$$A = \iint\limits_{R} f(x,\ y)\ dx\ dy = \int_{c}^{d} \left[\int_{\phi_1(y)}^{\phi_2(y)} f\ (x,\ y)\ dx \right] dy$$

II. Volume under a Surface

Let R be the region 1n the xy – plane. Then the volume inside the cylinder, enclosed by the surface

$$Z = f(x,\ y),\ Z > 0 \text{ is } V = \iint\limits_{R} f\ (x,\ y)\ dx\ dy$$

Example 1: Evaluate $\displaystyle\int_{0}^{1}\int_{0}^{1} \frac{dy\ dx}{\sqrt{\left(1-x^2\right)\left(1-y^2\right)}}$ (RGPV DEC 2005)

Solution: Let $\displaystyle\int_{0}^{1}\int_{0}^{1} \frac{dy\ dx}{\sqrt{\left(1-x^2\right)\left(1-y^2\right)}}$

$$= \int_{0}^{1}\left[\int_{0}^{1} \frac{1}{\sqrt{1-y^2}}\ dy \right] \frac{1}{\sqrt{1-x^2}}\ dx$$

$$= \int_{0}^{1}\left[\sin^{-1}y \right]_{0}^{1} \frac{1}{\sqrt{1-x^2}}\ dx$$

$$= \frac{\pi}{2} \int_0^1 \frac{1}{\sqrt{1-x^2}} \, dx$$

$$= \frac{\pi}{2} \left[\sin^{-1} x \right]_0^1$$

$$= \frac{\pi}{2} \times \frac{\pi}{2} = \frac{\pi^2}{4}$$

***Example* 2:**   Evaluate $\int_1^2 \int_0^x \frac{dx \, dy}{x^2 + y^2}$

***Solution*:**   Let $\int_1^2 \int_0^x \frac{dx \, dy}{x^2 + y^2} = \int_1^2 \left[\int_0^x \frac{dy}{x^2 + y^2} \right] dx$

$$= \int_1^2 \left[\frac{1}{x} \tan^{-1} \frac{y}{x} \right]_0^x dx$$

$$= \int_1^2 \left[\frac{1}{x} \tan^{-1} 1 - \frac{1}{x} \tan^{-1} 0 \right] dx$$

$$= \int_1^2 \frac{\pi}{4} \times \frac{1}{x} \, dx \qquad \left(\because \tan^{-1} 0 = 0 \right)$$

$$= \frac{\pi}{4} \left[\log x \right]_1^2$$

$$= \frac{\pi}{4} \left[\log 2 - \log 1 \right] = \frac{\pi}{4} \log 2 \quad \left(\because \log 1 = 0 \right)$$

***Example* 3:** Evaluate $\int_0^1 \int_0^{\sqrt{1+x^2}} \frac{dy \, dx}{1 + x^2 + y^2}$

***Solution*:** Let $\int_0^1 \int_0^{\sqrt{1+x^2}} \frac{dy \, dx}{1 + x^2 + y^2} = \int_0^1 \left[\int_0^{\sqrt{1+x^2}} \frac{dy}{1 + x^2 + y^2} \right] dx$

$$= \int_0^1 \left[\int_0^{\sqrt{1+x^2}} \frac{dy}{\left(\sqrt{1+x^2} \right)^2 + y^2} \right] dx$$

$$= \int_0^1 \left[\frac{1}{\sqrt{1+x^2}} \tan^{-1} \frac{y}{\sqrt{1+x^2}} \right]_0^{\sqrt{1+x^2}} dx$$

$$= \int_0^1 \frac{1}{\sqrt{1+x^2}} \left(\tan^{-1} 1 - \tan^{-1} 0 \right) dx$$

$$= \frac{\pi}{4} \int_0^1 \frac{1}{\sqrt{1+x^2}} \, dx$$

$$= \frac{\pi}{4} \left[\log \left(x + \sqrt{1+x^2} \right) \right]_0^1$$

$$= \frac{\pi}{4} \log \left(1 + \sqrt{2} \right)$$

***Example* 4:** Evaluate $\int_0^1 \int_0^{x^2} e^{y/x} \, dy \, dx$

***Solution*:** Let $\int_0^1 \int_0^{x^2} e^{y/x} \, dy \, dx \;= \int_0^1 \left[\int_0^{x^2} e^{y/x} \, dy \right] dx$

$$= \int_0^1 \left[xe^{y/x} \right]_0^{x^2} dx$$

$$= \int_0^1 \left[x\, e^x - x \right] dx \qquad \left(\because e^0 = 1 \right)$$

$$= \int_0^1 x \left(e^x - 1 \right) dx$$

$$= \left[x \left(e^x - x \right) \right]_0^1 - \int_0^1 \left(e^x - x \right) dx \qquad \text{(by parts)}$$

$$= \left(e - 1 \right) - \left(e^x - \frac{x^2}{2} \right)_0^1$$

$$= e - 1 - \left(e - \frac{1}{2} - 1 \right)$$

$$= \frac{1}{2}$$

Example 5: Evaluate $\int_0^2 \int_0^{\sqrt{2x-x^2}} x \; dx \; dy$

Solution: Let $\int_0^2 \int_0^{\sqrt{2x-x^2}} x \; dx \; dy = \int_0^2 x \left[\int_0^{\sqrt{2x-x^2}} dy \right] dx$

$$= \int_0^2 x \; [y]_0^{\sqrt{2x-x^2}} dx$$

$$= \int_0^2 x \; \sqrt{2x-x^2} \; dx$$

$$= \int_0^2 x \; \sqrt{1-(x-1)^2} \; dx$$

Putting $x - 1 = \sin \theta$, then $dx = \cos \theta \; d\theta$, we get

$$= \int_{-\pi/2}^{\pi/2} (1+\sin \theta) \sqrt{1-\sin^2 \theta} \times \cos \theta \; d\theta$$

$$= 2 \int_0^{\pi/2} (1+\sin \theta) \cos^2 \theta \; d\theta \qquad \text{(by property of definite integral)}$$

$$= 2 \int_0^{\pi/2} \cos^2 \theta \; d\theta \qquad [\because \sin \theta \cos^2 \theta \text{ is odd function}]$$

$$= 2 \times \frac{1}{2} \times \frac{\pi}{2} = \frac{\pi}{2}$$

Example 6: Evaluate $\iint_R xy \, (x+y) \, dx \, dy$ over the region r bounded by the curve $y = x^2$ and

$y = x$ $\qquad\qquad\qquad\qquad\qquad\qquad\qquad\qquad$ (RGPV Dec 2003, 2004)

Solution: The given curve be $y = x^2$ and $y - x$

We get $x^2 = x$

$\qquad \Rightarrow \; x \, (x-1) = 0$

$\qquad \Rightarrow \; x = 0, x = 1$

So that $y = 0, y = 1$

Therefore, points are (0,0) and (1,1). For limits $y = x^2$ to $y = x$ and x varies from

$\qquad x = 0$ to $x = 1$

$\therefore \qquad \iint_R xy \, (x+y) \, dx \, dy = \int_{x=0}^{1} \int_{y=x^2}^{x} \left(x^2 y + xy^2\right) dy \; dx$

$$= \int_0^1 \left(\frac{x^2 y^2}{2} + \frac{xy^3}{3} \right)_{x^2}^{x} dx$$

$$= \int_0^1 \left(\frac{x^4}{2} + \frac{x^4}{3} \right) - \left(\frac{x^6}{2} + \frac{x^7}{3} \right) dx$$

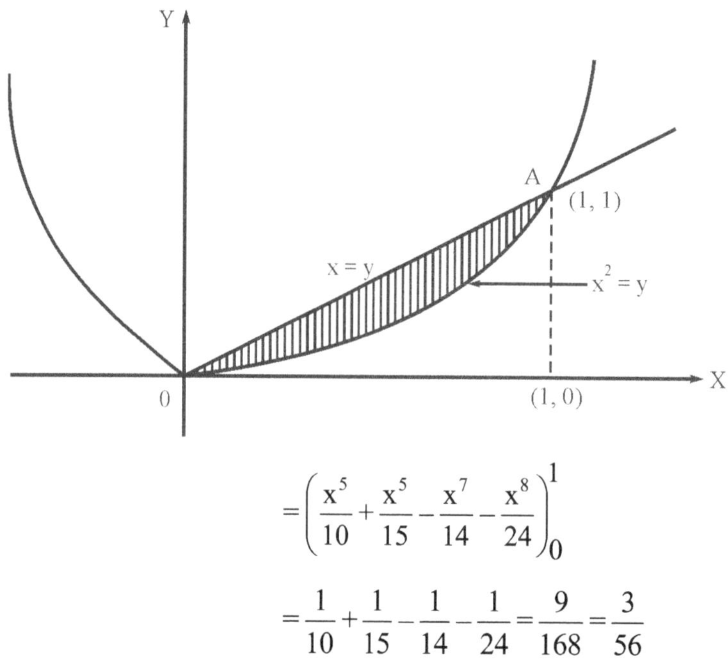

$$= \left(\frac{x^5}{10} + \frac{x^5}{15} - \frac{x^7}{14} - \frac{x^8}{24} \right)_0^1$$

$$= \frac{1}{10} + \frac{1}{15} - \frac{1}{14} - \frac{1}{24} = \frac{9}{168} = \frac{3}{56}$$

***Example* 7:** Evaluate $\iint\limits_R (x + y)^2 \, dx \, dy$, over the region r bounded by the ellipse $\dfrac{x^2}{a^2} + \dfrac{y^2}{b^2} = 1$

(RGPV JAN 2006)

***Solution*:** For the ellipse $\dfrac{x^2}{a^2} + \dfrac{y^2}{b^2} = 1$

$$\Rightarrow \quad \frac{y}{b} = \pm \sqrt{1 - \frac{x^2}{a^2}}$$

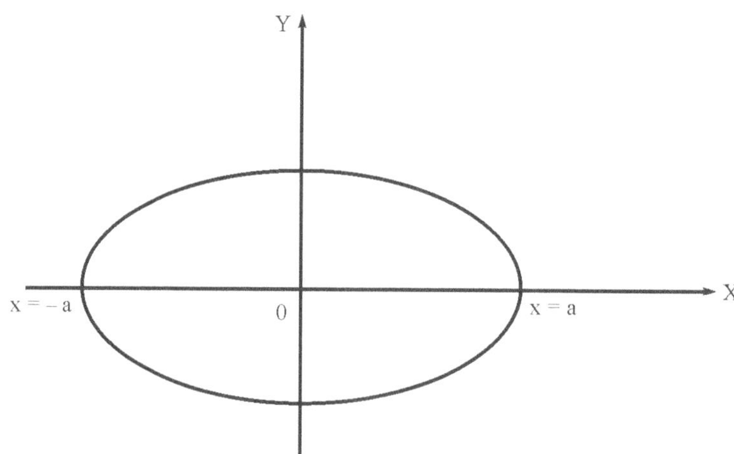

So the region of integration can be considered as bounded by the curves

$$y = -b\sqrt{1 - \frac{x^2}{a^2}} \text{ to } y = b\sqrt{1 - \frac{x^2}{a^2}} \text{ and x varies as } x = -a \text{ to } x = a$$

$$\therefore \iint\limits_{R} (x+y)^2 \, dx \, dy = \int_{-a}^{a} \int_{-b\sqrt{1-\frac{x^2}{a^2}}}^{b\sqrt{1-\frac{x^2}{a^2}}} (x^2 + 2xy + y^2) \, dy \, dx$$

$$= 2\int_{-a}^{a} \int_{0}^{b\sqrt{1-\frac{x^2}{a^2}}} \left(x^2 + y^2\right) dy \, dx$$

$$(\therefore xy \text{ is an odd function})$$

$$= 2\int_{-a}^{a} \left[x^2 y + \frac{y^3}{3} \right]_{0}^{b\sqrt{1-\frac{x^2}{a^2}}} dx$$

$$= 2\int_{-a}^{a} \left[x^2 b\sqrt{1 - \frac{x^2}{a^2}} + \frac{1}{3}b^3\left(1 - \frac{x^2}{a^2}\right)^{\frac{3}{2}} \right] dx$$

$$= 4b\int_{0}^{a} \left[x^2\sqrt{1 - \frac{x^2}{a^2}} + \frac{1}{3}b\left(1 - \frac{x^2}{a^2}\right)^{\frac{3}{2}} \right] dx$$

Putting $x = a \sin\theta \Rightarrow dx = a \cos\theta \, d\theta$, we get

$$= 4b\int_{0}^{\frac{\pi}{2}} \left(a^2 \sin^2\theta \cos\theta + \frac{1}{3}b^2 \cos^3\theta \right) a \cos\theta \, d\theta$$

$$= 4ab\int_{0}^{\frac{\pi}{2}} \left(a^2 \sin^2\theta \cos^2\theta + \frac{1}{3}b^2 \cos^4\theta \right) d\theta$$

$$= 4ab \left[a^2 \frac{\overline{3/2}\ \overline{3/2}}{2\overline{3}} + \frac{1}{3}b^2 \frac{\overline{3/2} \times \overline{1/2}}{2\overline{3}} \right]$$

$$= 4ab \left[a^2 \times \frac{\frac{1}{2}\sqrt{\pi} \times \frac{1}{2}\sqrt{\pi}}{2 \times 2 \times 1} + \frac{1}{3}b^2 \times \frac{\frac{3}{2} \times \frac{1}{2}\sqrt{\pi} \times \sqrt{\pi}}{2 \times 2 \times 1} \right]$$

$$= 4ab \left[\frac{1}{16}\pi a^2 + \frac{1}{16}\pi b^2 \right]$$

$$= \frac{1}{4}\pi \, ab \left(a^2 + b^2\right)$$

***Example* 8:** Evaluate $\iint\limits_{R} x^2\, y^2\, dx\, dy$, where R is the region bounded by $x = 0$, $y = 0$ and $x^2 + y^2 = 1$, $x \geq 0$, $y \geq 0$.   (RGPV Jan 2007)

Solution: Let the curve be $x = 0$, $y = 0$, $x^2 + y^2 = 1$

Limits are $y = 0$ to $y = \sqrt{1-x^2}$

and $x = 0$ to $x = 1$

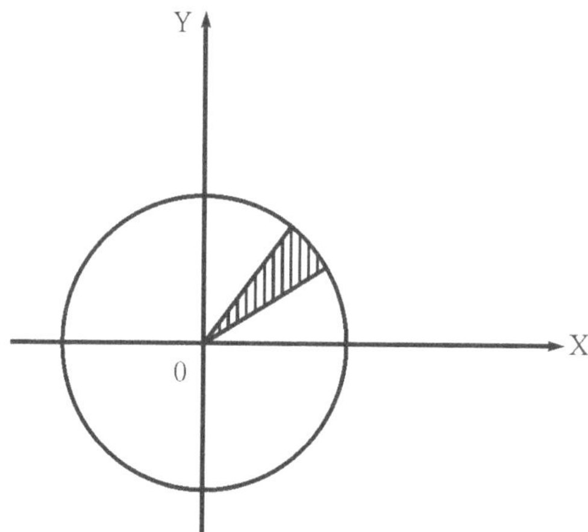

$$\therefore \iint\limits_{R} x^2\, y^2\, dx\, dy = \int_0^1 \int_0^{\sqrt{1-x^2}} x^2\, y^2\, dy\, dx$$

$$= \int_0^1 \left[\frac{x^2\, y^3}{3}\right]_0^{\sqrt{1-x^2}} dx$$

$$= \frac{1}{3} \int_0^1 x^2 \left(1-x^2\right)^{\frac{3}{2}} dx$$

Putting $x = \sin\theta \Rightarrow dx = \cos\theta\, d\theta$ and θ varies as $\theta = 0$ to $\theta = \dfrac{\pi}{2}$, then

$$= \frac{1}{3} \int_0^{\frac{\pi}{2}} \sin^2\theta \, \cos^3\theta \, \cos\theta \, d\theta$$

$$= \frac{1}{3} \int_0^{\frac{\pi}{2}} \sin^2\theta \, \cos^4\theta \, d\theta$$

$$= \frac{1}{3} \times \frac{\left\lfloor \frac{3}{2} \right. \left\lfloor \frac{5}{2} \right.}{2\left\lfloor 4 \right.}$$

$$= \frac{1}{3} \times \frac{\frac{1}{2}\sqrt{\pi} \times \frac{3}{2} \times \frac{1}{2}\sqrt{\pi}}{2 \times 3 \times 2 \times 1} = \frac{\pi}{96}$$

***Example* 9:** Evaluate $\iint\limits_{A} xy\ dx\ dy$ over the region in the positive quadrant for which $x + y \leq 1$.

Solution: The region of integration is bounded by the two axes and the line $x + y = 1$. Thus, limits are

$$y = 0 \text{ to } y = 1 - x \quad \text{and} \quad x = 0 \text{ to } x = 1$$

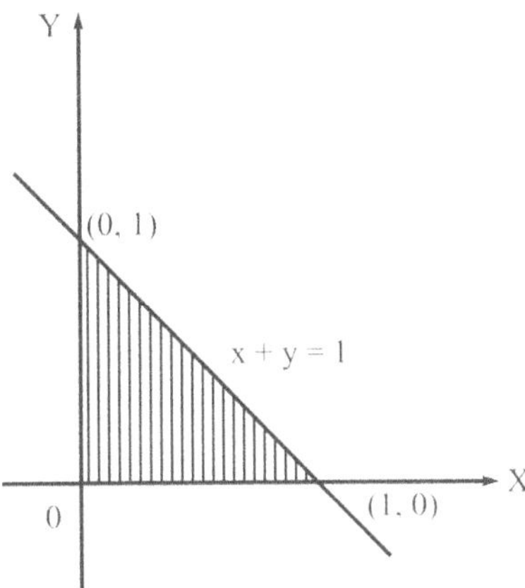

Therefore,

$$\iint\limits_{R} xy\ dx\ dy = \int_0^1 \int_0^{1-x} xy\ dy\ dx$$

$$= \int_0^1 \left[\frac{xy^2}{2} \right]_0^{1-x} dx$$

$$= \int_0^1 \frac{x}{2}(1-x)^2\ dx$$

$$= \frac{1}{2} \int_0^1 x(1 - 2x + x^2)\ dx$$

$$= \frac{1}{2} \left[\frac{x^2}{2} - \frac{2x^3}{3} + \frac{x^4}{4} \right]_0^1$$

$$= \frac{1}{2} \left[\frac{1}{2} - \frac{2}{3} + \frac{1}{4} \right] = \frac{1}{24}$$

***Example* 10:** Evaluate $\iint\limits_{A} y\ dx\ dy$, where r is the region bounded by the parabolas $y^2 = 4x$ and $x^2 = 4y$

Solution: Let the curve be $y^2 = 4x$ and $x^2 = 4y$

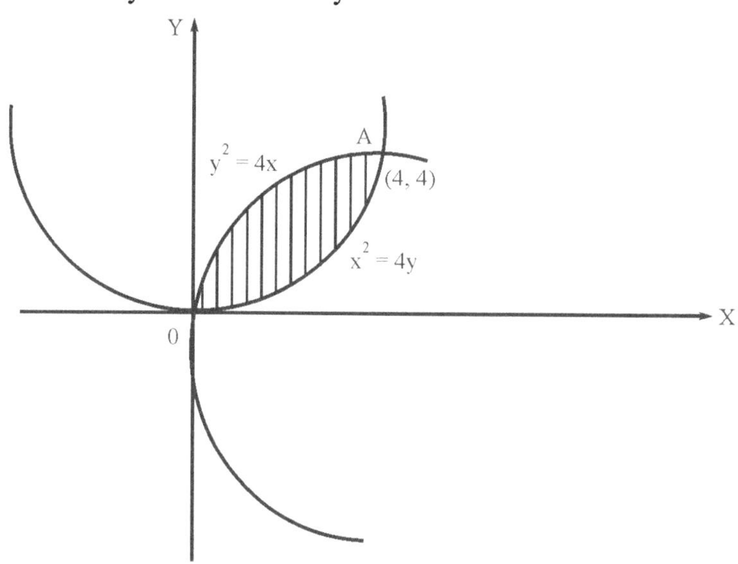

By solving $\left(\dfrac{x^2}{4}\right)^2 = 4x$

$\Rightarrow \qquad x(x^3 - 64) = 0 \;\Rightarrow\; x = 0, 4$

and $\qquad y = 0, 4$

$\therefore$ points are $(0, 0)$ and $(4, 4)$

Limits are $\;\; y = \dfrac{x^2}{4}\;$ to $\;y = 2\sqrt{x}\;$ and x varies as $x = 0$ to $x = 4$.

$\therefore \qquad\qquad \iint_R y \; dx \; dy = \int_0^4 \int_{x^2/4}^{2\sqrt{x}} y \; dy \; dx$

$$= \int_0^4 \left[\frac{y^2}{2}\right]_{x^2/4}^{2\sqrt{x}} dx$$

$$= \frac{1}{2} \int_0^4 \left[4x - \frac{x^4}{16}\right] dx$$

$$= \frac{1}{2} \left[\frac{4x^2}{2} - \frac{x^5}{80}\right]_0^4$$

$$= \frac{1}{2} \left[32 - \frac{64}{5}\right] = \frac{48}{5}$$

***Example* 11:** Let R be the region between the parabola $y = x^2$ and straight line $y = x + 6$, then evaluate $\iint_x y \, dx \, dy$.

***Solution*:** Let $y = x^2$ and $y = x + 6$

    By solving $x^2 = x + 6 \Rightarrow x = -2, 3$

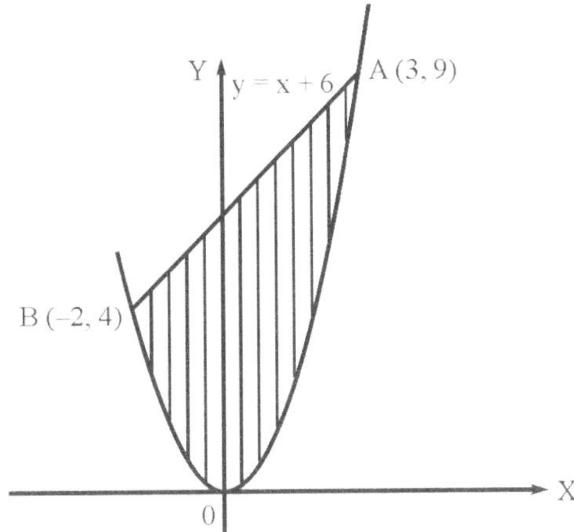

The points are $(-2, 4)$ and $(3, 9)$

Therefore y varies from $y = x^2$ to $y = x + 6$

and x varies from $x = -2$ to $x = 3$.

$$\therefore \qquad \iint_R x \, dx \, dy = \int_{-2}^{3} \int_{x^2}^{x+6} x \, dy \, dx$$

$$= \int_{-2}^{3} [xy]_{x^2}^{x+6} dx$$

$$= \int_{-2}^{3} [x(x+6) - x^3] \, dx$$

$$= \left[\frac{x^3}{3} + 6\frac{x^2}{2} - \frac{x^4}{4} \right]_{-2}^{3}$$

$$= \left(9 + 27 - \frac{81}{4} \right) - \left(-\frac{8}{3} + 12 - 4 \right) = \frac{125}{12}$$

***Example* 12:** Evaluate $\iint \dfrac{xy}{\sqrt{a^2 - y^2}} \, dx \, dy$ over the positive quadrant of the circle $x^2 + y^2 = a^2$.

***Solution*:** Let given curve be $x^2 + y^2 = a^2$

$$\Rightarrow \qquad y = \sqrt{a^2 - x^2}$$

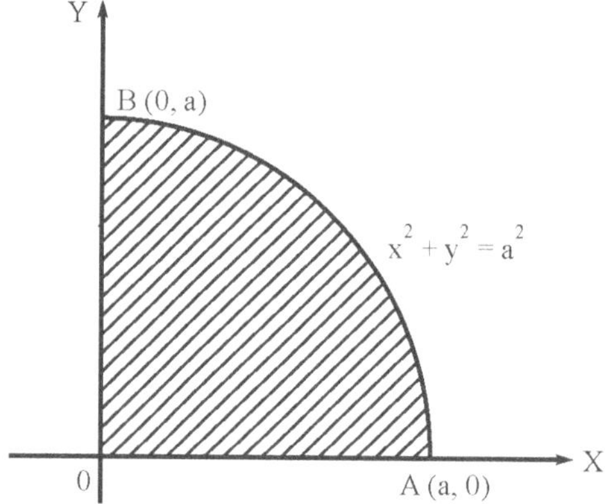

Thus y varies from $y = 0$ to $y = \sqrt{a^2 - x^2}$

and x varies from $x = 0$ to $x = a$. in the positive quadrant.

$$\therefore \quad \iint \frac{xy}{\sqrt{a^2 - y^2}}\, dxdy = \int_0^a \int_0^{\sqrt{a^2-x^2}} \frac{xy}{\sqrt{a^2 - y^2}}\, dy\, dx$$

$$= \int_0^a x\left[-\sqrt{a^2 - y^2}\,\right]_0^{\sqrt{a^2-x^2}} dx$$

$$= \int_0^a x\left[-\sqrt{a^2 - a^2 + x^2} + a\right] dx$$

$$= \int_0^a x\,(-x + a)\, dx$$

$$= \left[-\frac{x^3}{3} + \frac{ax^2}{2}\right]_0^a = -\frac{a^3}{3} + \frac{a^3}{2} = \frac{a^3}{6}$$

14.3   Double Integral in Polar Coordinates

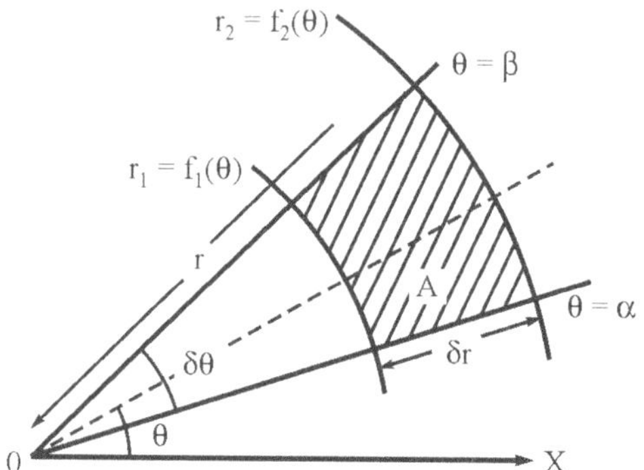

Let r, θ be the polar coordinates of the function f(r, θ), which is to be integrated over a certain area A whose boundary is also given in polar coordinates. If the inner and outer boundaries of the radial strips are given by curves

$$r_1 = f_1\,(\theta) \qquad \text{and} \qquad r_2 = f_2(\theta)$$

in which θ is constant during integration

$$\text{Thus} \qquad \iint_A f(r,\theta)dA = \int_\alpha^\beta \int_{f_1(\theta)}^{f_2(\theta)} f(r,\theta)\, rdr\, d\theta$$

***Example* 13:**   Evaluate $\displaystyle\int_0^\pi \int_0^{a(1+\cos\theta)} r^3 \sin\theta \; d\theta \; dr$

***Solution*: Let**

$$I = \int_0^\pi \int_0^{a(1+\cos\theta)} r^3 \sin\theta \; d\theta \; dr$$

$$= \int_0^\pi \sin\theta \left[\int_0^{a(1+\cos\theta)} r^3 dr \right] d\theta$$

$$= \int_0^\pi \sin\theta \left[\frac{r^4}{4} \right]_0^{a(1+\cos\theta)} d\theta$$

$$= \frac{1}{4} \int_0^\pi \sin\theta \cdot a^4 (1+\cos\theta)^4 \; d\theta$$

Putting $1 + \cos\theta = t \Rightarrow -\sin\theta \; d\theta = dt$, we get

$$I = \frac{a^4}{4} \int_2^0 t^4 \; (-dt)$$

$$= \frac{a^4}{4} \int_0^2 t^4 \; dt$$

$$= \frac{a^4}{4} \left[\frac{t^5}{5} \right]_0^2 = \frac{a^4}{4} \cdot \frac{2^5}{5} = \frac{8a^4}{5}$$

***Example* 14:** Evaluate $\displaystyle\iint r^2 d\theta \; dr$ over the area of the circle $r = a\cos\theta$.   (RGPV June 2006)

***Solution*:** Limits are $r = 0$ to $r = a\cos\theta$

and $\theta = -\dfrac{\pi}{2}$ to $\theta = \dfrac{\pi}{2}$

Thu, the given integral

$$= \int_{-\frac{\pi}{2}}^{\frac{\pi}{2}} \int_0^{a\cos\theta} r^2 \; dr \; d\theta$$

$$= \int_{-\frac{\pi}{2}}^{\frac{\pi}{2}} \left[\frac{r^3}{3} \right]_0^{a\cos\theta} d\theta$$

$$= \frac{1}{3} \int_{-\frac{\pi}{2}}^{\frac{\pi}{2}} a^3 \cos^3\theta \; d\theta$$

$$= \frac{2a^3}{3} \int_0^{\frac{\pi}{2}} \cos^3\theta \; d\theta$$

$$= \frac{2a^3}{3} \cdot \frac{\left| \! \frac{1}{2} \right. \cdot \left| \! 2 \right.}{2 \left| \! \frac{5}{2} \right.} = \frac{2a^3}{3} \cdot \frac{\sqrt{\pi}}{2 \cdot \frac{3}{2} \cdot \frac{1}{2} \sqrt{\pi}} = \frac{4}{9} a^3$$

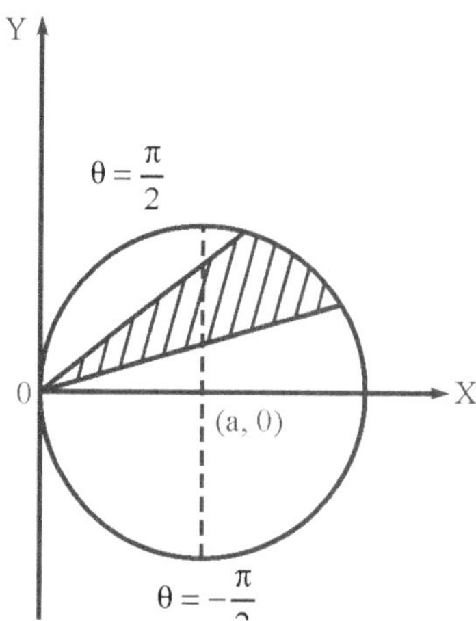

***Example* 15:** Evaluate $\displaystyle\int_0^\infty \int_0^\infty e^{-(x^2+y^2)}\, dx\, dy$ $\hspace{3cm}$ (RGPV June 2005)

***Solution*:** Given limits of x and y are 0 to ∞ (given)

Transform into polar coordinates

Substituting $x = r\cos\theta,\; y = r\sin\theta,\; dx\, dy = r\, dr\, d\theta$

Then limits r varies from 0 to ∞ and θ varies from 0 to $\dfrac{\pi}{2}$.

$$\therefore \qquad \int_0^\infty \int_0^\infty e^{-\left(x^2+y^2\right)} dx\, dy = \int_0^{\pi/2}\int_0^\infty e^{-r^2}.dr\, d\theta$$

Putting $\qquad r^2 = t \Rightarrow 2r\, dr = dt$

$$\therefore \qquad = \int_0^{\frac{\pi}{2}} \int_0^\infty e^{-t}.\frac{dt}{2}\,.\, d\theta$$

$$= \frac{-1}{2} \int_0^{\frac{\pi}{2}} \left[e^{-t} \right]_0^\infty d\theta$$

$$= \frac{-1}{2} \int_0^{\frac{\pi}{2}} \left[e^{-\infty} - e^0 \right] d\theta$$

$$= \frac{-1}{2} \int_0^{\frac{\pi}{2}} (-1)\, d\theta \hspace{2cm} \left(\because e^{-\infty} = 0 \right)$$

$$= \frac{1}{2} \left[\theta \right]_0^{\pi/2} = \frac{\pi}{4}$$

***Example* 16:** By changing to polar coordinates, evaluate $\displaystyle\int_0^2 \int_0^{\sqrt{2x-x^2}} \frac{x}{\sqrt{x^2+y^2}}\, dx\, dy$

***Solution*:** The given limits of integration be

$$y = 0, \ y = \sqrt{2x - x^2} \ \text{ and } \ x = 0, x = 2$$

Thus, the region of integration is upper half of the circle $x^2 + y^2 - 2x = 0$ in xy-plane.

Now, changing into polar coordinates, by substituting $x = r\cos\theta$, $y = r\sin\theta$, $dxdy = r\, d\theta\, dr$

Limits varies as $r = 0$ to $r = 2\cos\theta$ and θ varies as $\theta = 0$ to $\theta = \dfrac{\pi}{2}$.

Hence

$$\int_0^2 \int_0^{\sqrt{2x-x^2}} \frac{x}{\sqrt{x^2+y^2}} = \int_0^{\frac{\pi}{2}} \int_0^{2\cos\theta} \frac{r\cos\theta}{r} \cdot r\, dr\, d\theta$$

$$= \int_0^{\frac{\pi}{2}} \cos\theta \left[\int_0^{2\cos\theta} r\, dr \right] d\theta$$

$$= \int_0^{\frac{\pi}{2}} \cos\theta \left[\frac{r^2}{2} \right]_0^{2\cos\theta} d\theta$$

$$= \int_0^{\frac{\pi}{2}} \cos\theta \cdot 2\cos^2\theta\, d\theta$$

$$= 2 \int_0^{\frac{\pi}{2}} \cos^3\theta\, d\theta$$

$$= 2 \cdot \frac{\left|\frac{1}{2}\right. \left|\overline{2}\right.}{2\left|\overline{\frac{3}{2}}\right.} = \frac{\sqrt{\pi}}{\frac{3}{2}\cdot\frac{1}{2}\cdot\sqrt{\pi}} = \frac{4}{3}$$

***Example* 17:** Integrate $r\sin\theta$ over the area of the cardioid $r = a\,(1 + \cos\theta)$ above the initial line.

***Solution*:** The region of integration is the portion of cardioid above the initial line which can be bounded by the curves

$$r = 0, r = a\,(1 + \cos\theta) \quad \text{and} \quad \theta = 0, \theta = \pi$$

Hence,

$$\int_0^{\pi} \int_0^{a(1+\cos\theta)} r\sin\theta \cdot r\, d\theta\, dr$$

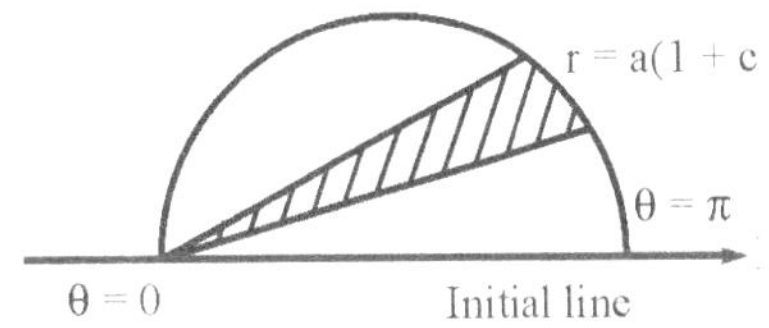

$$= \int_0^\pi \sin\theta \left[\int_0^{a(1+\cos\theta)} r^2 \ dr \right] d\theta$$

$$= \int_0^\pi \sin\theta \left[\frac{r^3}{3} \right]_0^{a(1+\cos\theta)} d\theta$$

$$= \frac{a^3}{3} \int_0^\pi \sin\theta \ (1+\cos\theta)^3 \ d\theta$$

$$= \frac{a^3}{3} \int_0^\pi 2 \sin\frac{\theta}{2} \cos\frac{\theta}{2} \left(2\cos^2\frac{\theta}{2} \right)^3 d\theta$$

$$= \frac{16}{3} a^3 \int_0^\pi \sin\frac{\theta}{2} \cos^7\frac{\theta}{2} \ d\theta$$

Putting $\quad \dfrac{\theta}{2} = \phi \ \Rightarrow \ d\theta = 2d\phi$, we get

$$= \frac{16}{3} a^3 \int_0^{\pi/2} \sin\phi \ \cos^7\phi \ . \ 2d\theta$$

$$= \frac{32}{3} a^3 \ \frac{\lfloor 1 \ \lfloor 4}{2 \lfloor 5}$$

$$= \frac{32}{3} a^3 \ . \ \frac{3.2.1}{2.4.3.2.1} = \frac{4}{3} a^3$$

Example **18:** Transform the integral $\displaystyle\int_0^a \int_0^{\sqrt{a^2-x^2}} y^2 \sqrt{x^2+y^2} \ dx \ dy$ by changing to polar coordinates and hence evaluate it.

Solution: The given limits of integration be $y = 0$, $y = \sqrt{a^2-x^2}$

and $x = 0$, $x = a$

Then the region of integration be a circle $x^2 + y^2 = a^2$ in the first quadrant.

Now, transform into polar coordinate $x = r \cos\theta$, $y = r \sin\theta$, dx dy $= r \ d\theta \ dr$

Hence $r^2 = a^2 \Rightarrow r = a$

The limits of r varies from $r = 0$ to $r = a$ and $\theta = 0$ to $\theta = \dfrac{\pi}{2}$

$$\therefore \quad \int_0^a \int_0^{\sqrt{a^2-x^2}} y^2 \sqrt{x^2+y^2} \ dx \ dy = \int_0^{\pi/2} \int_0^a r^2 \sin^2\theta . r . r \ d\theta \ dr$$

$$= \int_0^{\pi/2} \sin^2\theta \left[\int_0^a r^4 dr \right] d\theta$$

$$= \int_0^{\pi/2} \sin^2 \theta \left[\frac{r^5}{5} \right] d\theta$$

$$= \frac{a^5}{5} \int_0^{\pi/2} \sin^2 \theta \, d\theta$$

$$= \frac{a^5}{5} \cdot \frac{\dfrac{1}{2} \left| \dfrac{1}{2} \cdot \right| \dfrac{1}{2}}{\left| 3 \right.}$$

$$= \frac{a^5}{5} \cdot \frac{\dfrac{1}{2} \cdot \sqrt{\pi} \cdot \sqrt{\pi}}{2.1} = \frac{1}{20} \pi a^5$$

14.4   Evaluation of Area by Double Integral

***Example* 19:** Find the area between the curves $y^2 = x^3$ and $y + x^2$.   (RGPV 2002)

***Solution*:** Let the curves $y = x^3$ and $y = x^2$

By solving $\qquad x^3 + x^2 = 0$

$\Rightarrow \qquad\qquad x^2 (x - 1) = 0$

$\Rightarrow \qquad\qquad x = 0, 1 \quad$ and $\quad y = 0, 1$

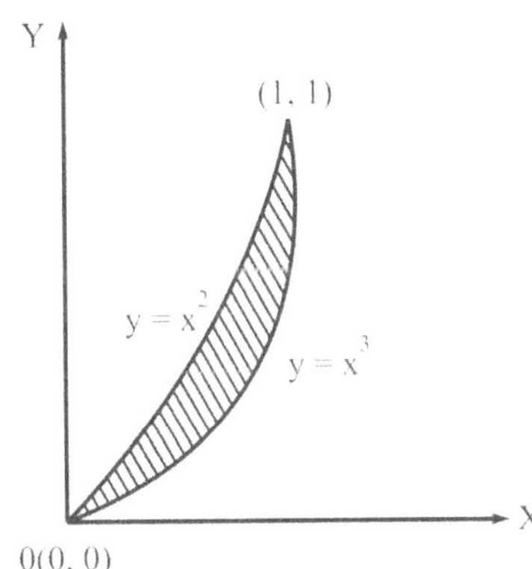

$\therefore$ points are $(0, 0)$ $(1,1)$

Limits of y varies from $y = x^3$ to $y = x^2$

and x varies from $x = 0$ to $x = 1$

$$\therefore \qquad\qquad \text{Area} = \int_0^1 \int_{x^3}^{x^2} dx \, dy$$

$$= \int_0^1 \left[y \right]_{x^3}^{x^2} dx$$

$$= \int_0^1 (x^2 - x^3) \, dx$$

$$= \left[\frac{x^3}{3} - \frac{x^4}{4}\right]_0^1 = \frac{1}{3} - \frac{1}{4} = \frac{1}{12} \text{ square unit.}$$

***Example* 20:** Express the area between the curves $x^2 + y^2 = a^2$ and $x + y = a$ as a double integral and evaluate it.

(RGPV June 2003)

***Solution*:** Let the curves $x^2 + y^2 = a^3$ and $x + y = a$

By solving
$$x^2 + (a - x)^2 = a^2$$
$$\Rightarrow \quad x^2 + a^2 - 2ax + x^2 = a^2$$
$$\Rightarrow \quad 2x^2 - 2ax = 0$$
$$\Rightarrow \quad 2x\,(x - a) = 0$$
$$\Rightarrow \quad x = 0, a$$

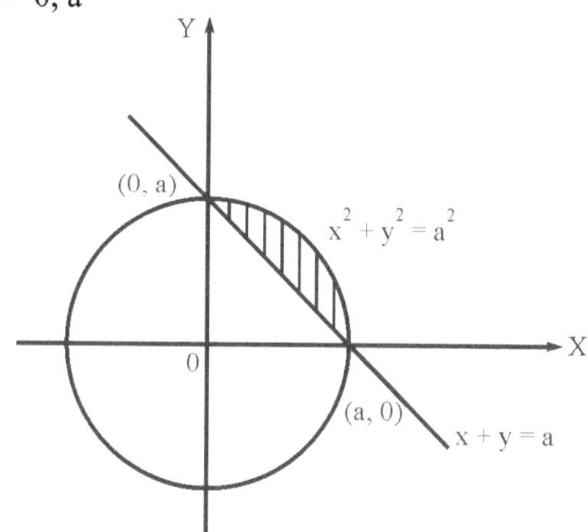

$\therefore$ points are $(0, a)$ and $(a, 0)$

$\therefore$ limits of y varies from $y = a - x$ to $y = \sqrt{a^2 - x^2}$

and x varies from $x = 0$ to $x = a$

$$\therefore \qquad \text{Area} = \int_0^a \int_{a-x}^{\sqrt{a^2-x^2}} dy\ dx$$

$$= \int_0^a \left[y\right]_{a-x}^{\sqrt{a^2-x^2}} dx$$

$$= \int_0^a \left[\sqrt{a^2 - x^2} - (a - x)\right] dx$$

$$= \left[\frac{x}{2}\sqrt{a^2 - x^2} + \frac{a^2}{2}\sin^{-1}\frac{x}{a} - ax + \frac{x^2}{2}\right]_0^a$$

$$= \frac{1}{2} a^2 \sin^{-1} 1 - a^2 + \frac{a^2}{2}$$

$$= \frac{\pi}{2} \frac{a^2}{2} - \frac{a^2}{2} = \frac{a^2}{4} (\pi - 2) \text{ square unit.}$$

***Example* 21:** Find by double integral, the area lying between the parabola $y = 4x - x^2$ and the line $y = x$.

(RGPV June 2008)

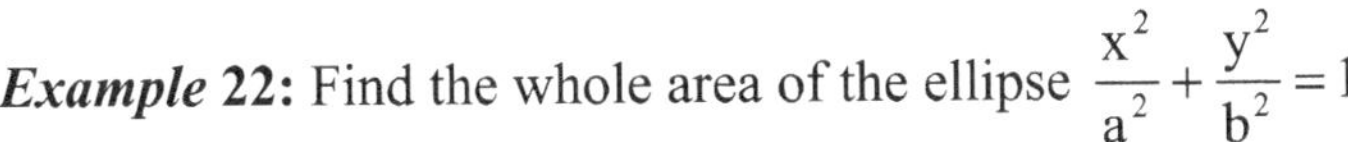

***Solution*:** Let the curves $y = 4x - x^2$ and $y = x$

By solving $x = 4x - x^2$

$\Rightarrow \qquad x^2 - 3x = 0$

$\Rightarrow \qquad x(x - 3) = 0$

$\Rightarrow \qquad x = 0, 3 \qquad$ and $\qquad y = 0, 3$

points are $(0, 0)$ and $(3, 3)$

$\therefore$ limits of y varies from $y = x$ to

$\qquad y = 4x - x^2$

and x varies from $x = 0$ to $x = 3$

$$\therefore \quad \text{Area} = \int_0^3 \int_x^{4x-x^2} dy\, dx$$

$$= \int_0^3 \left[y \right]_x^{4x-x^2} dx$$

$$= \int_0^3 \left[4x - x^2 - x \right] dx$$

$$= \int_0^3 \left[3x - x^2 \right] dx$$

$$= \left[\frac{3x^2}{2} - \frac{x^3}{3} \right]_0^3 = \frac{27}{2} - \frac{27}{3} = \frac{9}{2} \text{ square unit.}$$

***Example* 22:** Find the whole area of the ellipse $\dfrac{x^2}{a^2} + \dfrac{y^2}{b^2} = 1$

or

Compute the value of $\iint_R dx\, dy$ when R is the ellipse $\dfrac{x^2}{a^2} + \dfrac{y^2}{b^2} = 1$ of positive quadrant.

(RGPV April 2009)

Solution: Given equation of ellipse $\dfrac{x^2}{a^2} + \dfrac{y^2}{b^2} = 1$

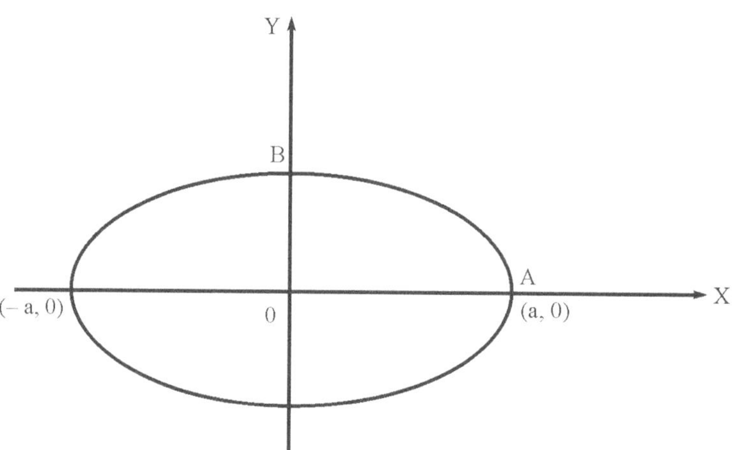

$$\Rightarrow \frac{y^2}{b^2} = 1 - \frac{x^2}{a^2}$$

$$\Rightarrow y = b\sqrt{1 - \frac{x^2}{a^2}}$$

In the first quadrant, limits of y varies from $y = 0$ to $y = \dfrac{b}{a}\sqrt{1 - \dfrac{x^2}{a^2}}$ and x varies from $x = 0$ to $x = a$.

$$\therefore \text{Area (OAB)} = \int_0^a \int_0^{b\sqrt{1-\frac{x^2}{a^2}}} dy\, dx$$

$$= \int_0^a [y]_0^{\frac{b}{a}\sqrt{a^2-x^2}}\, dx$$

$$= \int_0^a \frac{b}{a}\sqrt{a^2 - x^2}\, dx$$

$$= \frac{b}{a}\left[\frac{x}{2}\sqrt{a^2 - x^2} + \frac{a^2}{2}\sin^{-1}\frac{x}{a}\right]_0^a$$

$$= \frac{b}{a}\left[\frac{a^2}{2}\sin^{-1}1 - 0\right]$$

$$= \frac{ab}{2}\cdot\frac{\pi}{2} = \frac{\pi ab}{4}\ \text{square unit}$$

Now, whole area of ellipse = 4. area (OAB)

$$= 4\cdot\frac{\pi ab}{4}$$

$$= \pi ab \text{ square unit.}$$

Example 23: Find the area of the cardioid $r = a\,(1 + \cos\theta)$

Solution: Let the curve be $r = a\,(1 + \cos\theta)$

The limits are $r = 0$ to $r = a\,(1 + \cos\theta)$ and $\theta = 0$ to $\theta = \pi$

Hence the required are $= 2 \times$ area of OABO

$$= 2 \int_0^\pi \int_0^{a(1+\cos\theta)} r\ dr\ d\theta$$

$$= 2 \int_0^\pi \left[\frac{r^2}{2}\right]_0^{a(1+\cos\theta)} d\theta$$

$$= \int_0^\pi a^2 (1 + \cos\theta)^2\ d\theta$$

$$= 4a^2 \int_0^\pi \cos^4 \frac{\theta}{2}\ d\theta$$

Putting $\dfrac{\theta}{2} = \phi \Rightarrow d\theta = 2d\phi$, we get

$$\text{Area} = 4a^2 \int_0^{\frac{\pi}{2}} \cos^4 \phi \ . \ 2d\phi$$

$$= 8a^2\ \frac{\left\lfloor\dfrac{5}{2}\right. \left\lfloor\dfrac{1}{2}\right.}{2\left\lfloor 3\right.}$$

$$= 8a^2\ \frac{\dfrac{3}{2}.\dfrac{1}{2}\sqrt{\pi}.\sqrt{\pi}}{2.2.1} = \frac{3}{2}a^2\pi.$$

Example 24: Find the whole area of the asteroid $x^{\frac{2}{3}} + y^{\frac{2}{3}} = a^{\frac{2}{3}}$

or

Find the complete are of the asteroid $x = a\cos^3 t,\ y = a\sin^3 t$

Solution: The given curve be $x^{\frac{2}{3}} + y^{\frac{2}{3}} = a^{\frac{2}{3}}$

$$\Rightarrow \qquad y = \left(a^{\frac{2}{3}} - x^{\frac{2}{3}}\right)^{\frac{3}{2}}$$

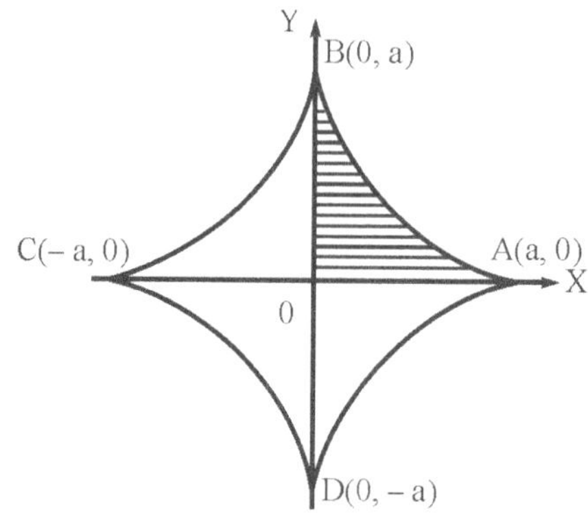

The limits of y are $y = 0$ to $y = \left(a^{\frac{2}{3}} - x^{\frac{2}{3}} \right)^{\frac{3}{2}}$

and x from $x = 0$ to $x = a$

$\therefore$ The whole area $= 4 \times$ area of OAB

$$= 4 \int_0^a \int_0^{\left(a^{\frac{2}{3}} - x^{\frac{2}{3}} \right)^{\frac{3}{2}}} dy \ dx$$

$$= 4 \int_0^a \left(a^{\frac{2}{3}} - x^{\frac{2}{3}} \right)^{\frac{3}{2}} dx$$

Putting $x = a \sin^3\theta \Rightarrow dx = 3a \sin^2\theta \cos\theta \ d\theta$

$$= 4 \int_0^{\frac{\pi}{2}} a\left(1 - \sin^2\theta \right)^{\frac{3}{2}} . \ 3a \sin^2\theta \cos\theta \ d\theta$$

$$= 12a^2 \int_0^{\frac{\pi}{2}} \sin^2\theta \cos^4\theta \ d\theta$$

$$= 12a^2 \frac{\left\lfloor \frac{3}{2} \right. \left\lfloor \frac{5}{2} \right.}{2 . \left\lfloor 4 \right.}$$

$$= 12a^2 \frac{\frac{1}{2}\sqrt{\pi} . \frac{3}{2} . \frac{1}{2}\sqrt{\pi}}{2.3.2.1}$$

$$= \frac{3}{8}\pi a^2$$

***Example* 25:** Find the area of the Leminscate $r^2 = a^2 \cos2\theta$

Solution: The given curve be $r^2 = a^2 \cos2\theta$

Putting $r = 0$, $\cos2\theta = 0 \Rightarrow 2\theta = \pm \frac{\pi}{2}$

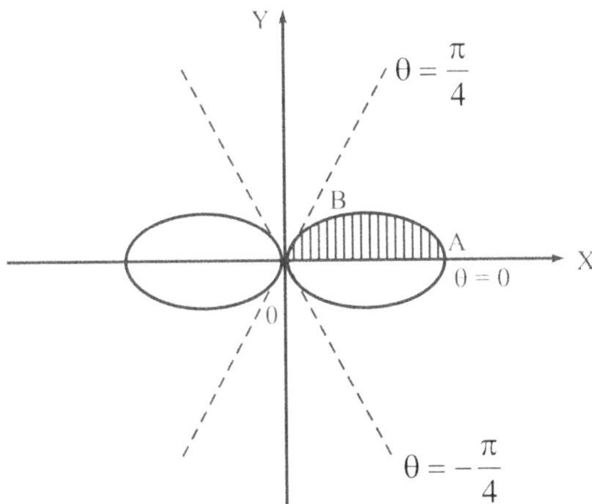

hence $\theta = \pm \dfrac{\pi}{4}$

The limits for one loop are $r = 0$ to $r = a\sqrt{\cos 2\theta}$ and $\theta = 0$ to $\theta = \dfrac{\pi}{4}$

$\therefore$ Area of one loop $= 2 \times$ area of OABO

$$= 2\int_0^{\frac{\pi}{4}} \int_0^{a\sqrt{\cos 2\theta}} r \, dr \, d\theta$$

$$= 2\int_0^{\frac{\pi}{4}} \left[\frac{r^2}{2}\right]_0^{a\sqrt{\cos 2\theta}} d\theta$$

$$= \int_0^{\frac{\pi}{4}} a^2 \cos 2\theta \, d\theta$$

$$= a^2 \left[\frac{\sin 2\theta}{2}\right]_0^{\frac{\pi}{4}}$$

$$= \frac{a^2}{2}\left[\sin\frac{\pi}{2} - 0\right] = \frac{a^2}{2}$$

$\therefore$ Required area $= 2 \times$ area of one loop

$$= 2 \times \frac{a^2}{2}$$

$$= a^2.$$

14.5 Triple Integrals

Let f(x, y, z) be a function of the three independent variables x, y, z defined in a bounded region V of the three dimensional space. We divide the region V in to n sub divisions $\delta v_1, \delta v_2, \ldots \delta v_n$.

i.e. δv_r (say). Then the limit of the sum $= \sum_{r=1}^{n} f(x_r, y_r, z_r)\, \delta v_r$.

Taking limit as $n \to \infty$ and sub-division tend to zero is known as triple integral and is defined as $\iiint_v f(x, y, z)\, dv$

14.6 Evaluation of Triple Integrals

Let the surfaces z = Ψ_1, (x, y) and z = Ψ_2 (x, y) and y = $\phi_1(x)$, y = $\phi_2(x)$, x = a and x = b.

Then

$$\iiint_v f(x,y,z)\, dv = \int_a^b \left[\int_{\phi_1(x)}^{\phi_2(x)} \left\{ \int_{\Psi_1(x)}^{\Psi_2(x)} f(x,y,z)\, dz \right\} dy \right] dx$$

In order to evaluate a triple integral, first we integrate with respect to z taking x and y are constant and the compute the double integral of this function over the region as has already been done.

***Example* 26:** Evaluate $\displaystyle\int_2^4 \int_0^x \int_0^{x+y} z\, dz\, dy\, dx$
$\hfill$ (RGPV June 2003)

***Solution*:** Let $\quad I = \displaystyle\int_2^4 \int_0^x \left(\int_0^{x+y} z\, dz \right) dy\, dx$

$$= \int_2^4 \int_0^x \left(\frac{z^2}{2} \right)_0^{x+y} dy\, dx$$

$$= \frac{1}{2} \int_2^4 \left(\int_0^x (x+y)^2\, dy \right) dx$$

$$= \frac{1}{2} \int_2^4 \left(\int_0^x x^2 + y^2 + 2xy\, dy \right) dx$$

$$= \frac{1}{2} \int_2^4 \left(x^2 y + \frac{y^3}{3} + xy^2 \right)_0^x dx$$

$$= \frac{1}{2} \int_2^4 \frac{7}{3} x^3\, dx$$

$$= \frac{7}{6} \left[\frac{x^4}{4} \right]_2^4$$

$$= \frac{7}{24} [256 - 16] = \frac{7}{24} \times 240 = 70$$

***Example* 27**: Find the value of $\displaystyle\int_1^e \int_0^{\log y} \int_1^{e^x} \log z \, dz \, dy \, dx$

Solution: Let $\quad I = \displaystyle\int_1^e \int_0^{\log y} \left(\int_1^{e^x} \log z \, dz \right) dx \, dy$

$$= \int_1^e \int_0^{\log y} \left(z \log z - z \right)_1^{e^x} dx \, dy$$

$$= \int_1^e \left[\int_0^{\log y} \left(e^x \cdot x - e^x + 1 \right) dx \right] dy$$

$$= \int_1^e \left[x e^x - e^x - e^x + x \right]_0^{\log y} dy$$

$$= \int_1^e \left[y \, \log y - 2y + \log y + 2 \right] dy$$

$$= \int_1^e \left[(1+y) \, \log y + 2 \, (1-y) \right] dy$$

$$= \left[\log y \left(y + \frac{y^2}{2} \right) \right]_1^e - \int_1^e \frac{1}{y} \left(y + \frac{y^2}{2} \right) dy + \left[2y - y^2 \right]_1^e$$

$$= e + \frac{e^2}{2} - \int_1^e \left(1 + \frac{y}{2} \right) dy + 2e - e^2 - 1$$

$$= -\frac{e^2}{2} + 3e - 1 + \left[y + \frac{y^2}{4} \right]_1^e$$

$$= -\frac{e^2}{2} + 3e - 1 - \frac{e^2}{4} - e + \frac{1}{4} + 1$$

$$= -\frac{3}{4} e^2 + 2e + \frac{1}{4}$$

***Example* 28**: Evaluate $\displaystyle\int_0^1 \int_0^{\sqrt{1-x^2}} \int_{\sqrt{x^2+y^2}}^1 \frac{dx \, dy \, dz}{\sqrt{x^2 + y^2 + z^2}}$ (RGPV June 2007)

Solution: Let $\quad I = \displaystyle\int_0^1 \int_0^{\sqrt{1-x^2}} \left(\int_{\sqrt{x^2+y^2}}^1 \frac{1}{\sqrt{x^2 + y^2 + z^2}} \, dz \right) dy \, dx$

We change to spherical polar coordinates (r, θ, ϕ), so that $x = r \sin\theta \cos\phi$, $y = r \sin\theta \sin\phi$, $z = r \cos\theta$, and $x^2 + y^2 + z^2 = r^2$, $dx \, dy \, dz = r^2 \sin\theta \, dr \, d\theta \, d\phi$.

Hence, the limits for $z^2 = x^2 + y^2$ (cone) and $x^2 + y^2 = 1$ (cylinder) bounded by the plane $z = 1$ in the first quardrant i.e. r varies from $r = 0$ to $r = \sec\theta$, θ varies from $\theta = 0$ to $\theta = \dfrac{\pi}{4}$ and ϕ varies from $\phi = 0$ to $\phi = \dfrac{\pi}{2}$. Therefore

$$I = \int_{\phi=0}^{\frac{\pi}{2}} \int_{\theta=0}^{\frac{\pi}{4}} \int_{r=0}^{\sec\theta} \frac{1}{r} \cdot r^2 \sin\theta \; dr \; d\theta \; d\phi$$

$$= \int_0^{\frac{\pi}{2}} \int_0^{\frac{\pi}{4}} \left[\int_0^{\sec\theta} r \; dr \right] \sin\theta \; d\theta \; d\phi$$

$$= \int_0^{\frac{\pi}{2}} \int_0^{\frac{\pi}{4}} \left[\frac{r^2}{2} \right]_0^{\sec\theta} \sin\theta \; d\theta \; d\phi$$

$$= \frac{1}{2} \int_0^{\frac{\pi}{2}} \int_0^{\frac{\pi}{4}} \sec^2\theta \, \sin\theta \; d\theta \; d\phi$$

$$= \frac{1}{2} \int_0^{\frac{\pi}{2}} \left[\int_0^{\frac{\pi}{4}} \sec\theta \, \tan\theta \; d\theta \right] d\phi$$

$$= \frac{1}{2} \int_0^{\frac{\pi}{2}} \left[\sec\theta \right]_0^{\frac{\pi}{4}} d\phi$$

$$= \frac{1}{2} \left(\sqrt{2} - 1 \right) \int_0^{\pi/2} d\phi$$

$$= \frac{1}{2} \left(\sqrt{2} - 1 \right) \left[\phi \right]_0^{\pi/2} = \frac{\left(\sqrt{2} - 1 \right)\pi}{4}$$

***Example* 29:** Evaluate $\displaystyle\int_0^a \int_0^x \int_0^{x+y} e^{x+y+z} \; dz \; dy \; dx$ (RGPV April 2010)

Solution: Let

$$I = \int_0^a \int_0^x \left(\int_0^{x+y} e^{x+y+z} \; dz \right) dy \; dx$$

$$= \int_0^a \int_0^x \left(e^{x+y+z} \right)_0^{x+y} dy \; dx$$

$$= \int_0^a \left[\int_0^x \left(e^{2x+2y} - e^{x+y} \right) dy \right] dx$$

$$= \int_0^a \left[\frac{e^{2x+2y}}{2} - e^{x+y} \right]_0^x dx$$

$$= \int_0^a \left[\frac{e^{4x} - e^{2x}}{2} - e^{2x} + e^x \right] dx$$

$$= \left[\frac{1}{2} \left\{ \frac{e^{4x}}{4} - \frac{e^{2x}}{2} \right\} - \frac{e^{2x}}{2} + e^x \right]_0^a$$

$$= \frac{1}{8} e^{4a} - \frac{1}{4} e^{2a} - \frac{e^{2a}}{2} + e^a - \frac{1}{8} + \frac{1}{4} + \frac{1}{2} - 1$$

$$= \frac{1}{8} e^{4a} - \frac{3}{4} e^{2a} + e^a - \frac{3}{8}$$

***Example* 31:** Evaluate $\iiint_R (x - 2y + z)\, dx\, dy\, dz$, where R is the region determined by

$$0 \le x \le 1, \, 0 \le y \le x^2, \, 0 \le z \le x + y. \hspace{2cm} \text{(RGPV Dec 2003)}$$

***Solution*: Let**

$$I = \int_0^1 \int_0^{x^2} \left[\int_0^{x+y} (x - 2y + z)\, dz \right] dy\, dx$$

$$= \int_0^1 \int_0^{x^2} \left[(x - 2y)\, z + \frac{z^2}{2} \right]_0^{x+y} dy\, dx$$

$$= \int_0^1 \int_0^{x^2} \left[(x - 2y)\, (x + y) + \frac{(x + y)^2}{2} \right] dy\, dx$$

$$= \int_0^1 \int_0^{x^2} \left[\frac{3}{2} x^2 - \frac{3}{2} y^2 \right] dy\, dx$$

$$= \frac{3}{2} \int_0^1 \left[x^2 y - \frac{y^3}{3} \right]_0^{x^2} dx$$

$$= \frac{3}{2} \int_0^1 \left[x^4 - \frac{x^6}{3} \right] dx$$

$$= \frac{3}{2} \left[\frac{x^5}{5} - \frac{x^7}{21} \right]_0^1 = \frac{3}{2} \left[\frac{1}{5} - \frac{1}{21} \right] = \frac{3}{2} \cdot \frac{16}{105} = \frac{8}{35}.$$

***Example* 31:** Evaluate $\iiint_v z\, dx\, dy\, dz$, where the region of integration v is cylinder, which is bounded by the surfaces $z = 0$, $z = 1$, $x^2 + y^2 = 4$.

***Solution*:** It is clear that x varies from $z = 0$ to $z = 1$, y varies from $y = -\sqrt{4 - x^2}$ to $y = \sqrt{4 - x^2}$ and x varies from $x = -2$ to $x = 2$.

Hence $\iiint_v z\ dx\ dy\ dz = \int_{-2}^{2}\int_{-\sqrt{4-x^2}}^{\sqrt{4-x^2}}\int_{0}^{1} z\ dz.\ dy\ dx$

$$= \int_{-2}^{2}\int_{-\sqrt{4-x^2}}^{\sqrt{4-x^2}}\left[\frac{z^2}{2}\right]_0^1 dy\ dx$$

$$= \frac{1}{2}\int_{-2}^{2}\int_{-\sqrt{4-x^2}}^{\sqrt{4-x^2}} dy\ dx$$

$$= \frac{1}{2}\int_{-2}^{2}\left[y\right]_{-\sqrt{4-x^2}}^{\sqrt{4-x^2}} dx$$

$$= \int_{-2}^{2}\sqrt{4-x^2}\ dx$$

$$= 2\int_{0}^{2}\sqrt{4-x^2}\ dx$$

Putting $x = 2\sin\theta \Rightarrow dx = 2\cos\theta\ d\theta$

$$\therefore \qquad = 2\int_{0}^{\pi/2} 2\sqrt{1-\sin^2\theta}\ .\ 2\cos\theta\ d\theta$$

$$= 8\int_{0}^{\pi/2}\cos^2\theta\ d\theta$$

$$= 8\ .\ \frac{1}{2}\ .\ \frac{\pi}{2} = 2\pi$$

***Example* 32:** Find the value of $\iiint (x+z)\ dx\ dy\ dz$, when the region of integration is defined by $x^2 + y^2 + z^2 \leq 1$, $x \geq 0$, $y \geq 0$ and $z \geq 0$.

Solution: Since the region is first octant of the sphere $x^2 + y^2 + z^2 = 1$.

Therefore limits of z varies from $z = 0$ to $z = \sqrt{1-x^2-y^2}$, y varies from $y = 0$ to $y = \sqrt{1-x^2}$ and x varies from $x = 0$ to $x = 1$. Hence

$$\iiint (x+z)\ dx\ dy\ dz = \int_{0}^{1}\int_{0}^{\sqrt{1-x^2}}\int_{0}^{\sqrt{1-x^2-y^2}} (x+z)\ dz.\ dy\ dx$$

$$= \int_{0}^{1}\int_{0}^{\sqrt{1-x^2}}\left[xz + \frac{z^2}{2}\right]_0^{\sqrt{1-x^2-y^2}} dy\ dx$$

$$= \int_{0}^{1}\int_{0}^{\sqrt{1-x^2}}\left[x\sqrt{1-x^2-y^2} + \frac{1-x^2-y^2}{2}\right] dy\ dx$$

$$= \int_{0}^{1}\left[x.\left(\frac{y}{2}\sqrt{1-x^2-y^2} + \frac{1-x^2}{2}\sin^{-1}\frac{y}{\sqrt{1-x^2}}\right) + \frac{(1-x^2)y}{2} - \frac{y^3}{6}\right]_0^{\sqrt{1-x^2}} dx$$

$$= \int_0^1 \left[x \cdot \left(\frac{1-x^2}{2}\right) \frac{\pi}{2} + \frac{\left(1-x^2\right)^{\frac{3}{2}}}{2} - \frac{\left(1-x^2\right)^{\frac{3}{2}}}{6} \right] dx$$

$$= \frac{\pi}{4} \int_0^1 (x - x^3) \, dx + \frac{1}{3} \int_0^1 \left(1-x^2\right)^{\frac{3}{2}} dx$$

Putting $x = \sin\theta \Rightarrow dx = \cos\theta \, d\theta$ in the second integral

$$= \frac{\pi}{4} \left(\frac{x^2}{2} - \frac{x^4}{4}\right)_0^1 + \frac{1}{3} \int_0^{\pi/2} \cos^4 \theta \, d\theta$$

$$= \frac{\pi}{16} + \frac{1}{3} \cdot \frac{3}{4} \cdot \frac{1}{2} \cdot \frac{\pi}{2} = \frac{\pi}{16} + \frac{\pi}{16} = \frac{\pi}{8}$$

Practice Problems

1. Evaluate the following double integrals:

(a) $\int_0^a \int_0^b (x^2 + y^2) \, dx \, dy$

(b) $\int_0^1 \int_{y^2}^y (1 + xy^2) \, dx \, dy$

(c) $\int_1^2 \int_0^x \frac{dx \, dy}{x^2 + y^2}$

(d) $\int_0^a \int_0^{\sqrt{a^2-y^2}} (a^2 - x^2 - y^2) \, dy \, dx$

(e) $\int_0^1 \int_0^{\sqrt{1/2\,(1-y)^2}} \frac{dy \, dx}{\sqrt{1-x^2-y^2}}$

$$\left[\textbf{Ans:} \ (a) \ \frac{1}{3} \, ab \, (a^2 + b^2) \ (b) \ \frac{41}{210} \ (c) \ \frac{\pi}{4} \log 2 \ (d) \ \frac{\pi}{4} \right]$$

2. Evaluate $\iint x^2 y^2 dx \, dy$ over the region $x^2 + y^2 \le 1$.

$$\left[\textbf{Ans:} \ (a) \ \frac{\pi}{24} \right]$$

3. Evaluate $\iint \frac{xy}{\sqrt{1-y^2}}$ over the positive quadrant of the circle $x^2 + y^2 = 1$.

$$\left[\textbf{Ans:} \ \frac{1}{6} \right]$$

4. Find by double integration the area of the region enclosed by the following curves:
    (a) $x^2 + y^2 = a^2$ and $x + y = a$ (in the first quadrant)

$$\left[\textbf{Ans:} \ \frac{(\pi - 2)}{4} a^2 \right]$$

    (b) $(x^2 + 4a^2) \, y = 8a^3$, $2y = x$ and $x = 0$

$$[\textbf{Ans:} \ (\pi - 1) \, a^2]$$

5. Find the area common to the circle $x^2 + y^2 = a^2$ and $x^2 + y^2 = 2ax$.

$$\left[\mathbf{Ans:}\ \left(\frac{\pi}{3} - \frac{\sqrt{3}}{4}\right)a^2\right]$$

6. Find by double integration the area of the region bounded by the circle $x^2 + y^2 = a^2$.

$$[\mathbf{Ans:}\ \pi a^2]$$

7. Show that the area of the region included between the cardioids $r = a\,(1 + \cos\theta)$ and $r = a\,(1 - \cos\theta)$ is $\dfrac{a^2}{2}(3\pi - 8)$.

8. Find the area bounded by the curves $y = x - 2$ and $y^2 = 2x + 4$.

$$[\mathbf{Ans:}\ 18]$$

9. Find the area between the parabolas $y^2 = 4ax$ and $x^2 = 4ay$.

$$\left[\mathbf{Ans:}\ \frac{16}{3}a^2\right]$$

10. Evaluate $\iint r^3 dr\, d\theta$ over the area bounded between the circles $r = 2\cos\theta$ and $r = 4\cos\theta$.

11. Find the whole area of the curve $a^2 x^2 = y^3(2a - y)$.

$$[\mathbf{Ans:}\ \pi a^2]$$

12. Evaluate $\iint e^{-(x^2 + y^2)}\, dx\, dy$, the integration being over the circle $x^2 + y^2 = a^2$.

$$\left[\mathbf{Ans:}\ \pi\left(1 - e^{-a^2}\right)\right]$$

13. Evaluate $\displaystyle\int_{-c}^{c}\int_{-b}^{b}\int_{-a}^{a}(x^2 + y^2 + z^2)\, dz\, dy\, dx$

$$\left[\mathbf{Ans:}\ \frac{8abc}{3}(a^2 + b^2 + c^2)\right]$$

14. Find the value of $\displaystyle\int_{0}^{1}\int_{0}^{\sqrt{1-x^2}}\int_{0}^{\sqrt{1-x^2-y^2}} xyz\, dx\, dy\, dz$.

$$\left[\mathbf{Ans:}\ \frac{1}{48}\right]$$

15. Evaluate $\displaystyle\int_{0}^{2}\int_{0}^{x}\int_{0}^{x+y} e^x(y + 2z)\, dx\, dy\, dz$.

$$\left[\mathbf{Ans:}\ 19\left(1 + \frac{e^2}{3}\right)\right]$$

16. If the region V is bounded by the planes $x + y + z = 1$, $x = 0$, $y = 0$, $z = 0$, then find the value of $\displaystyle\iiint_{v}\frac{dx\, dy\, dz}{(1 + x + y + z)^3}$.

$$\left[\mathbf{Ans:}\ \frac{1}{2}\left(\log 2 - \frac{5}{8}\right)\right]$$

17. Find the value of the integral $\iiint_v 2z\, dx\, dy\, dz$, where v is the region of integration bounded by solid cone $x^2 + y^2 = z^2$ and plane $z = 1$.

$$\left[\text{Ans}: \frac{\pi}{2}\right]$$

18. Evaluate triple integral of the function $f(x, y, z) = x^2$ over the region v enclosed by planes $x = 0$, $y = 0$, $z = 0$ and $x + y + z = a$

$$\left[\text{Ans}: \frac{a^5}{60}\right]$$

19. Find the value of $\iiint x^2\, dx\, dy\, dz$, taken throughout the ellipsoid $\dfrac{x^2}{a^2} + \dfrac{y^2}{b^2} + \dfrac{z^2}{c^2} = 1$.

$$\left[\text{Ans}: \frac{4}{15}\, \pi a^3\right]$$

20. Evaluate $\iiint_v (x^2 + y^2 + z^2)\, dz\, dy\, dx$, where v is the volume of the cube bounded by the planes and $x = y = z = a$.

$$[\text{Ans}: a^5]$$

21. Evaluate $\displaystyle\int_0^{\pi/2} \int_0^{a\cos\theta} \int_0^{\sqrt{a^2 - r^2}} r\, dz\, dr\, d\theta$

$$\left[\text{Ans}: \frac{a^3}{3}\left(\frac{\pi}{2} - \frac{2}{3}\right)\right]$$

22. Evaluate $\displaystyle\int_0^{\pi} \int_0^{\pi} \int_0^{\pi} \cos\,(x + y + z)\, dx\, dy\, dz$

$$[\text{Ans}: 0]$$

23. Evaluate $\iiint_v (2x + y)\, dz\, dy\, dx$, where v is the closed region bounded by the plane $z = 4 - x^2$ and the planes $x = 0$, $y = 0$, $y = 2$, $z = 0$.

$$\left[\text{Ans}: \frac{80}{3}\right]$$

14.7   Change of Order of Integration

Case I

The process of evaluating a double integral, if limits of integration with respect to both the variables are constant, than c we can always change the order of integration i.e.

$$\int_a^b \int_c^d f(x, y)\, dx\, dy = \int_c^d \int_a^b f(x, y)\, dy\, dx, \quad \text{where } a \le x \le b,\ c \le y \le d$$

Case II

When the limits of integration in a double integral, with respect to y are the functions of x and the limits of integration with respect to x are constant, then we shall integrate first with respect to y (x as constant) and then with respect to x. i.e.

$$\int_a^b dx \int_{f_1(x)}^{f_2(x)} f(x, y)\, dy$$

when it is required to change the order of integration in above integral, we first of all knowing the region of integration, then we can put in the limits for integration in the reverse order i.e.

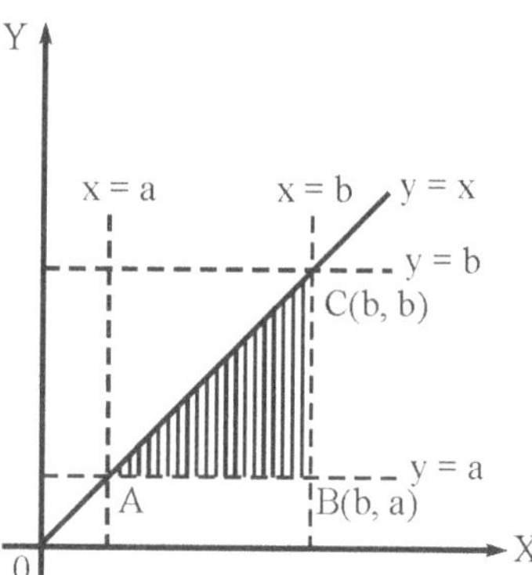

The new limits of x as a function of y and the limits of y as constant, which is defined in the form.

$$\int_c^d dy \int_{\phi_1(y)}^{\phi_2(y)} f(x, y)\, dx$$

Example 33: Show that $\int_a^b \int_a^x f(x, y)\, dx\, dy = \int_a^b \int_y^b f(x, y)\, dy\, dx$

Solution: In the double integration in the left hand side, the region of integration that y varies from the line y = a to the line y = x and x varies as x = a to x = b.

If we reverse the order of integration, then we have x varies from the line x = y to the line x = b and y varies as y = a to y = b.

Hence $\int_a^b \int_a^x f(x, y)\, dx\, dy = \int_a^b \int_y^b f(x, y)\, dy\, dx$

Example 34: Change the order of integration in $\int_0^a \int_y^a \dfrac{x\, dx\, dy}{(x^2 + y^2)}$, and hence evaluate.

(RGPV June 2003)

Solution: Here, integration is bounded by the following lines.

$$x = y, \quad x = a, \quad y = 0, \quad y = a$$

For   x = 0,   y = 0,   x = a,   y = a

∴ points are (0, 0) (a, 0) and (a, a)

∴ region OAB is bounded by these points.

For the change of order of integration, the   limits of y varies from y = 0 to y = x and x varies from x = 0 to x = a.

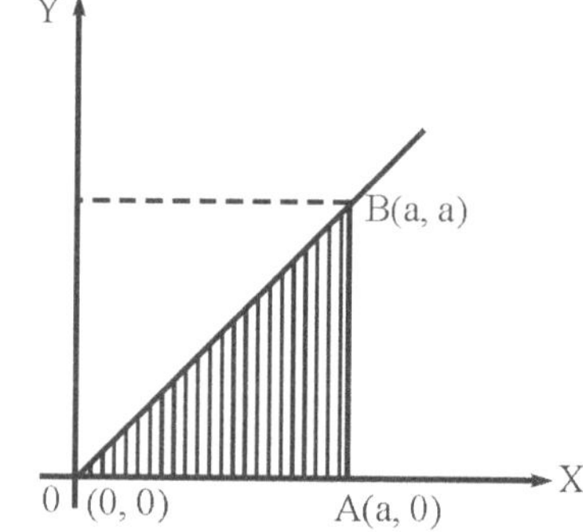

Hence $\displaystyle\int_{y=0}^{a}\int_{x=y}^{a}\frac{x\,dx\,dy}{(x^2+y^2)}=\int_{x=0}^{a}\left[\int_{y=0}^{y=x}\frac{x}{x^2+y^2}dy\right]dx$

$$=\int_{0}^{a}x\left[\frac{1}{x}\,\tan^{-1}\frac{y}{x}\right]_{0}^{x}dx$$

$$=\int_{0}^{a}(\tan^{-1}1-\tan^{-1}0)\,dx$$

$$=\frac{\pi}{4}\int_{0}^{a}dx=\frac{\pi}{4}\left[x\right]_{0}^{a}=\frac{a\pi}{4}.$$

***Example* 35:** Change the order of integration $\displaystyle\int_{0}^{1}\int_{x^2}^{2-x}xy\,dx\,dy$ and evaluate.

(RGPV Jan 2008)

***Solution*:** Let $I=\displaystyle\int_{0}^{1}\int_{x^2}^{2-x}xy\,dx\,dy$

The region of integration in the given integral is defined as following:

$y=x^2$ (parabola)   .....(i)

$y=2-x\Rightarrow x+y=2$ (line)   .....(ii)

$x=0\Rightarrow$ y axis   .....(iii)

$x=1$ (line)   .....(iv)

at parabola $\rightarrow$ $x=0,\ y=0$ and $x=1,\ y=1$

at line $\rightarrow$ $x=0,\ y=2$ and $x=1,\ y=1$

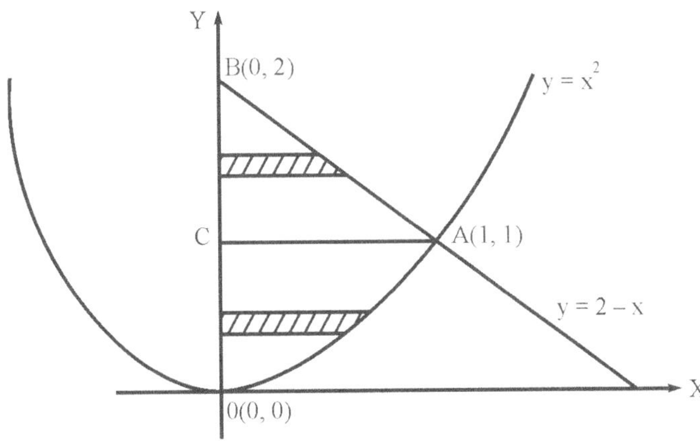

$\therefore$ points are $(0, 0)$, $(1, 1)$ for parabola and $(0, 2)$, $(1, 1)$ for line and the point of intersection be $(1, 1)$ for parabola and the line. We draw a line AC from A is parallel to x-axis which divides the region in two parts OAC and CAB.

For the region OAC: x varies from x = 0 to x = $\sqrt{y}$ and y varies from y = 0 to y = 1. Hence the change of order of integration becomes

$$\int_0^1 \int_0^{\sqrt{y}} xy \ dy \ dx$$

For the region CAB: x varies from x = 0 to x = 2 – y and y varies from y = 1 to y = 2. Hence the change of order of integration.

becomes $\qquad \int_1^2 \int_0^{2-y} xy \ dy \ dx$

$$\therefore \qquad I = \int_0^1 \int_0^{\sqrt{y}} xy \ dy \ dx + \int_1^2 \int_0^{2-y} xy \ dy \ dx$$

$$= \int_0^1 \left[\frac{x^2}{2}\right]_0^{\sqrt{y}} y \ dy + \int_1^2 \left[\frac{x^2}{2}\right]_0^{2-y} y \ dy$$

$$= \frac{1}{2} \int_0^1 y^2 \ dy + \frac{1}{2} \int_0^2 y (2-y)^2 \ dy$$

$$= \frac{1}{2} \left[\frac{y^3}{3}\right]_0^1 + \frac{1}{2} \int_1^2 (4y - 4y^2 + y^3) \ dy$$

$$= \frac{1}{6} + \frac{1}{2} \left[2y^2 - \frac{4}{3}y^3 + \frac{y^4}{4}\right]_1^2$$

$$= \frac{1}{6} + \frac{1}{2} \left[\left(8 - \frac{32}{3} + 4\right) - \left(2 - \frac{4}{3} + \frac{1}{4}\right)\right] = \frac{3}{8}.$$

Example 36: Evaluate the following integral by changing the order of integration

$$\int_0^1 \int_x^{\sqrt{2-x^2}} \frac{x \ dy \ dx}{\sqrt{x^2 + y^2}} \qquad\qquad \text{(RGPV June 2008)}$$

Solution: Here the limits of integration are

$$y = x, \quad y = \sqrt{2 - x^2} \qquad \text{and} \quad x = 0, \quad x = 1$$

i.e., $x^2 + y^2 = 2$ is a circle with centre (0, 0) and radius $\sqrt{2}$.

The straight line y = x intersect the circle $x^2 + y^2 = 2$ at (–1, –1) and (1, 1)

Now, when x = 0, y = 0 and x = 1, y = 1

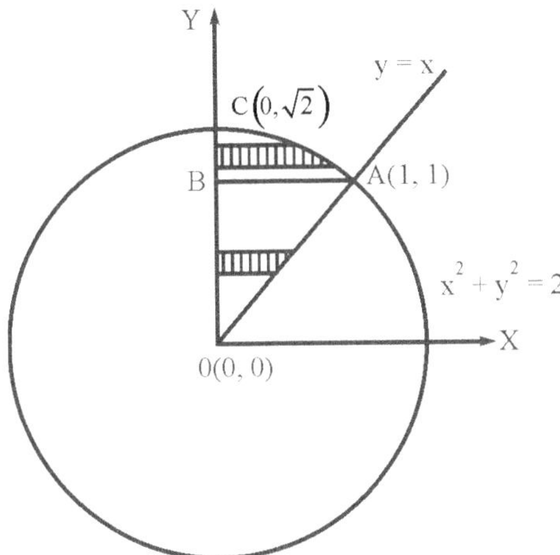

i.e., points are (0, 0) and (1, 1) on the line and

when $x = 0$, $y = \sqrt{2}$ and $x = 1$, $y = 1$

i.e., points are $\left(0, \sqrt{2}\right)$ and $(1, 1)$ on the circle.

Hence the integral region is OACBO and the strips change this has point at A(1, 1) parallel to x-axis, which divides the region into two parts i.e., OAB and BAC.

For region OAB: y varies form $y = 0$ to $y = 1$ and x varies from $x = 0$ to $x = y$. i.e.

$$\int_0^1 \int_0^y \frac{x\ dy\ dx}{\sqrt{x^2 + y^2}}$$

For region BAC: y varies from $y = 1$ to $y = \sqrt{2}$ and x varies from $x = 0$ to $x = \sqrt{2 - y^2}$ i.e.

$$\int_1^{\sqrt{2}} \int_0^{\sqrt{2-y^2}} \frac{x\ dy\ dx}{\sqrt{x^2 + y^2}}$$

Hence, the change of order of integration will be

$$\int_0^1 \int_x^{\sqrt{2-x^2}} \frac{x\ dy\ dx}{\sqrt{x^2 + y^2}} = \int_0^1 \int_0^y \frac{x\ dy\ dx}{\sqrt{x^2 + y^2}} + \int_1^{\sqrt{2}} \int_0^{\sqrt{2-y^2}} \frac{x\ dy\ dx}{\sqrt{x^2 + y^2}}$$

$$= \int_0^1 \left[\sqrt{x^2 + y^2}\right]_0^y dy + \int_1^{\sqrt{2}} \left[\sqrt{x^2 + y^2}\right]_0^{\sqrt{2-y^2}} dy$$

$$= \int_0^1 \left(\sqrt{2}y - y\right) dy + \int_1^{\sqrt{2}} \left(\sqrt{2} - y\right) dy$$

$$= \left(\sqrt{2} - 1\right) \left[\frac{y^2}{2}\right]_0^1 + \left[\sqrt{2}y - \frac{y^2}{2}\right]_1^{\sqrt{2}}$$

$$= \frac{\sqrt{2}-1}{2} + 2 - 1 - \sqrt{2} + \frac{1}{2}$$

$$= 1 - \frac{1}{\sqrt{2}}$$

***Example* 37**: Change the order of integration $\displaystyle\int_0^\infty \int_0^x xe^{\frac{-x^2}{y}} \, dx \, dy$ and evaluate.

(RGPV Dec 2008)

Solution: The limits of integration are $y = 0$, $y = x$ and $x = 0$, $x = \infty$

The region of integration are OAB exists in the positive quadrant.

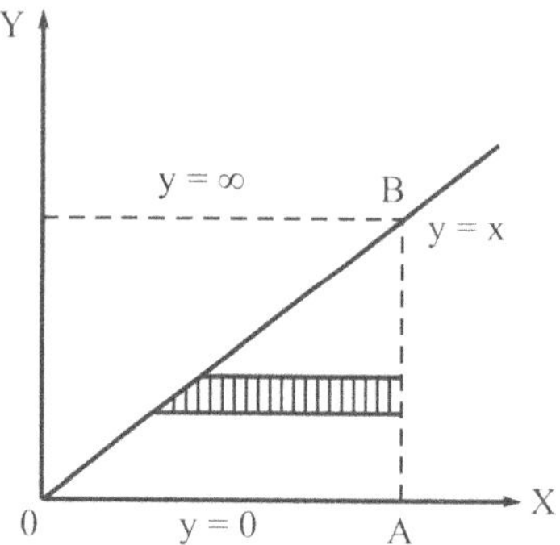

The limits of x varies from $x = y$ to $x = \infty$ and y from $y = 0$ to $y = \infty$ for the change of order of integration. Hence

$$\int_0^\infty \int_0^x xe^{\frac{-x^2}{y}} \, dx \, dy = \int_0^\infty \int_y^\infty xe^{\frac{-x^2}{y}} \, dy \, dx$$

$$= \int_0^\infty \left[\frac{-y}{2} \, e^{\frac{-x^2}{y}} \right]_y^\infty dy$$

$$= \int_0^\infty \frac{-1}{2} \left[ye^{-\infty} - ye^{-y} \right] dy$$

$$= \frac{1}{2} \int_0^\infty ye^{-y} \, dy \qquad \left(\because e^{-\infty} = 0 \right)$$

$$= \frac{1}{2} \left[-ye^{-y} - e^{-y} \right]_0^\infty$$

$$= \frac{1}{2} \left[e^{-\infty} - e^0 \right] = \frac{1}{2} \left[\because \lim_{y\to\infty} ye^{-y} = 0 \right]$$

***Example* 38:** Find the value of the integral $\int_0^\infty \int_x^\infty \dfrac{e^{-y}}{y}\ dx\ dy$, by changing the order of integral

(RGPV June 2009)

***Solution*:** The limits of integration are $y = x$, $y = \infty$ and $x = 0$, $x = \infty$

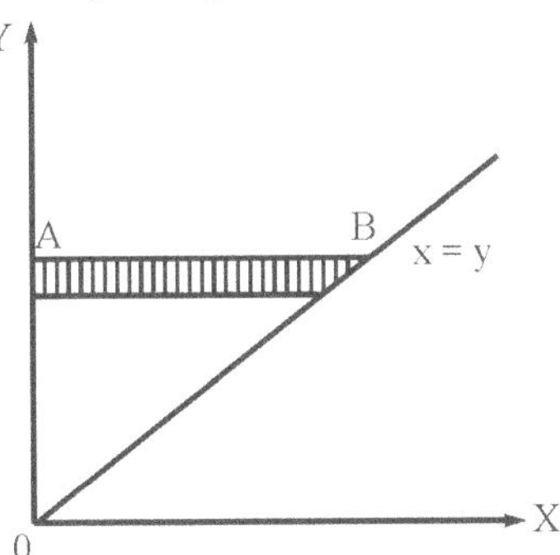

For the change of order of integration, the given integral is integrate first with respect to x then with respect to y. So the strip AB is parallel to x-axis.

The limits of x varies from $x = 0$ to $x = y$ and y varies from $y = 0$ to $y = \infty$. i.e.

$$\int_0^\infty \int_x^\infty \frac{e^{-y}}{y}\ dx\ dy = \int_0^\infty \int_0^y \frac{e^{-y}}{y}\ dy\ dx$$

$$= \int_0^\infty \frac{e^{-y}}{y}\ \big[x\big]_0^y\ dy$$

$$= \int_0^\infty e^{-y}\ dy = \Big[-e^{-y}\Big]_0^\infty = -\Big[e^{-\infty} - e^0\Big] = 1 \qquad \Big(\because e^{-\infty} = 0\Big)$$

***Example* 39:** Change the order of integration $\int_0^1 \int_0^{2-x} xy\ dx\ dy$ and hence evaluate.

(RGPV April 2010)

***Solution*:** The limits of given integration are $y = 0$, $y = 2 - x$ and $x = 0$, $x = 1$

when $x = 0$, $y = 2$ i.e. points of the line are $(1, 1)$ and $(0, 2)$ and $x = 1$, $y = 1$.

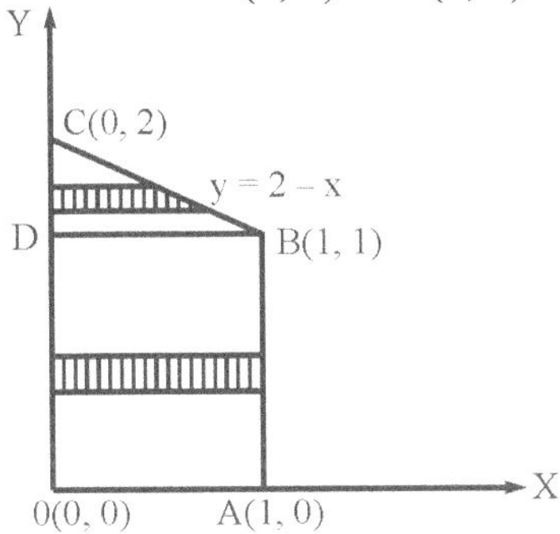

The points are $(0, 0)$ $(0, 2)$ $(1, 0)$ and $(1, 1)$

The region of integration OABCDO, which is divided in two parts i.e. OABD and DBC.

For the region OABD: x varies from x = 0 to x = 1 and y varies from y = 0 to y = 1. i.e.

$$\int_0^1 \int_0^1 xy \, dy \, dx$$

For the region DBC: x varies from x = 0 to x = 2 – y and y varies from y = 1 to y = 2 i.e.

$$\int_1^2 \int_0^{2-y} xy \, dy \, dx$$

Hence, the change of integration

$$\int_0^1 \int_0^{2-x} xy \, dx \, dy = \int_0^1 \int_0^1 xy \, dy \, dx + \int_1^2 \int_0^{2-y} xy \, dy \, dx$$

$$= \int_0^1 \left[\frac{x^2}{2}\right]_0^1 y \, dy + \int_1^2 \left[\frac{x^2}{2}\right]_0^{2-y} y \, dy$$

$$= \frac{1}{2} \int_0^1 y \, dy + \frac{1}{2} \int_1^2 \left[2-y\right]^2 y \, dy$$

$$= \frac{1}{2} \left[\frac{y^2}{2}\right]_0^1 + \frac{1}{2} \left[\frac{4y^2}{2} + \frac{y^4}{4} - \frac{4y^3}{3}\right]_1^2$$

$$= \frac{1}{4} + \frac{1}{2} \left[8 + 4 - \frac{32}{3}\right] = \frac{11}{24}.$$

Practice Problems

1. Change the order of integration in $\int_0^3 \int_1^{\sqrt{4-y}} (x + y) \, dx \, dy$.

$$\left[\textbf{Ans :} \int_1^2 \int_0^{4-x^2} (x + y) \, dy \, dx \right]$$

2. Change the order of integration in $\int_a^b \int_0^x f(x,y) \, dy \, dx$.

$$\left[\textbf{Ans :} \int_a^b \int_y^b f(x,y) \, dx \, dy \right]$$

3. Change the order of integration in $\int_0^a \int_{a-\sqrt{a^2-y^2}}^{a+\sqrt{a^2-y^2}} dx \, dy$ and hence evaluate it.

$$\left[\textbf{Ans :} \int_0^{2a} \int_0^{\sqrt{2ax-x^2}} dy \, dx, \frac{\pi}{2} a^2 \right]$$

4. Change the order of integration in $\int_0^a \int_{mx}^{lx} f(x, y) \, dx \, dy$

$$\left[\textbf{Ans :} \int_0^{am} \int_{y/l}^{y/m} f(x,y) \, dy \, dx + \int_{am}^{al} \int_{y/l}^{a} f(x,y) \, dy \, dx \right]$$

5. Change the order of integration in the double integral $\int_0^{a\cos\alpha} \int_{x\tan\alpha}^{\sqrt{a^2-x^2}} f(x,y)\, dx\, dy$.

$$\left[\mathbf{Ans:}\ \int_0^{a\sin\alpha} \int_0^{y\cot\alpha} f(x,y)\, dy\, dx + \int_{a\sin\alpha}^{a} \int_0^{\sqrt{a^2-y^2}} f(x,y)\, dy\, dx \right]$$

6. Change the order of integration in $\int_0^{2a} \int_{\sqrt{2ax-x^2}}^{\sqrt{2ax}} v\, dx\, dy$.

$$\left[\mathbf{Ans:}\ \int_0^{a} \int_{y^2/2a}^{a-\sqrt{a^2-y^2}} v\, dy\, dx + \int_0^{a} \int_{a+\sqrt{a^2-y^2}}^{a} v\, dy\, dx + \int_0^{2a} \int_{y^2/2a}^{2a} v\, dy\, dx \right]$$

7. Change the order of integration in $\int_0^1 \int_{\sqrt{x}}^1 e^{x/y}\, dx\, dy$.

$$\left[\mathbf{Ans:}\ \int_0^1 \int_0^{y^2} e^{x/y}\, dy\, dx,\ \frac{1}{2}\right]$$

CHAPTER – 15

Volume and Surface of Solid

15.1 Volume by Double Integral

If f(x, y) be a function over a region R in the xy-plane, then volume under the surface $Z = f(x, y)$ is

$$\iint_R f(x,y)\, dA \quad \text{or} \quad \iint_R z\, dA \quad \text{or} \quad \int_a^b \int_{\phi_1(x)}^{\phi_2(x)} z\, dxdy$$

15.2 Volume by Triple Integral

If f(x, y, z) be a function over a region R then volume is

$$\iiint_V dv \quad \text{or} \quad \int_a^b \int_{\phi_1(x)}^{\phi_2(x)} \int_{\psi_1(x,y)}^{\psi_2(x,y)} dxdydz$$

15.3 Volume of solids of Revolution

A solid generated by revolving a plane area about a line in the plane is known as solid of revolution and the line is said to be the axis of revolution.

(i) **_Revolution about X-axis:_** The volume of solid generated by the revolution of the area bounded by the curve y = f(x), the ordinates x = a and x = b and the x-axis is

$$V = \int_a^b \pi y^2 dx$$

(ii) **_Revolution about Y-axis:_** The volume of solid generated by revolution of the area bounded by the curve x = ϕ(y), the ordinates y = c and y = d and the y-axis is

$$V = \int_c^d \pi x^2 dy$$

(iii) The volume of solid generated by revolution about the initial line of the area bounded by the curve r = f(θ), and the radii vector $\theta = \alpha$ and $\theta = \beta$ is

$$V = \frac{2}{3} \int_{\alpha}^{\beta} \pi r^3 \sin\theta \, d\theta$$

***Example* 1:** Find the volume generated by revolving the ellipse $\dfrac{x^2}{a^2} + \dfrac{y^2}{b^2} = 1$, about X-axis.

***Solution*:** Let the curve be $\qquad \dfrac{x^2}{a^2} + \dfrac{y^2}{b^2} = 1$

$$\Rightarrow \qquad y^2 = \frac{b^2}{a^2}(a^2 - x^2)$$

and limit of x varies as x = –a to x = a (but y = 0)

$$\therefore \quad \text{volume} = \int_{-a}^{a} \pi y^2 dx$$

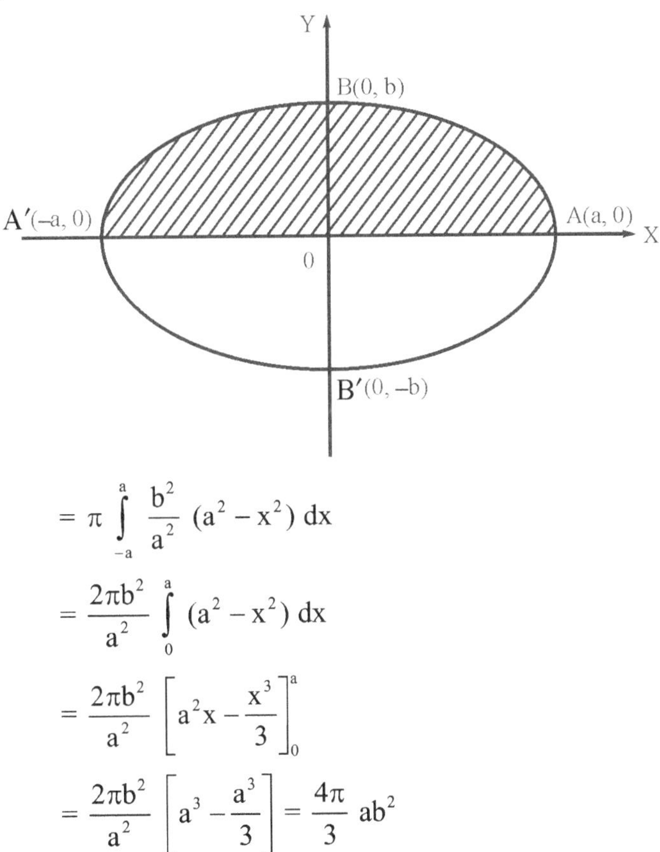

$$= \pi \int_{-a}^{a} \frac{b^2}{a^2}(a^2 - x^2)\, dx$$

$$= \frac{2\pi b^2}{a^2} \int_{0}^{a} (a^2 - x^2)\, dx$$

$$= \frac{2\pi b^2}{a^2} \left[a^2 x - \frac{x^3}{3} \right]_{0}^{a}$$

$$= \frac{2\pi b^2}{a^2} \left[a^3 - \frac{a^3}{3} \right] = \frac{4\pi}{3} ab^2$$

***Example* 2:** Find the volume of the solid generated by the revolution of the curve $X = a \cos^3 t$, $y = b \sin^3 t$ or

$$\left(\frac{x}{a}\right)^{2/3} + \left(\frac{y}{b}\right)^{2/3} = 1, \quad \text{about} \quad X - \text{axis.}$$

***Solution*:** Given curve be $x = a \cos^3 t$, $y = b \sin^3 t$

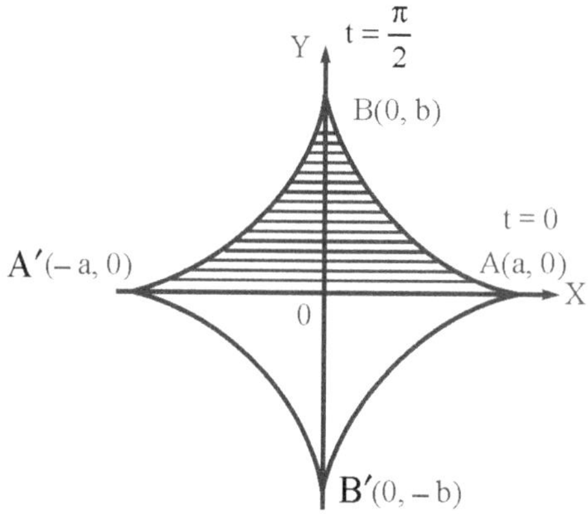

volume generated by the revolution of area about X-axis

$$= 2 \int_{\pi/2}^{0} \pi y^2 \, dx$$

$$= 2\pi \int_{\pi/2}^{0} b^2 \, \sin^6 t \left(\frac{dx}{dt}\right) dt$$

$$= 2\pi b^2 \int_{\pi/2}^{0} \sin^6 + 3a \cos^2 t \, (- \sin t) \, dt$$

$$= 6\pi a b^2 \int_{0}^{\pi/2} \sin^7 t \, \cos^2 t \, dt$$

$$= 6\pi a b^2 \, \frac{\overline{|4} \cdot \overline{\left|\dfrac{3}{2}\right.}}{2 \cdot \overline{\left|\dfrac{11}{2}\right.}}$$

$$= 3\pi a b^2 \, \frac{3.2.1 \, \dfrac{1}{2} \, \overline{\left|\dfrac{1}{2}\right.}}{\dfrac{9}{2} \cdot \dfrac{7}{2} \cdot \dfrac{5}{2} \cdot \dfrac{3}{2} \cdot \dfrac{1}{2} \, \overline{\left|\dfrac{1}{2}\right.}} = \frac{32}{105} \, \pi a b^2$$

***Example* 3:** Find the volume generated by revoluting the area in the first octant bounded by the parabola $y^2 = 0x$ and its latus rectum about X-axis.

***Solution*:** Given curve be $y^2 = 8x$

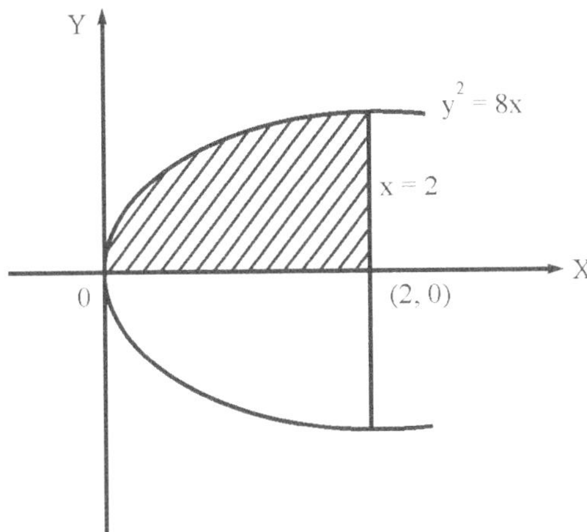

whose vertex $(0, 0)$ and focus $= (2, 0)$ and its latus rectum is $x = 2$.

$$\therefore \text{ volume } = \int_0^2 \pi y^2 dx$$

$$= \pi \int_0^2 8x \ dx$$

$$= 8\pi \left[\frac{x^2}{2}\right]_0^2 = 16\pi$$

***Example* 4:** Find the volume of the parabola generated by the revolution of the parabola $y^2 = 4ax$ about X-axis from $x = 0$ to $x = h$.

***Solution*:**

The given curve be $y^2 = 4ax$

$$\therefore \quad \text{volume} = \int_0^h \pi y^2 dx$$

$$= \pi \int_0^h 4ax \ dx$$

$$= 4a\pi \left[\frac{x^2}{2}\right]_0^h = 2\pi ah^2$$

***Example* 5:** Find the volume bounded by the paraboloid $x^2 + 4y^2 + z = 4$ and the xy-plane

***Solution*:**

The equation of paraboloid be $x^2 + 4y^2 + z = 4$

$$\Rightarrow \qquad z = 4 - x^2 - 4y^2$$

Taking z = 0, $\qquad x^2 + 4y^2 = 4 \qquad \Rightarrow y = \pm\frac{1}{2}\sqrt{4-x^2}$ and $\qquad x = \pm 2$

$\therefore$ Required volume

$$v = \iint_R z \, dA$$

$$= \int_{-2}^{2} \int_{-1/2\sqrt{4-x^2}}^{1/2\sqrt{4-x^2}} (4 - x^2 - 4y^2) \, dx \, dy$$

$$= \int_{-2}^{2} \left[(4-x^2)y - \frac{4}{3}y^3 \right]_{-1/2\sqrt{4-x^2}}^{1/2\sqrt{4-x^2}} dx$$

$$= \int_{-2}^{2} \left[(4-x^2)\frac{1}{2}\sqrt{4-x^2} - \frac{4}{3}\cdot\frac{1}{8}(4-x^2)^{3/2} + (4-x^2)\frac{1}{2}\sqrt{4-x^2} - \frac{4}{3}\cdot\frac{1}{8}(4-x^2)^{3/2} \right] dx$$

$$= \frac{2}{3} \int_{-2}^{2} (4-x^2)^{3/2} \, dx$$

Putting $x = 2 \sin\theta \Rightarrow dx = 2\cos\theta \, d\theta$, we have

$$V = \frac{2}{3} \int_{-\pi/2}^{\pi/2} 16 \cos^4\theta \, d\theta$$

$$= \frac{32}{3} \cdot 2 \int_{0}^{\pi/2} \cos^4\theta \, d\theta = \frac{64}{3} \cdot \frac{\left|\frac{5}{2}\right.\left|\frac{1}{2}\right.}{2\left|3\right.} = 4\pi$$

***Example* 6:** Find the volume under the plane $x + y + z = 6$ and above the traiangle in the xy-plane bounded by $2x = 3y$, $y = 0$, $x = 3$.

***Solution*:** The given curve be $x + y + z = 6$

$$\Rightarrow z = 6 - x - y$$

$\therefore$ Required volume

$$V = \int_0^3 \int_0^{2x/3} z\, dA$$

$$= \int_0^3 \int_0^{2x/3} (6-x-y)\, dx\, dy$$

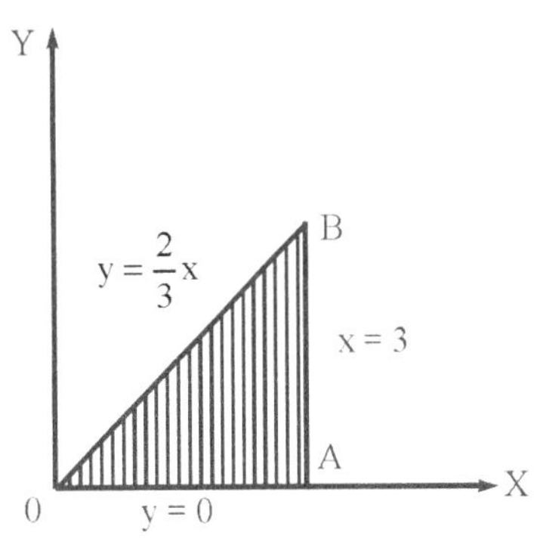

$$= \int_0^3 \left[6y - \frac{xy}{2} - \frac{y^2}{2} \right]_0^{2x/3} dx$$

$$= \int_0^3 \left[4x - \frac{2}{3}x^2 - \frac{2}{9}x^2 \right] dx$$

$$= \int_0^3 \left(4x - \frac{8}{9}x^2 \right) dx$$

$$= \left[2x^2 - \frac{8}{27}x^3 \right]_0^3 = 18 - 8 = 10$$

***Example* 7:** Find the volume in the positive octant of the ellipsoid

$$\frac{x^2}{a^2} + \frac{y^2}{b^2} + \frac{z^2}{c^2} = 1$$

***Solution*:** Let the ellipsoid be $\dfrac{x^2}{a^2} + \dfrac{y^2}{b^2} + \dfrac{z^2}{c^2} = 1$

$$\Rightarrow \qquad z = c\sqrt{1 - \frac{x^2}{a^2} - \frac{y^2}{b^2}}$$

and the plane xoy, z = 0, then

$$\frac{x^2}{a^2} + \frac{y^2}{b^2} = 1$$

$\therefore$ limits of y varies as $y = 0$ to $y = b\sqrt{1 - \dfrac{x^2}{a^2}}$ and $x = 0$ to $x = a$

Hence, the required volume is $V = \iint_R z\, dA$

$$V = \int_0^a \int_0^{b\sqrt{1-\frac{x^2}{a^2}}} c\sqrt{1 - \frac{x^2}{a^2} - \frac{y^2}{b^2}}\, dx\, dy$$

$$= \frac{c}{b} \int_0^a \left[\frac{1}{2} y \sqrt{4^2 - y^2} + \frac{1}{2} 4^2 \sin^{-1} \frac{y}{4} \right]_0^4 dx$$

$$\left(\text{put } \sqrt{1 - \frac{x^2}{a^2}} = 4 \right)$$

$$= \frac{c}{b} \int_0^a \frac{1}{2} 4^2 . \frac{\pi}{2} dx$$

$$= \frac{\pi c}{4b} \int_0^a b^2 \left(1 - \frac{x^3}{a^2} \right) dx$$

$$= \frac{\pi bc}{4} \left[x - \frac{x^3}{3a^2} \right]_0^a = \frac{1}{6} \pi abc$$

Example 8: Find the triple integration the volume of paraboloid of revolution $x^2 + y^2 = 4z$ cut off by the plane $z = 4$.

Solution: Given curve be $x^2 + y^2 = 4z$ and plane $z = 4$

The limits of z varies as $z = \dfrac{x^2 + y^2}{4}$ to $z = 4$, and put $z = 4$, $x^2 + y^2 = 16$ i.e., circle

Hence, the limits of y varies from $y = \pm \sqrt{16 - x^2}$ and $x = \pm 2$.

$\therefore$ Required volume $V = \iiint_R dxdydz$

$$= \int_{-2}^{2} \int_{-\sqrt{16-x^2}}^{\sqrt{16-x^2}} \int_{\frac{x^2+y^2}{4}}^{4} dzdydx$$

$$= \int_{-2}^{2} \int_{-\sqrt{16-x^2}}^{\sqrt{16-x^2}} [z]_{\frac{x^2+y^2}{4}}^{4} \, dydx$$

$$= \int_{-2}^{2} \int_{-\sqrt{16-x^2}}^{\sqrt{16-x^2}} \left(4 - \frac{x^2 + y^2}{4} \right) dydx$$

$$= \frac{1}{4} \int_{-2}^{2} \int_{-\sqrt{16-x^2}}^{\sqrt{16-x^2}} \left[16 - (x^2 + y^2) \right] dy \, dx$$

putting $x = r \cos \theta$, $y = r \sin\theta$, $dxdy = rdr \, d\theta$. We get

$$V = \int\limits_{0}^{2\pi} \int\limits_{0}^{4} \left[\frac{16 - r^2}{4} \right] r \, dr \, d\theta$$

$$= \frac{1}{4} \int\limits_{0}^{2\pi} \left[\frac{16r^2}{2} - \frac{r^4}{4} \right]_{0}^{4} d\theta$$

$$= \frac{1}{4} \int\limits_{0}^{2\pi} 64 \, d\theta$$

$$= \frac{64}{4} \left[\theta \right]_{0}^{2\pi} = 32\pi$$

***Example* 9:** Find the volume bounded by the cylinder $x^2 + y^2 = a^2$ and the cone $x^2 + y^2 = z^2$.

***Solution*:** Given curves are $x^2 + y^2 = a^2$ and $x^2 + y^2 = z^2$

The limits of z varies from $z = \pm\sqrt{x^2 + y^2}$

The required volume $V = \iiint\limits_{v} dx\,dy\,dz$

$$= \iint\limits_{s} \int\limits_{-\sqrt{x^2+y^2}}^{\sqrt{x^2+y^2}} dz\,dy\,dx$$

$$= \iint\limits_{s} [z]_{-\sqrt{x^2+y^2}}^{\sqrt{x^2+y^2}} dy\,dx$$

$$= 2 \iint\limits_{s} \sqrt{x^2 + y^2} \, dy\,dx$$

Now, for $x^2 + y^2 = a^2$, which is a circle,

Putting $x = r\cos 0$, $y = r\sin\theta$, $dx\,dy = r\,dr\,d\theta$, we get

$$V = 2 \int\limits_{0}^{2\pi} \int\limits_{0}^{a} r^2 \, dr \, d\theta$$

$$= 2 \int\limits_{0}^{2\pi} \left[\frac{r^3}{3} \right]_{0}^{a} d\theta$$

$$= \frac{2}{3} a^3 \int\limits_{0}^{2\pi} d\theta = \frac{2}{3} a^3 \left[\theta \right]_{0}^{2\pi} = \frac{4}{3} \pi a^3$$

***Example* 10:** Find the volume bounded by the cylinder $x^2 + y^2 = 4$, and the planes $y + z = 3$ and $z = 0$

***Solution*:** Given curves be $x^2 + y^2 = 4$ and $y + z = 3$

The limits of z varies as $z = 0$ to $z = 3 - y$

y varies as $y = \pm\sqrt{4-x^2}$ and $x = \pm 2$

$\therefore$ Required volume $= \iiint_v dx\,dy\,dz$

$$= \int_{-2}^{2} \int_{-\sqrt{4-x^2}}^{\sqrt{4-x^2}} \int_{0}^{3-y} dz\,dy\,dx$$

$$= \int_{-2}^{2} \int_{-\sqrt{4-x^2}}^{\sqrt{4-x^2}} \left[Z\right]_0^{3-y} dy\,dx$$

$$= \int_{-2}^{2} \int_{-\sqrt{4-x^2}}^{\sqrt{4-x^2}} \left[3-y\right] dy\,dx$$

$$= \int_{-2}^{2} \left[3y - \frac{y^2}{2}\right]_{-\sqrt{4-x^2}}^{\sqrt{4-x^2}} dx$$

$$= \int_{-2}^{2} \left[3\sqrt{4-x^2} - \frac{(4-x^2)}{2} + 3\sqrt{4-x^2} + \frac{(4-x^2)}{2}\right] dx$$

$$= 6 \int_{-2}^{2} \sqrt{4-x^2}\ dx$$

$$= 6 \left[\frac{x}{2}\sqrt{4-x^2} + \frac{4}{2}\ \sin^{-1}\frac{x}{2}\right]_{-2}^{2}$$

$$= 6 \left[2 - \sin^{-1}1 - 2\sin^{-1}(-1)\right] = 6 \left[2.\frac{\pi}{2} + 2.\frac{\pi}{2}\right] = 12\pi$$

Example 11: By triple integration find the volume of a hemisphere of radius a.

Solution: Let the equation of sphere be $x^2 + y^2 + z^2 = a^2$

The limits of z varies as $z = 0$ to $z = \sqrt{a^2 - x^2 - y^2}$

$\therefore$ The required volume $\qquad V = \iiint_v dx\ dy\ dz$

$$= \iint_s \int_0^{\sqrt{a^2-x^2-y^2}} dz\ dy\ dx$$

$$= \iint_s \sqrt{a^2 - x^2 - y^2}\, dy\ dx$$

Since $x^2 + y^2 + a^2$ be a circle, then

$x = r \cos \theta$, $y = r \sin \theta$, $dxdy = rdr\, d\theta$, we get

$$V = \int_0^{2\pi} \int_0^a \sqrt{a^2 - r^2}\ r\ dr\ d\theta$$

putting $a^2 - r^2 = t \Rightarrow dt = -2rdr$, then

$$V = \int_0^{2\pi} \left[-\frac{1}{2}\ \frac{(a^2 - r^2)^{\frac{3}{2}}}{\frac{3}{2}} \right]_0^a d\theta$$

$$= -\frac{1}{3} \int_0^{2\pi} \left(-a^3\right) d\theta$$

$$= \frac{a^3}{3} \left[\theta\right]_0^{2\pi} = \frac{2}{3}\ \pi a^3$$

***Example* 12:** Find the volume enclosed between the cylinder $x^2 + y^2 = 2ax$, $z^2 = 2ax$.

***Solution*:** Given curves be $x^2 + y^2 = 2ax$ and $z^2 = 2ax$

The limits of z from $z = -2ax$ to $z = 2ax$, y from $y = -\sqrt{2ax - x^2}$ to

$y = \sqrt{2ax - x^2}$ and x from $x = 0$ to $x = 2a$. Then

The required volume $V = \iiint_v dx\ dy\ dz$

$$= \int_0^{2a} \int_{-\sqrt{2ax-x^2}}^{\sqrt{2ax-x^2}} \int_{-2ax}^{2ax} dz\ dy\ dx$$

$$= 2 \int_0^{2a} \int_{-\sqrt{2ax-x^2}}^{\sqrt{2ax-x^2}} \sqrt{2ax}\ dy\ dx$$

$$= 4 \int_0^{2a} \int_0^{\sqrt{2ax-x^2}} \sqrt{2ax}\ dy\ dx$$

$$= 4 \int_0^{2a} \left[\sqrt{2ax}\ . y \right]_0^{\sqrt{2ax-x^2}} dx$$

$$= 4 \int_0^{2a} \sqrt{2ax}\ \sqrt{2ax - x^2}\ dx$$

$$= 4\sqrt{2a} \int_0^{2a} x\sqrt{2a - x}\ dx$$

Putting $x = 2a \sin^2\theta \Rightarrow dx = 4a \sin\theta \cos\theta\, d\theta$, we get

$$V = 4\sqrt{2a} \int_0^{\pi/2} 2a \sin^2\theta \cdot \sqrt{2a} \cos\theta \cdot 4a \sin\theta \cos\theta\, d\theta$$

$$= 64a^3 \int_0^{\pi/2} \sin^3\theta \cos^2\theta\, d\theta$$

$$= 64a^3 \; \frac{\overline{\left|2\right.}\dfrac{\overline{\left|\dfrac{3}{2}\right.}}{}}{2\,\dfrac{\overline{\left|7\right.}}{2}}$$

$$= 64a^3 \; \frac{1 \cdot \dfrac{1}{2}\,\overline{\left|\dfrac{1}{2}\right.}}{2 \cdot \dfrac{5}{2} \cdot \dfrac{3}{2} \cdot \dfrac{1}{2}\dfrac{\overline{\left|1\right.}}{2}} = \frac{128}{15}\,a^3.$$

***Example* 13:** Find the volume enclosed bounded by the paraboloid $x^2 + y^2 = az$, the cylinder $x^2 + y^2 = 2ay$ and the plane $z = 0$.

***Solution*:** Let the given curves be $x^2 + y^2 = az$, $x^2 + y^2 = 2ay$ and $z = 0$

The limits of z varies as $z = 0$ to $z = \dfrac{x^2 + y^2}{a}$

$\therefore$ the required volume $V = \iiint_v dx\, dy\, dz$

$$= \iint_s \int_0^{\frac{x^2+y^2}{a}} dz\, dy\, dx$$

$$= \iint_s \frac{x^2 + y^2}{a}\, dy\, dx$$

Since $x^2 + y^2 = 2ay$ be a circle, then

$x = r \cos\theta$, $y = r \sin\theta$, $dx\, dy = r\, dr d\theta$, we get

limits of r as $\qquad\qquad x^2 + y^2 = 2ay$

$\qquad\qquad \Rightarrow \qquad r^2 = 2ar \sin\theta$

$\qquad\qquad \Rightarrow \qquad r\,(r - 2a \sin\theta) = 0$

$\qquad\qquad \Rightarrow \qquad r = 0$ to $r = 2a \sin\theta$

$\qquad$ and $\qquad\qquad \theta = 0$ to $\theta = \pi$

Hence
$$V = \int_0^\pi \int_0^{2a\sin\theta} \frac{r^2}{a} \cdot r \, dr \, d\theta$$

$$= \frac{1}{a} \int_0^\pi \left[\frac{r^4}{4}\right]_0^{2a\sin\theta} d\theta$$

$$= \frac{1}{4a} \int_0^\pi 16a^4 \sin^4\theta \, d\theta$$

$$= 4a^3 \cdot 2 \int_0^{\pi/2} \sin^4\theta \, d\theta$$

$$= 8a^3 \frac{\left|\frac{5}{2}\right. \left|\frac{1}{2}\right.}{2\left|\overline{3}\right.}$$

$$= 8a^3 \frac{\frac{3}{2} \cdot \frac{1}{2} \left|\overline{\frac{1}{2}}\right. \left|\overline{\frac{1}{2}}\right.}{2.2.1} = \frac{3}{2}\pi a^3$$

Example 14: Find the volume common to the cylinder $x^2 + y^2 = a^2$ and $x^2 + z^2 = a^2$

Solution: The given curves be $x^2 + y^2 + a^2$ and $x^2 + z^2 = a^2$

The limits of z from $z = \pm\sqrt{a^2 - x^2}$, y from $y = \pm\sqrt{a^2 - x^2}$ and $x = \pm a$

The required volume
$$v = \iiint_v dx \, dy \, dz$$

$$= \int_{-a}^{a} \int_{-\sqrt{a^2-x^2}}^{\sqrt{a^2-x^2}} \int_{-\sqrt{a^2-x^2}}^{\sqrt{a^2-x^2}} dz \, dy \, dx$$

$$= 2\int_{-a}^{a} \int_{-\sqrt{a^2-x^2}}^{\sqrt{a^2-x^2}} \sqrt{a^2 - x^2} \, dy \, dx$$

$$= 2\int_{-a}^{a} \sqrt{a^2 - x^2} \left[y\right]_{-\sqrt{a^2-x^2}}^{\sqrt{a^2-x^2}} dx$$

$$= 4\int_{-a}^{a} \left(a^2 - x^2\right) dx$$

$$= 4\left(a^2x - \frac{x^3}{3}\right)_{-a}^{a}$$

$$= 4\left[a^3 - \frac{a^3}{3} + a^3 - \frac{a^3}{3}\right] = \frac{16}{3}a^3.$$

***Example* 15:** Find the triple integral, the volume of the sphere $x^2 + y^2 + z^2 = a^2$

***Solution*:** Proceeding as example 11.

The volume of hemisphere $= \dfrac{2}{3}\,\pi a^3$.

$\therefore$ The volume of shhere $= 2 \times$ volume of hemisphere

$$= 2 \times \frac{2}{3}\pi a^3$$

$$= \frac{4}{3}\pi a^3.$$

***Example* 16:** Find the volume enclosed between the cylinder $x^2 + y^2 = ax$ and $z^2 = ax$.

***Solution*:** Given curves be $x^2 + y^2 = ax$ and $z^2 = ax$

The limits of z are $z = \pm\sqrt{ax}$, $y = \pm\sqrt{ax - x^2}$, $x = 0$ to $x = a$

The required volume

$$V = \iiint_v dx\ dy\ dz$$

$$= \int_0^a \int_{-\sqrt{ax-x^2}}^{\sqrt{ax-x^2}} \int_{-\sqrt{ax}}^{\sqrt{ax}} dz\ dy\ dx$$

$$= \int_0^a \int_{-\sqrt{ax-x^2}}^{\sqrt{ax-x^2}} [Z]_{-\sqrt{ax}}^{\sqrt{ax}}\ dy\ dx$$

$$= 2\sqrt{a} \int_0^a \int_{-\sqrt{ax-x^2}}^{\sqrt{ax-x^2}} \sqrt{x}\ dy\ dx$$

$$= 2\sqrt{a} \int_0^a \sqrt{x}\,[y]_{-\sqrt{ax-x^2}}^{\sqrt{ax-x^2}}\ dy\ dx$$

$$= 4\sqrt{a} \int_0^a \sqrt{x}\ \sqrt{ax^2 - x^2}\ dx$$

Putting $x = a\sin^2\theta \Rightarrow dx = 2a\sin\theta\cos\theta\ d\theta$, we get

$$V = 4\sqrt{a} \int_0^{\pi/2} \sqrt{a}\sin\theta \cdot a\sin\theta\cos\theta \cdot 2a\sin\theta\cos\theta\ d\theta$$

$$= 8a^3 \int_0^{\pi/2} \sin^3\theta \, \cos^2\theta \, d\theta$$

$$= 8a^3 \cdot \frac{\overline{\left|2\right.} \, \dfrac{\overline{\left|3\right.}}{2}}{2 \cdot \dfrac{\overline{\left|7\right.}}{2}} = \frac{16}{15} a^3.$$

***Example* 17:** Find the volume of the region bounded by the surface $y = x^2$, $x = y^2$ and the planes $z = 0$, $z = 3$.

***Solution*:** Given curves be $y = x^2$, $x = y^2$, $z = 0$ and $z = 3$

The limits of z are $z = 0$ to $z = 3$

$$y \text{ are } y = x^2 \text{ to } y = \sqrt{x}$$

for $x \rightarrow x = x^4 \Rightarrow x^4 - x = 0 \Rightarrow x(x^3 - 1) = 0 \Rightarrow x = 0 \text{ to } x = 1$

$\therefore$ The required volume $= \iiint_v dx \, dy \, dz$

$$= \int_0^1 \int_{x^2}^{\sqrt{x}} \int_0^3 dz \, dy \, dx$$

$$= \int_0^1 \int_{x^2}^{\sqrt{x}} [Z]_0^3 \, dy \, dx$$

$$= 3\int_0^1 [y]_{x^2}^{\sqrt{x}} \, dx$$

$$= 3\int_0^1 \left(\sqrt{x} - x^2\right) dx$$

$$= 3\left[\frac{2}{3} x^{\frac{3}{2}} - \frac{x^3}{3}\right]_0^1$$

$$= 3\left[\frac{2}{3} - \frac{1}{3}\right] = 3 \cdot \frac{1}{3} = 1$$

15.4 Surface of Solid by Revolution

(a) The area of the surface of the solid generated by revolving about the x-axis, the curve $y = f(x)$ and the ordinates $x = a$ to $x = b$ is

$$S = \int_{x=a}^{b} 2\pi y \; ds = \int_{a}^{b} 2\pi y \; \frac{ds}{dx} \cdot dx$$

where $$\frac{ds}{dx} = \sqrt{1 + \left(\frac{dy}{dx}\right)^2}$$

(b) The area of the surface of the solid generated by revolving about the y-axis, the curve $x = f(y)$ and the abscissae $y = c$ to $y = d$ is

$$S = \int_{y=c}^{d} 2\pi x \; ds = \int_{c}^{d} 2\pi x \; \frac{ds}{dy} \; dy$$

where $$\frac{ds}{dx} = \sqrt{1 + \left(\frac{dx}{dy}\right)^2}$$

(c) The surface of the solid generated by the revolution of the parametric curve $x = f(t)$, $y = g(t)$ and $t = t_1$ to $t = t_2$ is

$$S = \int_{t=t_1}^{t_2} 2\pi y \; \frac{ds}{dt} \; dt$$

where $$\frac{ds}{dx} = \sqrt{\left(\frac{dx}{dt}\right)^2 + \left(\frac{dy}{dt}\right)^2}$$

(d) The area of the surface of the solid generated by revolving about the initial line, the curve $r = f(\theta)$ and $\theta = \alpha$ to $\theta = \beta$ is

$$S = \int_{\theta=\alpha}^{\beta} 2\pi y \; \frac{ds}{d\theta} \; d\theta \qquad \text{or} \qquad S = \int_{r=r_1}^{r_2} 2\pi y \; \frac{ds}{dr} \; dr$$

where $$\frac{ds}{d\theta} = \sqrt{r^2 + \left(\frac{dr}{d\theta}\right)^2} \qquad \text{or} \qquad \frac{ds}{dr} = \sqrt{1 + \left(\frac{r \; d\theta}{dr}\right)^2}$$

(e) The curved surface of the solid generated by revolving the arc of the curve about any line (i.e. the axis of rotation is neither x-axis nor y-axis) is $\int 2\pi x$ (foot of pertendicular from the point on the given line) ds.

***Example* 18:** Find the surface area of the solid generated by the revolution of the arc of the parabola $y^2 = 4ax$ bounded by its latus rectum about the x-axis of x.

***Solution*:** The equation of generated curve be

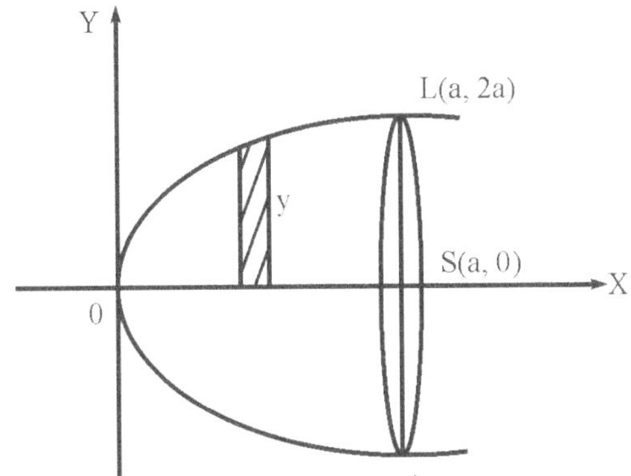

$$y^2 = 4ax$$

$$\therefore \quad \frac{dy}{dx} = \frac{2a}{y} 4$$

$$\therefore \quad \frac{ds}{dx} = \sqrt{1+\left(\frac{dy}{dx}\right)^2} = \sqrt{1+\frac{4a^2}{y^2}} = \sqrt{\frac{x+a}{x}}$$

Let S be the required surgace area generated by revolving the arc OL for latus rectum $x = a$ and for vertex $x = 0$. Then

$$S = \int_0^a 2\pi y \, \frac{ds}{dx} \, dx$$

$$= 2\pi \int_0^a \sqrt{4ax} \, \sqrt{\frac{x+a}{x}} \, dx$$

$$= 4\pi\sqrt{a} \int_0^a \sqrt{x+a} \, dx$$

$$= 4\pi\sqrt{a}.\frac{2}{3} \left[(x+a)^{\frac{3}{2}} \right]_0^a$$

$$= \frac{8\pi\sqrt{a}}{3} \left[(2a)^{\frac{3}{2}} - a^{\frac{3}{2}} \right] = \frac{8\pi a^2}{3} (2\sqrt{2} - 1)$$

***Example* 19:** Find the surface of the solid generated by the revolution of the astroid $x^{\frac{2}{3}} + y^{\frac{2}{3}} = a^{\frac{2}{3}}$ or $x = a\cos^3 t$, $y = a\sin^3 t$ about the x-axis.

***Solution*:** The parametric equations of the curve be

$$x = a\cos^3 t, \qquad\qquad y = a\sin^3 t$$

$$\frac{dx}{dt} = -3a \cos^2 t \sin t,$$

$$\frac{dy}{dt} = 3a \sin^2 t + \cos t$$

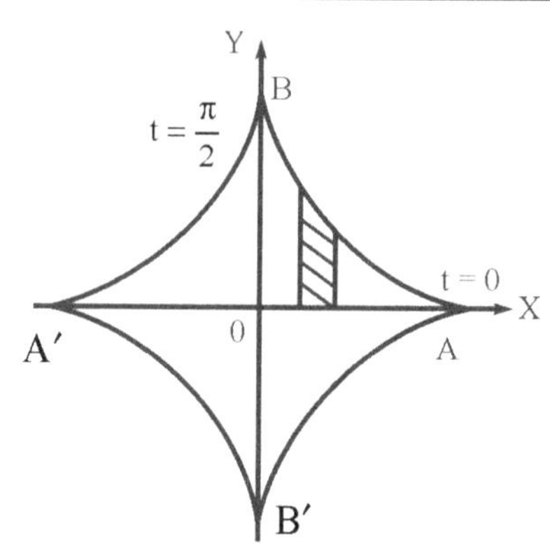

$$\therefore \quad \frac{ds}{dt} = \sqrt{\left(\frac{dx}{dt}\right)^2 + \left(\frac{dy}{dt}\right)^2}$$

$$= \sqrt{9a^2 \cos^4 t \sin^2 t + 9a^2 \sin^4 t \cos^2 t}$$

$$= 3a \sin t \cos t$$

$\therefore$ The required surface area = 2 × (area of surface generated by revoling of the arc AB about x-axis)

$$= 2 \int_0^{\pi/2} 2\pi y \, \frac{ds}{dt} \cdot dt$$

$$= 4\pi \int_0^{\pi/2} a \sin^3 t \cdot 3a \sin t \cos t \, dt$$

$$= 12\pi a^2 \int_0^{\pi/2} \sin^4 t \, \cos t \, dt$$

$$= 12\pi a^2 \, \frac{\left|\dfrac{5}{2}\right. \cdot \left|1\right.}{2 \cdot \left|\dfrac{7}{2}\right.}$$

$$= 12\pi a^2 \, \frac{\dfrac{3}{2} \cdot \dfrac{1}{2} \left|\dfrac{1}{2}\right. \cdot 1}{2 \cdot \dfrac{5}{2} \cdot \dfrac{3}{2} \cdot \dfrac{1}{2} \left|\dfrac{1}{2}\right.} = \frac{12\pi a^2}{5}$$

Example 20: Find the surface area of the sphere of radius a.

Solution: Let the equation of sphere be

$$x^2 + y^2 = a^2$$

$$\therefore \quad 2x + 2y \, \frac{dy}{dx} = 0$$

$$\text{or} \qquad \frac{dy}{dx} = \frac{-x}{y}$$

$$\therefore \qquad \frac{ds}{dx} = \sqrt{1 + \left(\frac{dy}{dx}\right)^2} = \sqrt{1 + \frac{x^2}{y^2}} = \sqrt{\frac{x^2 + y^2}{y^2}} = \sqrt{\frac{a^2}{y^2}} = \frac{a}{y}$$

$\therefore$ The required surface area $= \int_{-a}^{a} 2\pi y \, \frac{ds}{dx} \cdot dx$

$$= 2\pi \int_{-a}^{a} y \cdot \frac{a}{y} \cdot dx$$

$$= 2\pi a \left[x\right]_{-a}^{a} = 4\pi a^2$$

Example 21: Find the area of the surface of the solid generated by the revolution of the ellipse $x^2 + 4y^2 = 16$ about major axis.

Solution: Let the equation of ellipse be $x^2 + 4y^2 = 16$

$$\therefore \qquad \frac{dy}{dx} = \frac{-x}{4y}$$

The required area surface $= 2\int_{0}^{4} 2\pi y \, \frac{ds}{dx} \, dx$

$$= 4\pi \int_{0}^{4} y \cdot \sqrt{1 + \left(\frac{dy}{dx}\right)^2} \, dx$$

$$= 4\pi \int_{0}^{4} y \sqrt{1 + \frac{x^2}{16y^2}} \, dx$$

$$= 4\pi \int_0^4 y \sqrt{\frac{16y^2 + x^2}{4y}}\ dx$$

$$= \pi \int_0^4 \sqrt{16y^2 + x^2}\ dx$$

$$= \pi \int_0^4 \sqrt{64 - 4x^2 + x^2}\ dx$$

$$= \pi \int_0^4 \sqrt{64 - 3x^2}\ dx$$

$$= \pi\sqrt{3} \int_0^4 \sqrt{\frac{64}{3} - x^2}\ dx$$

$$= \pi\sqrt{3} \left[\frac{x}{2} \sqrt{\frac{64}{3} - x^2} + \frac{1}{2} \cdot \frac{64}{3} \sin^{-1} \frac{x}{\frac{8}{\sqrt{3}}} \right]_0^4$$

$$= 8\pi \left[1 + \frac{4\pi\sqrt{3}}{9} \right]$$

Example 22: Find the area of the surface of revolution formed by revolving curve $r = 2a\cos\theta$ about the initial line.

Solution: Let the given curve be $r = 2a\cos\theta$

$$\therefore \qquad \frac{dr}{d\theta} = -2a\sin\theta$$

$$\therefore \qquad \frac{ds}{d\theta} = \sqrt{r^2 + \left(\frac{dr}{d\theta}\right)^2} = \sqrt{4a^2\cos^2\theta + 4a^2\sin^2\theta} = 2a$$

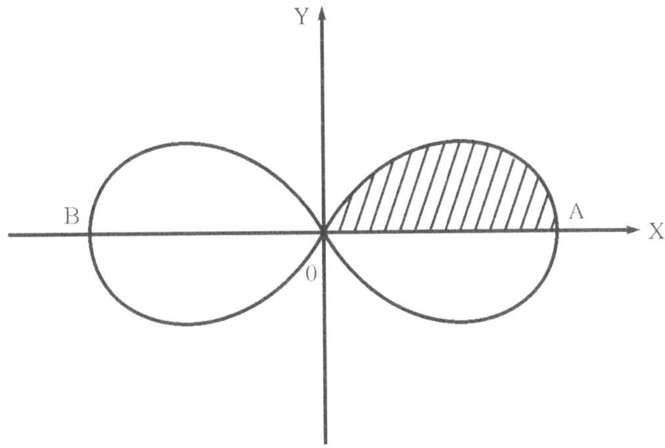

The required surface $= 2 \int\limits_{0}^{\pi/4} 2\pi y \cdot \dfrac{ds}{d\theta} \cdot d\theta$

$$= 4\pi \int\limits_{0}^{\pi/4} r\sin\theta \cdot 2a \, d\theta \qquad (\because y = \sin\theta)$$

$$= 16\pi a^{2} \int\limits_{0}^{\pi/4} \cos\theta \, \sin\theta \, d\theta$$

$$= 16\pi a^{2} \int\limits_{0}^{1/\sqrt{2}} t \, dt \qquad (\text{put } \sin\theta = t)$$

$$= 16\pi a^{2} \left(\dfrac{t^{2}}{2}\right)_{0}^{1/\sqrt{2}} = 16\pi a^{2} \cdot \dfrac{1}{4} = 4\pi a^{2}$$

***Example* 23:** Find the surface generated by the revolution of the cycloid $x = a\,(\theta - \sin\theta)$, $y = a\,(1 - \cos\theta)$ about the x-axis.

Solution: Let the curve be $x = a\,(a - \sin\theta)$, $y = a\,(1 - \cos\theta)$

$$\dfrac{dx}{d\theta} = a(1 - \cos\theta), \quad \dfrac{dy}{d\theta} = a\sin\theta$$

$$\dfrac{ds}{d\theta} = \sqrt{\left(\dfrac{dx}{d\theta}\right)^{2} + \left(\dfrac{dy}{d\theta}\right)^{2}} = \sqrt{a^{2}(1 - \cos\theta)^{2} + a^{2}\sin^{2}\theta}$$

$$= a\sqrt{1 + \cos^{2}\theta - 2\cos\theta + \sin^{2}\theta}$$

$$= a\sqrt{2 - 2\cos\theta}$$

$$= a\sqrt{2\,(1-\cos\theta)}$$

$$= 2a\,\sin\frac{\theta}{2}$$

$$\text{The required surface} = 2\int_{a}^{\pi} 2\pi y.\,\frac{ds}{d\theta}\,d\theta$$

$$= 4\pi\int_{a}^{\pi} a(1-\cos\theta)\,.\,2a\sin\frac{\theta}{2}\,d\theta$$

$$= 16\pi a^{2}\int_{0}^{\pi}\sin^{3}t\,d\theta$$

Putting $\qquad \dfrac{\theta}{2}=t \;\Rightarrow\; d\theta = 2dt$

$$= 32\pi a^{2}\int_{0}^{\pi/2}\sin^{3}t\,dt$$

$$= 32\pi a^{2}\;\frac{\sqrt{2}\left|\dfrac{1}{2}\right.}{2\left|\dfrac{5}{2}\right.} = \frac{64}{3}\pi a^{2}$$

Example 24: Find the surface of a hemisphere of radius a.

Solution: Let the curve be $x^{2}+y^{2}=a^{2}$

$$\therefore \qquad \frac{dy}{dx}=\frac{-x}{y}$$

$$\therefore \qquad \frac{ds}{dx}=\sqrt{1+\left(\frac{dy}{dx}\right)^{2}}=\sqrt{1+\frac{x^{2}}{y^{2}}}=\sqrt{\frac{x^{2}+y^{2}}{y^{2}}}=\frac{a}{y}$$

$$\therefore \text{ The required surface} = \int_{0}^{a} 2\pi y\,.\,\frac{ds}{dx}\,.\,dx$$

$$= 2\pi\int_{0}^{a} y\,.\,\frac{a}{y}\,dx$$

$$= 2\pi a\int_{0}^{a} dx = 2\pi a^{2}$$

Practice Problems

1. Find the volume of the wedge intercepted between the cylinder $x^2 + y^2 = 2ax$ and the planes $z = x$, $z = 2x$.

   $$[\textbf{Ans: } \pi a^3]$$

2. Find the volume of the tetrahedron bounded by the coordinate planes and the plane $x + y + z = 1$

   $$\left[\textbf{Ans: } \frac{1}{6}\right]$$

3. Find the volume cut from the paraboloid $4z = x^2 + y^2$ by the plane $x = 4$.

   $$[\textbf{Ans: } 32\pi]$$

4. Find the volume bounded by the xy-plane, the paraboloid $2z = x^2 + y^2$ and the cylinder $x^2 + y^2 = 4$

   $$[\textbf{Ans: } 4\pi]$$

5. Show that the volume of sphere of radius a is $\frac{4}{3}\pi a^3$.

6. Find the volume of the solid generated by revolving the ellipse $\dfrac{x^2}{a^2} + \dfrac{y^2}{b^2} = 1$ about the y-axis.

   $$\left[\textbf{Ans: } \frac{4}{3}\pi a^2 b\right]$$

7. Find the volume bounded by the surface $4z = 16 - 4x^2 - y^2$ and the plane $z = 0$.

   $$[\textbf{Ans: } 16\pi]$$

8. Find the volume enclosed by the surfaces $x^2 + y^2 = cz$, $x^2 + y^2 = 2ax$, $z = 0$

   $$\left[\textbf{Ans: } \frac{3\pi a^4}{2c}\right]$$

9. The axes of two right circular cylinder of the same radius a intersect at right angles. Prove that the volume which is inside both the cylinder is $\frac{16}{3}a^3$.

10. Find the volume bounded by $y^2 + z^2 = 4ax$, $y^2 = ax$, $x = 3a$.

    $$[\textbf{Ans: } a^2[6\pi + 9\sqrt{3}]]$$

11. Find the volume of the portion cut off from the cylinder which is determined by $2x^2 + y^2 = 29x$, and the planes $z = mx$ and $z = nx$.

$$\left[\textbf{Ans:}\ \frac{\pi\sqrt{2}a^3(n-m)}{\theta}\right]$$

12. Find the volume of the solid generated by the revolution of the curve $r = 2a\cos\theta$ about the initial line.

$$\left[\textbf{Ans:}\ \frac{4}{3}\pi a^3\right]$$

13. Find the volume of the solid generated by the revolution of the cardioid $r = a(1 - \cos\theta)$ about the initial line

$$\left[\textbf{Ans:}\ \frac{8}{3}\pi a^3\right]$$

14. Find the area of the surface of the sphere $x^2 + y^2 + z^2 = a^2$ which lies inside the cylinder $x^2 + y^2 = ax$.

$$[\textbf{Ans:}\ 2a^2(\pi - 2)]$$

15. Find the surface generated by the revolution of an arc of the catenary $y = c\cos h\left(\dfrac{x}{c}\right)$ about x-axis.

$$\left[\textbf{Ans:}\ \pi c\left[x + \frac{c}{2}\sinh\frac{2x}{c}\right]\right]$$

16. Find the surface area of the solid formed by the revolution, about the axis of y, of the part of the curve $ay^2 = x^3$ from $x = 0$ to $x = 4a$ which is above the x-axis.

$$\left[\textbf{Ans:}\ \frac{128\pi a^2}{1215}(125\sqrt{10} + 1)\right]$$

17. Prove that the surface area of the solid generated by the revolution of the loop of the curve $x = t^2,\ y = t - \dfrac{t^3}{3}$.

$$[\textbf{Ans:}\ 8\pi]$$

18. Find the surface generated by the revolution of the cycloid
    (i) $x = a(\theta + \sin\theta),\ y = a(1 - \cos\theta)$ about tangent at the vertex
    (ii) $x = a(\theta + \sin\theta),\ y = a(1 + \cos\theta)$ about its base

$$\left[\textbf{Ans:}\ \text{(i)}\ \frac{32\pi a^2}{3}\quad \text{(i)}\ \frac{64\pi a^2}{3}\right]$$

19. Find the surface of the solid formed by the revalution of the cardioid $r = a(1 - \cos \theta)$ about the initial line.

$$[\text{Ans: } \frac{32}{5}\pi a^2]$$

20. Find the surface of the cylinder $x^2 + z^2 = a^2$, that lies inside the cylinder $x^2 + y^2 = a^2$.

$$[\text{Ans: } 8a^2]$$